# CONTROL OF SOIL IN MILITARY CONSTRUCTION

DEPARTMENTS OF THE AIR FORCE AND THE ARMY

**Books for Business**
**New York - Hong Kong**

Control of Soils in Military Construction

by
Departments of the Air Force and the Army

ISBN: 0-89499-199-X

# CONTROL OF SOILS IN MILITARY CONSTRUCTION

# CHAPTER I
# INTRODUCTION

## 1. Purpose

This manual supplies engineer officers and noncommissioned officers with pertinent facts concerning the control of soils during military construction. It is designed as a guide to the military engineer for evaluating existing or potential soil conditions, for predicting soil behavior under varying conditions, and for assisting him to reach satisfactory solutions of the more simple problems dealing with foundations and earthwork constructions encountered in military operations.

## 2. Scope

This manual discusses briefly the formation and characteristics of soil and covers in detail the system used by the army to classify soils; classification systems used by other agencies also are presented. It discusses the movement of water through soils and frost action; the compressibility and shearing resistance of soils also receive attention. The bearing capacity of soils which serve as foundations is analyzed in some detail. The compaction and stabilization of soils used in military construction are discussed. The manual also deals with slopes, embankments, dikes, dams, and earth-retaining structures. The final chapter covers subsurface soil exploration. Emphasis is placed on construction throughout the manual. Specific testing procedures and design techniques are beyond the scope of this manual. Laboratory testing procedures and design of roads and airfields will be covered in separate technical manuals now under preparation. When printed, these manuals will be distributed automatically by The Adjutant General to organizations concerned.

## 3. References

Although the information contained in this manual should meet the needs of the military engineer for solving most of the common problems concerning control of soils, full use of all available references should be made where the importance of an assigned project requires more exact solution. Some additional information taken from other manuals has been included in the text. For a list of references see appendix I.

3

# CHAPTER 2

# FORMATION AND CHARACTERISTICS OF SOIL

## Section I.  FORMATION OF SOIL

### 4. Introduction

*a. Soil Defined.*  The engineering definition of the term *soil* is very broad.  *Soil* is all the earth material which covers the rock of the earth's crust.  Since both rock and soil are aggregates of minerals, the dividing line between them is subject to interpretation.  However, rock can best be visualized by the term *bedrock*, i. e. the mineral grains are premanently bonded together in massive formations broken only by occasional structural faults.  Soil, on the other hand is a natural aggregate of mineral grains ranging in particle size from large boulders to single mineral crystals of microscopic size which may be readily separated by mechanical action, such as gentle agitation in water.  Materials which are highly organic, such as the mud of a river bottom, and man-made fills are included in the definition of soil.

*b. Engineering Significance.*  An understanding of the natural processes by which soils are formed from the parent materials of the earth's crust is desirable as an aid in describing soils and in predicting soil behavior.  As soils are created they often acquire distinctive characteristics which are visible in the field and on maps and photographs.  When these characteristics are recognized and properly interpreted they may furnish much of the information which is needed in making a preliminary estimate of soil conditions.  While the detailed recognition and interpretation of these distinctive characteristics in some cases requires specialized training and experience which is beyond the scope of this manual, a general understanding of the processes of soil formation is nevertheless of importance to the military engineer.  Material which is contained in TM 5–545 is of importance in the understanding of the processes of soil formation.

### 5. Soil Formation

The tremendous forces of nature which operate ceaselessly about the face of the earth are or have been the primary agents in the formation of soils.  While it is the cumulative action of mechanical and chemical weathering which forms soil from rock, the nature of the soil-forming process has a marked effect upon the final material.

For the purposes of this manual and from the standpoint of the general processes responsible for their formation, soils may be divided into two groups, residual soils and transported soils.

## 6. Residual Soils

In the formation of a residual soil the rock material has been weathered in place. Here, while mechanical weathering may occur, chemical weathering is the dominant factor. Plant growth produces organic acids which are washed through the soil. Moisture in the zone of growth dissolves out minute quantities of material which the plant takes for food. The cumulative effect over the years causes a breakdown of the mineral particles. Water containing dissolved oxygen and carbon dioxide percolates through the soil, dissolves out the more soluble materials, and carries them away. As a result of this process and because of the assorted mineral structure of the rock material and other factors, the upper layers of the soil are usually fine-grained and relatively impervious to the flow of water. The imperviousness of the surface soils prevents the effect from reaching much more than fifty feet, even in tropical regions of heavy rainfall. Under this fine-grained material is a zone of the parent rock which is partially disintegrated; it may crumble easily and break down rapidly when exposed to loads, abrasions, or further weathering. The boundary line between soil and rock is usually not clearly defined. In general, these soils will present both drainage and foundation problems. Unless better transported soils are available nearby the engineer must go to the underlying parent rock for borrow materials to be used in construction processes. A typical residual soil is shown in figure 1. Residual soil deposits are characteristically erratic and variable in nature.

## 7. Transported Soils

By far the majority of the soils with which the civil engineer must deal are materials which have been transported and deposited at the place where he finds them. Three major forces, glacial ice, water, and wind, are the transporting agents. These forces have acted in a wide variety of ways and have produced an equally wide variety of soil deposits. Resulting foundation and construction problems are equally varied. These soils may be divided into glacial deposits, sedimentary or water-laid deposits, and aeolian or wind-laid deposits.

## 8. Glacial Deposits

The tremendous ice caps which at successive periods have covered much of the earth's surface had a very considerable effect on soil formation. The great ice caps were giant rock crushers of unbelievable force. As the ice crept forward in a plastic flow, under its own weight, it tore away the solid rock material to a great depth and

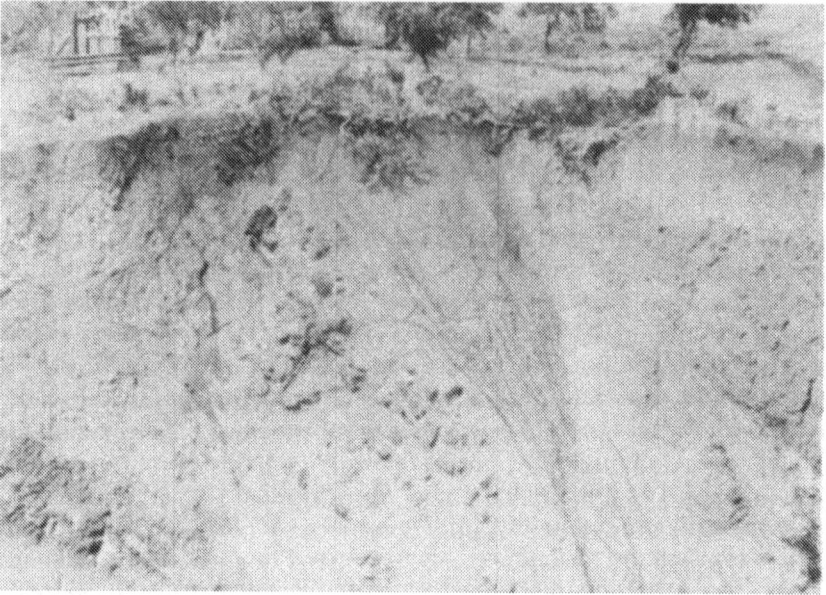

*Figure 1.   Residual soil forming from the in-place weathering of igneous rock.*

carried these materials forward.   In this grinding process particles were produced which vary in size from huge boulders to microscopic. Where these materials were deposited and not later sorted by water produced by melting of the ice, they were found as morainic deposits, sometimes called glacial till.   They are dense, compact deposits which frequently have a well-graded distribution of particle sizes from coarse to fine.   Many direct glacial deposits were later modified by the action of melt water; many sedimentary deposits were then

*Figure 2.   Stratified glacial drift consisting of interbedded sands and gravels.*

formed from materials originally transported by the moving ice sheets. Many glacial-laid sands and gravels have had most of the finer materials washed away by the melt waters. Included among deposits of this general nature are those which are called kames, eskers, and outwash plains. Figure 2 is an example of a glacial drift; note the absence of fines.

## 9. Sedimentary or Water-Laid Deposits

Water is one of the most active agents in the formation of soils. Soils which have been transported by or deposited in water are here divided into three principal groups: alluvial or stream-laid deposits, lacustrine or lake-laid deposits, and marine deposits.

*a. Alluvial Deposits.* Streams have a considerable power to carry soil materials. The capacity of a stream to transport material and the size of the material carried are functions of the velocity of the water. It is apparent that there will be a considerable variation in particle size which may be carried by a stream at different stages of flow and at different points along its length. As a stream begins in a mountainous area it generally has a high gradient and hence, a high velocity. At flood stages narrow deposits or terraces of gravel and sand may be formed in such a river. Lower down the valley of the stream may widen and the velocity of the water will be somewhat less. In such situations, deposits which are designated as alluvial fans may be formed; these also are likely to consist largely of sands and fine gravels. Finally, as the stream nears the sea its valley broadens out into a flood plain. It is in the flood plain, particularly during flood stages, that the river deposits most of its materials. As the velocity of the stream is reduced the coarser materials settle out first, followed by the finer ones. Such deposits are thus likely to be formed in layers (stratified). Particle size may vary from very fine sands to clay. These materials are in turn eroded away as the stream meanders back and forth across the valley. These meanders are continually being cut off into ox-bow lakes as the river in flood stage scours a new and shorter tract. As might be expected under such a variety of conditions stream-laid deposits are likely to be quite variable and somewhat unpredictable. In figure 3 is shown one type of deep water-laid gravel and sand.

*b. Lacustrine or Lake-Laid Deposits.* Lake deposits are usually fine-grained deposits. They may be of considerable extent or of limited area, depending on their original formation. For example, as the ice caps of the glacial age receded huge lakes were formed at the edge of the retreating ice. The melt water carrying the finer soil materials washed out into these lakes, which acted as giant settling basins. Thus great beds of clay and silt were formed, such as may be found on the Baltic Plain and the northern areas of European Russia, and in the Great Lakes region. Lesser lakes may be formed in the paths of

*Figure 3. A typical transported soil consisting of deep water-laid gravel and sand deposits.*

rivers and streams. In this case, and over long periods of time, these lakes have gradually filled with fine sediments until they have disappeared. In humid regions, such deposits frequently consist of alternate relatively thin layers (strata) of silt and clay; toward the top of the deposits the soil may contain much organic material, and the top layer may consist of peat or some other highly organic swamp soil. Ox-bow lakes, cut off from meandering rivers, receive new sediments each time the river floods until they likewise disappear.

   *c. Marine Deposits.* Marine soils are formed from materials carried into the sea by streams and by material eroded from the beaches by wave and tidal action. Some of the material is carried out into deep water, forming the so-called off-shore deposits, which are also sometimes characterized as deltaic sediments. Material which is carried out to deep water is sorted effectively and frequently deposited in rather uniform layers over the submerged plains bordering the coastline. Depending on the conditions of their laying, such deposits may be deep beds of either silt or clay; very frequently they consist of dense, impermeable, plastic clays. Shore deposits may be highly complex, owing to the mixing and transporting action of currents and waves. Typical beach deposits, however, commonly consist of sand and gravel of fairly uniform grain size. In figure 4 is shown an example of a marine deposit which has now been raised to a coastal plain.

*Figure 4. A typical coastal plain soil originating from marine deposits.*

Special mention may be made here of marine deposits formed by marine organisms. Marine organisms have two forms, but in either case the deposits are the shell materials of the dead organisms, which have persisted and precipitated into beds. One type has extracted silica from the sea water and has laid deposits known as *diatomaceous earths*. These deposits are usually fine-grained materials which act like clays. The other shell type is calcareous. The organism extracts calcium carbonate from the sea water. A variety of end products result, the best known of which is *coral*. Live coral is rockhard and, when crushed, is as good a material as a crushed limestone rock. Dead coral, on the other hand, has disintegrated into a fine-grained soft material.

## 10. Aeolian or Wind-Blown Soil Deposits

The sand dunes which are found in desert regions are typical wind-blown soil deposits. Wind is a selective agent in the transportation of soil particles, and its lifting force is limited to particles which are the size of very fine sand or smaller. Its effect is negligible on moist, cohesive soils. Vast deposits of wind-blown silt were formed when great winds, between the ice ages, swept across the desert lands, lifted, and transported the finer particles to be deposited many miles from the source. These aeolian soils are known as *loess*. Loess deposits are characterized by their uniform particle size, very loose structure, and pronounced vertical porosity and cleavage; they are generally yellow-brown in color. As erosion occurs in beds of loess, the banks usually stand vertical to a considerable height. Many loess deposits are hard because they contain cementing agents, such as calcium carbonate and iron oxide; when saturated with water they frequently become soft. Figure 5 is an example of a road cut in

9

loess, showing the high vertical banks these soils can maintain. It must be emphasized that such side slopes are not recommended in other silt soils; as will be indicated later, silt soils present many foundation and construction problems. Mention may also be made of volcanic ash as a wind-transported soil. In general, volcanic ash is light in weight and soaks up water readily; when it is partially decomposed and dried it forms a soft rock which is called *tuff*. Complete decomposition of the ash may result in the formation of highly plastic clays which are extremely compressible.

*Figure 5. Vertical cut in loess deposit. Banks cut from this wind-blown silt will usually stand vertically; such cuts are not recommended in other silt soils.*

## 11. The Soil Profile and Natural Variations

*a. Soil Profile Defined.* As time passes, the soil deposits which have been described in the preceding paragraphs undergo a maturing process. Every soil deposit then develops a characteristic profile near the surface of the ground because of weathering and the leaching action of water as it moves downward from the surface. The profile which is developed depends not only upon the nature of the deposit but upon such things as temperature, amount of rainfall and the type of vegetation. Under certain conditions complex profiles may be developed, particularly with old soils in humid regions; in dry regions the profile may be obscured.

*b. Soil Horizons.* Typical soil profiles have at least three layers as indicated in figure 6. The upper layer or A horizon is usually

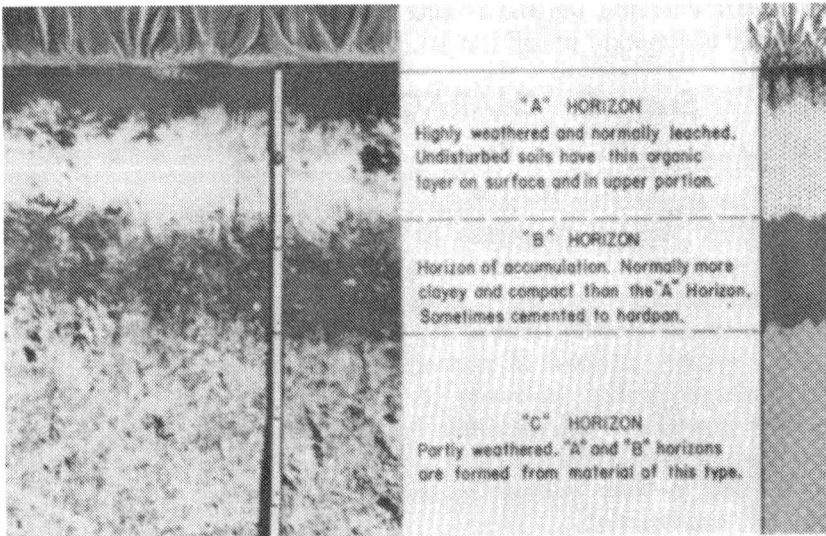

*Figure 6. Soil profile showing characteristic soil horizons.*

marked by a zone of accumulation of organic materials in its upper portion and a lower layer of lighter color from which the soil colloids and other soluble constituents have been removed. The B horizon represents the layer in which the soluble materials washed out of the A horizon have accumulated. This layer frequently contains a sizeable amount of clay and may be several feet thick. The C horizon is the parent material. As has been indicated, the development of a soil profile depends upon the downward movement of moisture. In arid and semi-arid regions the movement of moisture may be reversed and water brought to the surface because of evaporation; soluble salts may thus be brought to the surface and deposited.

*c. Pedology.* The study of the maturing of soils and the relationship of the soil profile to the parent material and its environment is called pedology. As will be explained later (par. 65) soils may be classified on the basis of their soil profiles; this approach is used by agricultural soil scientists and some engineering agencies. The system is of particular interest to engineers who are concerned with road and airfield problems.

## 12. Variable Nature of Soils

It is desirable at this point to emphasize the variable nature of soils and soil deposits. Not only are soils characteristically varied with depth, but several soil types can and often do exist within a relatively small area. This is of consequence to the engineer who must deal with soils, since the variations which do occur may be very important from an engineering standpoint. As will be emphasized in later discussions, the engineering properties of a soil are a function not only of the kind of soil but its condition.

## Section II. CHARACTERISTICS OF SOIL

## 13. Introduction

*a.* The engineering characteristics of soil vary greatly, depending upon such physical properties as particle size, gradation, particle shape, structure, density, and consistency. It is therefore desirable to define these properties, in most cases numerically, as a basis for the systematic classification of the many soil types. Such a classification system, utilized in connection with a common descriptive vocabulary, permits the ready identification of similar soils which may be expected to behave similarly in comparable situations. Properties which are discussed in this section are often important in themselves and in their relation to the control of soils encountered in military construction.

*b.* The nature of any given soil can be changed by manipulation. Vibration, for example, can change a loose sand into a dense one. Hence, the behavior of a soil in the field depends not only on the significant properties of the individual constituents of the soil mass, but also on those properties which are due to the arrangement of the particles within the mass.

*c.* Those properties which identify and classify soils, both as to categories and within categories, are known as index properties; the tests which determine these properties are frequently known as *classification tests.* Frequently, available field laboratory equipment or other considerations will permit the military engineer or his soil technician to determine only some of these index properties, and then only approximately. Often a hasty field identification will permit a sufficiently accurate evaluation for the problem at hand. However, neither the military engineer nor the soil technician can ever hope to develop his experience and judgment in estimating soil conditions or identifying soils to a point such that he can place safe reliance upon them, if he does not consistently make as detailed a determination of the index properties of all soils as the situation will permit, and subsequently correlate these identifying properties with the observed behavior of the soil.

## 14. Soil Particles

In a natural soil the soil particles or solids form a discontinuous mass with spaces or voids between the particles. These void spaces are normally filled with water and/or air. Organic material may be present in greater or lesser amounts. This portion of the discussion is concerned with the soil particles themselves. It should be noted that the terms *particle* and *grain* are used interchangeably herein and are taken to have the same meaning.

## 15. Particle Size and Gradation

As has been indicated previously, soils may be made up of grains which range in size or *diameter*, loosely speaking, from fairly large masses of detached rock to sub-microscopic materials. Important index properties, particularly for the coarser soils. Other properties are generally of greater importance for fine-grained soils.

## 16. Definitions in Terms of Grain Size

*a.* Soils may be divided into several different groups on the basis of the size of particles included in each group. Many different grain-size scales have been proposed and used. The scale used in the Unified Soil Classification System described in the next chapter is indicated in figure 9, and in the tabulation below:

|  | Sieve size | |
| --- | --- | --- |
|  | Passing | Retained on |
| Cobbles | -------- | 3 inch |
| Gravel | 3 inch | No. 4 |
| Sand | No. 4 | No. 200 |
| Fines | No. 200 | --------- |
| Organic material | (No size boundary) | |

Coarse gravel particles are comparable in size to a lemon, an egg, or a walnut, while fine gravel is about pea size  Sand particles range in size from that of rock salt, through table salt or granulated sugar, to powdered sugar. Below a No. 200 sieve, the particles (fines) are designated as *silt* or *clay*, depending on their plasticity characteristics (par. 53).

*b.* Other grain-size scales apply other limits of size to silt and clay. For example, many engineers in civil practice define silt as material which is less than 0.05 millimeters (mm.) in diameter and larger than

0.005 mm. Particles below 0.005 mm. are in the range of clay sizes. A particle 0.05 mm. in diameter is about as small as can be detected by the naked eye. It must be emphasized that below the No. 200 sieve particle size is relatively unimportant in most cases, other properties being of greater consequence. Particles below about 0.002 mm. (0.001 mm. in some grain-size scales) are frequently designated as *soil colloids*. Particles in this range of size have the properties of true colloids and may be expected to behave as such. The organic materials which may be present in a soil mass have no size boundaries.

## 17. Methods of Determining Grain Size (Mechanical Analysis)

Several methods may be employed to determine the size of soil particles contained in a soil mass and the distribution of particle sizes. Included among these are methods which may be designated as *sieve analysis, wet mechanical analysis,* and *combined mechanical analysis.* These methods are briefly described in this paragraph. Reference should be made to TB 5–253–1 for details of the test procedures.

*a. Sieve Analysis.* Separation of the soil into its fractions may be done by shaking the dry loose material through a nest of sieves of increasing fineness, i. e. successively smaller openings, as indicated in figure 7. The sieve analysis may be performed directly upon soils which contain little or no fines, e. g. a clean sand, or soils in which the fines may be readily separated from the coarser particles. Soils which have little dry strength and can be crushed easily in the fingers would generally fall in the latter category. If the character of the fines is such that the fine material adheres to the coarser particles and is not removed by dry sieving action, the sample is *prewashed* and the fine material removed. Material which is retained on the No. 200 sieve during the washing process is then dried and subjected to sieving as before. Sieves which are commonly used by the military engineer have square openings and are designated as 1-, ¾-, and ¼-inch sieves and U. S. Standard No. 4, 10, 40, 60, 100, and 200 sieves. Sieves which are designated as Nos. 20 and 80 also sometimes are used. The size of openings which corresponds to certain of the above sieve sizes is shown on the plots of figures 8 and 9. A No. 4 U. S. Standard sieve has 4 openings per lineal inch or 16 openings per square inch, etc.

*b. Wet Mechanical Analysis.* The practical lower limit for the use of sieves is the No. 200 sieve, which has openings which are 0.074 mm. square, and a total of 40,000 openings per square inch. However, it is sometimes desirable to determine the distribution of particle sizes below the No. 200 sieve. This may be determined by a process known as *wet mechanical analysis*, which employs the principle of sedimentation, i. e. that grains of different sizes fall through a liquid at different velocities. The wet mechanical analysis is not a normal

*Figure 7. Dry sieve analysis.*

field laboratory test and is not particularly important in military construction except, as will be seen later (par. 100), that the percentage of particles finer than 0.02 mm. has a direct bearing on the susceptibility of a soil to frost action. TB 5–253–1 gives a field method for performing a wet mechanical analysis for the determination of the percentage of material finer than 0.02 mm. The procedure therein described is called *decantation*.

*c. Combined Mechanical Analysis.* The procedures which have been described in *a* and *b* above are frequently combined to give a more complete picture of grain-size distribution. The procedure is then designated as a *combined mechanical analysis*.

*d. Other Methods.* Other methods based on sedimentation are frequently employed in soils laboratories, particularly when it is desired to determine the distribution of particles below the No. 200 sieve. One such method is the hydrometer analysis. If a complete picture of grain-size distribution is desired, it frequently is obtained by a combined sieve and hydrometer analysis. This method will not be described here.

## 18. Methods of Reporting the Results of Mechanical Analysis

The results of a mechanical analysis may be reported in one of two forms, tabular and graphical. The graphical method of presentation results in what is termed a *grain-size distribution curve*.

*a. Tabular Form.* The tabular forms which are illustrated in table I, below, are frequently used in reporting the results of a sieve analysis on a soil which is predominantly made up of coarse particles. Two methods of reporting data on the same soil are shown. Both methods are in common use, and the soils engineer should be familiar with both. The form in table I (Method A), in which the soil is divided into fractions, is frequently used when the soil is being checked for compliance with a standard specification, as for a gravel base or wearing course. Method B is more commonly used in reporting routine sieve analyses on natural soils, when a grain-size distribution curve is not plotted. Sieves other than those included in the table may be used.

*Table I. Tabular forms for reporting results of a sieve analysis on a coarse-grained soil.*

### METHOD A

| | Percent by weight |
|---|---|
| Retained on 1-inch sieve | 0. 0 |
| Passing 1-inch, retained on ¾-inch sieve | 4. 6 |
| Passing ¾-inch, retained on No. 4 sieve | 29. 3 |
| Passing No. 4, retained on No. 10 sieve | 28. 0 |
| Passing No. 10, retained on No. 40 sieve | 20. 0 |
| Passing No. 40, retained on Nc. 100 sieve | 8. 6 |
| Passing No. 100, retained on No. 200 sieve | 3. 5 |
| Passing No. 200 sieve | 6. 0 |
| Total | 100. 0 |

### METHOD B

| Sieve Size | Percent passing (finer than) by weight |
|---|---|
| 1-inch | 100. 0 |
| ¾-inch | 95. 4 |
| No. 4 | 66. 1 |
| No. 10 | 38. 1 |
| No. 40 | 18. 1 |
| No. 100 | 9. 5 |
| No. 200 | 6. 0 |

*b. Grain-size Distribution Curve.* The most commonly used method of presenting the results of a mechanical analysis graphically is shown in figure 8. Limiting particle diameters or sieve sizes are plotted horizontally on the logarithmic scale of a semi-log plot, while the percentage by weight of the material passing any given sieve size, or smaller (finer) than any given grain size, is plotted vertically on the arithmetic scale. The curve which is shown in figure 8 was obtained by plotting a point corresponding to each of the tabular values given in table I (Method B). These points were then connected to form a smooth curve, the grain-size distribution curve. The percentage corresponding to a particle size of 0.02 mm. was determined by wet mechanical analysis. This method of reporting has two principal advantages. First, it permits ready visualization of the distribution of particle sizes, or *gradation*. Second, it permits the determination of percentages which correspond to particle sizes which were not included in the original mechanical analysis.

## 19. Gradation and Associated Factors

As has been previously indicated, the distribution of particle sizes in a soil is known as its gradation. Considerations relative to gradation and associated factors are discussed in this paragraph. The discussion is primarily applicable to coarse-grained soils.

  *a. Effective Size and Uniformity Coefficient.* The grain size which corresponds to 10 percent on a grain-size distribution curve of the type of figure 8 is called Hazen's *effective size* and is designated by the symbol $D_{10}$. For the soil shown $D_{10}$ is 0.17 mm. As will be indicated later (par. 86), the effective size of clean sands and gravels can be related to their permeability. The *uniformity coefficient* is defined as the ratio between the grain diameter corresponding to 60 percent on the curve $(D_{60})$ and $D_{10}$. Hence, $C_u = D_{60}/D_{10}$. For the soil shown, $D_{60} = 4.0$ mm. and $C_u = 4.0/0.17 = 23.5$. The uniformity coefficient is used in judging gradation, as is indicated in the following discussion.

  *b. Coefficient of Gradation.* Another quantity which may be used to judge the gradation of a soil is the *coefficient of gradation*, $C_g$. It is given by the expression:

$$C_g = \frac{(D_{30})^2}{D_{60} \times D_{10}}, \text{ in which}$$

$D_{60}$ and $D_{10}$ have meanings previously given, and $D_{30}$ is the grain diameter corresponding to 30 percent on the grain size distribution curve. For the soil of figure 8, $D_{30} = 1.2$ mm. and $C_g = 2.1$.

  *c. Well-Graded Soils.* The grain-size curves for four different soils are shown in figure 9. Two of these soils, the ones which are labeled GW and SW, are well graded. The symbols GW, SW, GP, and SP shown on the drawing refer to groups within the Unified Soil Classification System, which is described in the next chapter. *Well-graded soils* are those which have a reasonably large spread between the largest and smallest particles, and have no marked deficiency in any one size. The grain-size curve of such a soil is marked by its smooth-

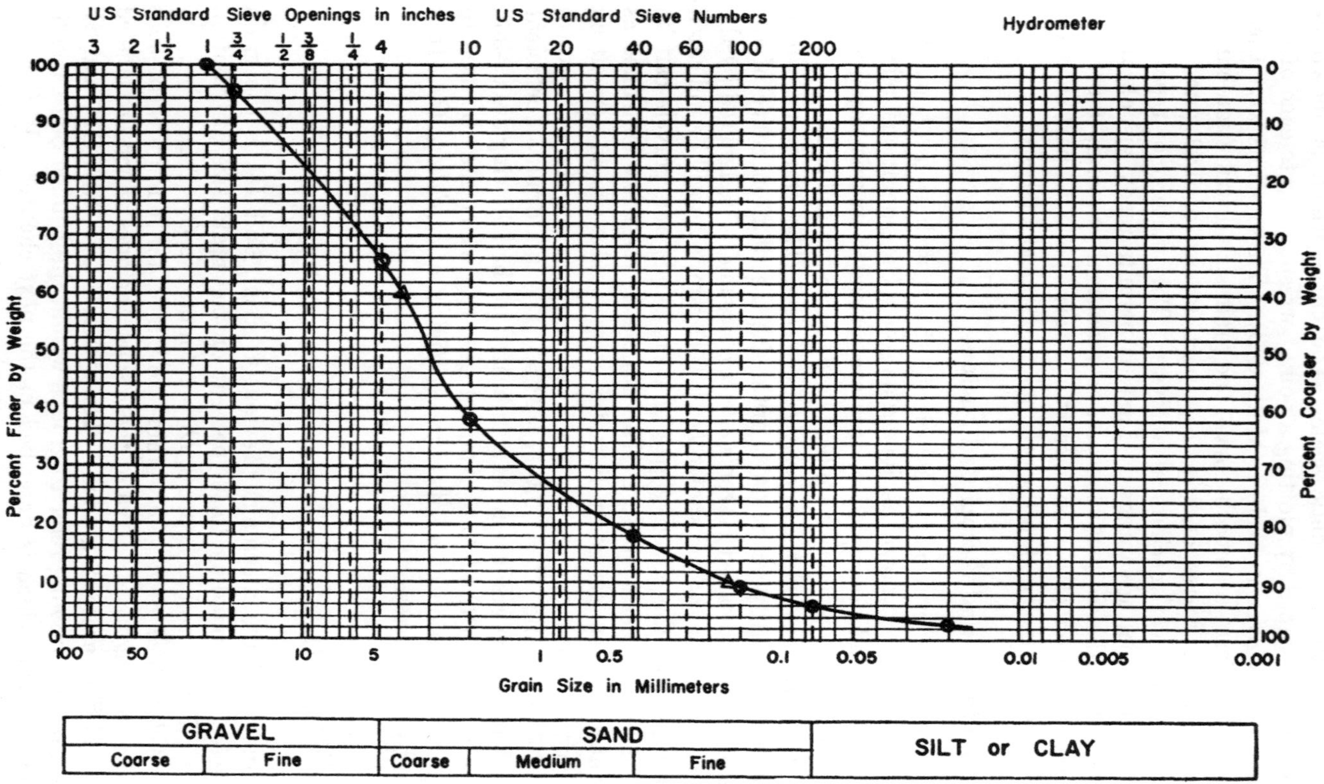

*Figure 8.  Grain-size distribution curve.*

ness and gradual changes in slope. In the Unified Soil Classification System, gravels must have a value of $C_u$ greater than 6; sands, greater than 4. Both gravels and sands must have a value of $C_g$ between 1 and 3 to be well graded. Gravels and sands which do not meet these criteria are termed *poorly graded*.

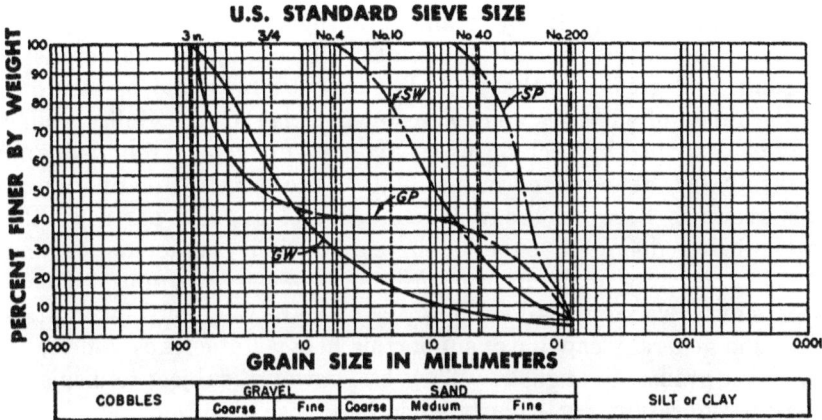

*Figure 9. Typical well-graded and poorly graded soils.*

*d. Poorly Graded Soils.* There are two general types of poorly graded coarse soils, *uniform* and *skip-graded* materials.

    (1) Uniform soils. The curve which is marked SP in figure 9 is poorly graded, uniform soil. By uniform is meant that the particles are nearly all the same size, i. e. the difference in size between the largest and smallest grains is small. The slope of the grain-size curve of such a soil is characteristically steep. Beach sands are typical of uniform soils.

    (2) Skip-graded soils. The soil which is represented by the symbol GP in figure 9 is also poorly graded. This material falls in the category of skip- or gap-graded soils. By this is meant that there is a deficiency in one size of particles. This gap in gradation produces a characteristic hump in the grain size curve. The coefficient of uniformity may not disclose a skip-graded soil.

## 20. Relation of Gradation to Density

Coarse materials which are well graded are generally preferable from an engineering standpoint, since good gradation is generally indicative of high density. This is because smaller particles are available to fill the void spaces between the larger grains. With soils of this type, high density is accompanied by high stability, or ability to support load.

## 21. Particle Shape

The shape of soil particles is also an important property, being related to their arrangement within the soil mass and to such characteristics as stability, compressibility and plasticity. Three principal shapes of soil grains have been recognized; these are bulky, flaky or scale-like, and needle-like. The last of these occurs only rarely and will not be further discussed here.

*a. Bulky Grains.* Bulky grains are those which have approximately equal length, height, and width. This shape is characteristic of sands and gravels, although bulky particles may also be included in the range of silt sizes and, occasionally, clay sizes. Bulky grains may be described by such terms as angular, subangular, rounded, and well rounded, as indicated in figure 10. The principal distinction in this group is between the well-rounded and angular particles, the latter being definitely superior in compaction characteristics and stability, because of the internal friction between the grains. Dense soils which are composed of bulky grains are relatively incompressible and generally capable of supporting heavy static loads, especially when the grains are angular. They are, however, subject to displacement by vibration, particularly when in a loose condition.

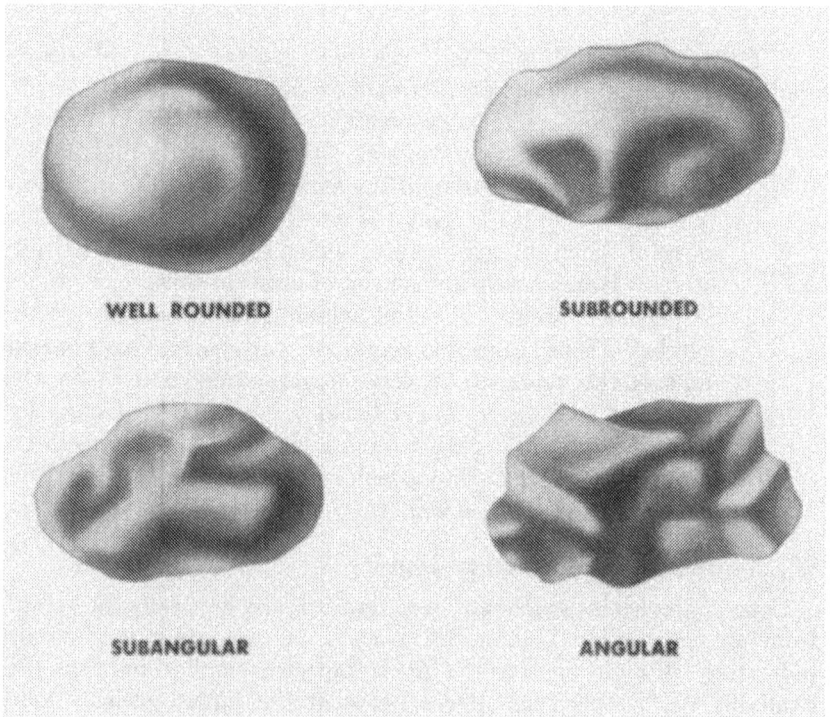

WELL ROUNDED          SUBROUNDED

SUBANGULAR          ANGULAR

*Figure 10. Bulky grains.*

*b. Flaky Grains.* Flaky grains are extremely thin as compared with their length and width. They have the general shape of a flake of mica, or a sheet of paper, as illustrated in figure 11. Some coarse particles, particularly those which have been formed by the mechanical breakdown of mica, are flaky or scale-like in shape. However, the most important fact is that the large majority of particles which fall in the range of clay sizes, including the so-called clay minerals, have this characteristic shape. As will be explained in more detail later, it is the presence of these extremely small flaky grains which is generally responsible for the plasticity of clay; this type of soil is also highly compressible under static load. The dimension which appears on the photograph of figure 11 is 1 micron, which is 0.001 mm

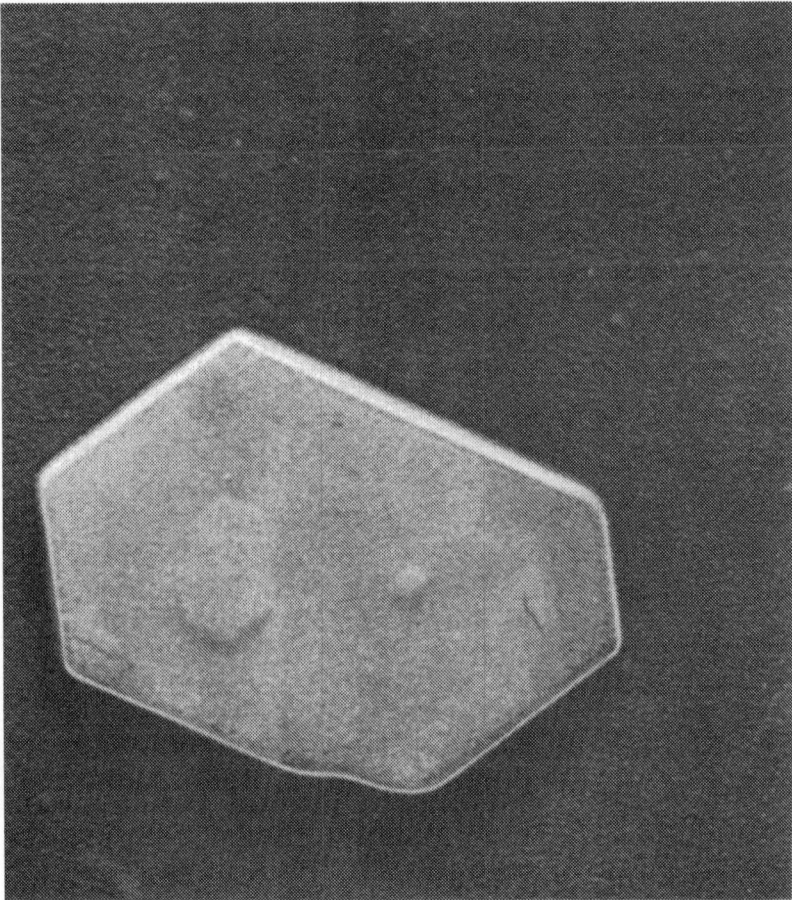

*Figure 11. Electron photomicrograph of a clay particle (kaolinite).*

## 22. Specific Gravity

The specific gravity of a homogeneous substance may be defined as the ratio between the weight per unit of volume of the material and the weight per unit of volume of water at 4° C. The specific gravity of the solids, i. e. the soil particles, which are contained in a soil mass is designated by the symbol G. The specific gravity of a soil generally is determined by measuring the volume of water displaced by a known dry weight of soil particles. The procedure is relatively simple for coarse soils, such as sand and gravel, and more difficult for fine-grained soils, like clay. Appropriate test procedures are contained in TB 5–253–1. The test is not too important in itself. It is necessary that the specific gravity of solids be known in order to make certain of the calculations which are presented in later portions of this section. The specific gravity of a given soil is largely dependent upon the specific gravity of the minerals from which it was derived. For most inorganic soils the specific gravity will be in the range of 2.60 to 2.80; many soils of this general nature have specific gravities between 2.65 and 2.75. The value of the specific gravity will be outside of this range if the soil was derived from either unusually light or unusually heavy minerals.

## 23. Volume and Weight Relationships

  a. *General.*
    (1) Inorganic soils commonly have what may be termed three-phase composition, the principal ingredients being the soil particles, water and air. In certain circumstances either the air or the water may be removed, or essentially so, as will be indicated in the following paragraphs. Organic material may be present in appreciable quantity in some soils.
    (2) The purpose here is to present certain volume and weight relationships which are in common use in soil engineering. In a sense, these relationships constitute an alphabet of soil engineering, since they are frequently used in presenting both qualitative and quantitative information relative to soil properties. Many examples of the usefulness of these concepts will be presented in later portions of the manual. For example, the curve of figure 37 shows the way in which the void ratio varies with the load applied to a clay soil. In figure 59 is shown the relationship between moisture content and dry unit weight for a typical soil subjected to dynamic compaction.

  b. *Simplified Sketch Showing Volumes and Weights.* Attention may now be given to the simplified sketch of figure 12. In this sketch the total volume is that occupied by an actual soil mass. The volume shown for each of the principal constituents is the same as that occupied by that material in the actual mass. Similarly, the weights are the same. The sketch does not attempt to show the arrangement

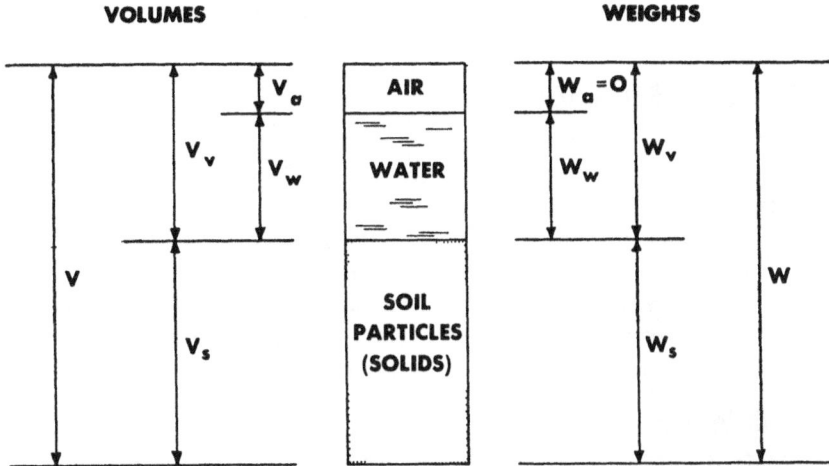

VOLUMES WEIGHTS

Figure 12. Simplified sketch indicating volumes and weights of materials in a typical inorganic soil.

of the three principal ingredients in the actual mass, only their weights and volumes.

## 24. Void Ratio and Porosity

The *void ratio, e,* is defined as the ratio between the volume of voids and the volume of solids. In terms of the sketch, $e = V_v/V_s$. It is always expressed as a decimal and may have a value greater than one. The numerical value of $e$ may range from as low as 0.30 or so for a very dense, well-graded granular material to 2.0 and higher for certain clay soils. *Porosity, n,* is the ratio, always expressed as a percentage, between the volume of voids and the total volume of the soil mass. From the sketch, $n = (V_v/V)$ 100. The porosity is much less frequently used than the void ratio.

## 25. Moisture Content

Moisture is an extremely important constituent of soils and may be present in several forms, as will be indicated as the discussion proceeds. The *moisture (or water) content, w,* is defined as the ratio expressed as a percentage, between the weight of water which is contained in a soil mass and the weight of soilds. In terms of the sketch of figure 12, $w = (W_w/W_s)100$. The moisture content may exceed 100 percent. By definition, when a soil mass is dried to constant weight in an oven maintained at a temperature between 100° C and 110° C, $w = 0\%$ and the soil is said to be oven-dry, or just *dry.* As defined here and as commonly used in soil engineering, the moisture content is expressed on a dry-weight basis. If a soil mass is air dry, as it might be if allowed to remain in an open pan in a laboratory for several hours or several days, the corresponding moisture content is called *hygroscopic moisture*. Suitable test procedures for determining the moisture content of a soil are given in TB 5–253–1.

## 26. Degree of Saturation

The *degree of saturation*, $S_r$, which is always expressed as a percentage, gives the relationship between the volume of water and the volume of voids. $S_r = (V_w/V_v)100$. In other words, it indicates the proportion of the voids in a soil mass which are filled with water. When $S_r = 100\%$, meaning that the voids are completely filled with water and no air is present in the soil, the soil is said to be *saturated*. A value of $S_r = 0\%$ corresponds to an oven-dry condition. If the value is between 0 and 100%, the soil is frequently said to be *partially saturated*.

## 27. Unit Weight

*a. Unit Weight* is the ratio between the weight of a soil mass and the corresponding volume. From figure 12, $\gamma = W/V$. In the English system unit weight is expressed in pounds per cubic foot; in the metric system, in grams per cubic centimeter. It is important to note that, in soils terminology, the terms unit weight and density are commonly used interchangeably. In the civil practice of highway engineering the term *wet unit weight* or *wet density* is applied if the moisture content is anything other than zero. If $w = 0\%$ (oven-dry) then the term *dry unit weight* or *dry density* is used; it is designated by the symbol $\gamma_d$. Referring to figure 12, $\gamma_d = W_s/V$. The general relationship between dry unit weight and wet unit weight is $\gamma = \gamma_d (100+w)/100$. This relationship is of some importance, since it is frequently desirable to change from dry unit weight to wet unit weight, and vice versa.

*b.* In foundation problems the density of a soil is usually expressed in terms of the wet unit weight, while in construction the dry unit weight basis is generally preferred. In some cases it may be more desirable to express density in terms of void ratio rather than unit weight. The term dry unit weight must not be confused with the unit weight of solids, $\gamma_s$. By unit weight of solids is meant the unit weight of the soil particles themselves, considered as a voidless mass. In figure 12, $\gamma_s = W_s/V_s$. In the metric system $\gamma_s$ is numerically equal to the specific gravity. In the English system $\gamma_s = G\gamma_w$, where $\gamma_w$ is the unit weight of water. The unit weight of (fresh) water is commonly taken to be 62.5 pounds per cubic foot. Test procedures for the determination of the wet and dry unit weights are contained in TB 5–253–1.

*c.* Wet unit weights of natural soils in place vary widely. For example, a sandy soil may have a unit weight of 115 to 135 pounds per cubic foot, depending on its gradation and whether it is in a loose or dense condition. Some very dense glacial tills have unit weights as high as 145 pounds per cubic foot. Most clays have unit weights ranging from 100 to 125 pounds per cubic foot in place, while soft organic soils may weigh less than 100 pounds per cubic foot. The density of most soils can be greatly increased during construction by compaction, as indicated in chapter 9.

## 28. Numerical Example of Volume-Weight Relationships

A simple example will serve to show the relationships among the various quantities defined above, and the way in which they may be calculated—

Assume that a soil has a wet unit weight of 125.0 pounds per cubic foot and a moisture content of 18.0%. The specific gravity of the soil particles is 2.65. Determine the dry unit weight, void ratio, porosity, and degree of saturation.

The quantities given in the problem are $\gamma=125$ pounds per cubic foot, $w=18.0\%$, and $G=2.65$. Assume that $V=1$ cubic foot. From paragraph 27 above,

$$\gamma_d = 100\gamma/(100+w) = 125.0/1.18$$

$\gamma_d = 105.9$ pounds per cubic foot

$$V_s = \frac{105.9}{2.65(62.5)} = 0.64 \text{ cubic foot}$$

$V_v = 1.00 - 0.64 = 0.36$ cubic foot

By definition, $e = \dfrac{0.36}{0.64} = 0.56$,

and $n = \dfrac{0.36(100)}{1} = 36.0\%$.

$$W_w = W - W_s - 0(W_a) = 125.0 - 105.9,$$

$W_w = 19.1$ pounds.

$$V_w = \frac{19.1}{62.5} = 0.31 \text{ cubic foot},$$

and $S_r = \dfrac{0.31}{0.36} \times 100 = 86\%$.

Thus, all the desired quantities have been solved for. In general, the approach which has been illustrated and which is based upon the fundamental definitions is the best and most usable, although various equations have been derived giving relationships among the various quantities.

## 29. Relative Density

As is discussed in the next paragraph, coarse-grained soils may exist in nature at a wide range of void ratios or densities. The value of the void ratio itself is not very effective in showing what the behavior of a soil will be, since one soil may be in a loose condition at a given value of the void ratio, while another may be dense. Much more useful in this respect is the term, *relative density*, which is given by the equation,

$$D_d = \frac{e_{max.} - e}{e_{max.} - e_{min.}},$$

where $e_{max}$ is the void ratio in the loosest possible condition, $e$ is the actual void ratio, and $e_{min.}$ is the void ratio in the most dense condition possible. Because of the difficulty of making a direct measurement of the void ratio of a given soil in the field, the relative density is usually measured indirectly. If necessary, laboratory procedures may be devised to determine $e_{max.}$ and $e_{min.}$, while $e$ may be determined by weight, volume and moisture content determinations upon an undisturbed sample of soil The relative density is of significance primarily for gravels and sands, while it is of much less importance for soils which contain appreciable amounts of silt and clay. The value of $D_d$ may be close to zero for a very dense soil and close to one for a very loose one.

## 30. Soil Structure

By *soil structure* is meant the arrangement of soil particles within the soil mass. There are three principal types of structure which are of consequence; single-grain, honeycomb and flocculent. Soils having each of these basic structures present somewhat different foundation and construction problems.

## 31. Single-Grain Structure

*a. General.* Soils which are composed of bulky particles, gravels, sands and most silts, have *single-grain* or *granular* structure. In general terms, this means that each soil grain is in contact with those around it, as illustrated in figure 13. Even though each grain is in contact with surrounding grains, still there may be considerable difference in density depending on the arrangement of the individual grains. This is illustrated by analogy in figure 13, in which spheres of uniform size are arranged in two different patterns. One arrangement represents the loosest one possible with all the spheres in contact with one another; the void ratio in this condition would be about 0.91. The other arrangement of spheres represents the most dense possible arrangement with uniform spheres; here the void ratio would be about 0.35. For most uniform soils, the void ratio in the dense condition is usually higher than 0.35 because of the angularity of the grains. Gradation has a great effect upon the density of the soils which have single-grain structure, since the smaller particles tend to fill the voids between the larger ones. This is also illustrated in figure 13. Well-graded soils have smaller minimum and maximum values of the void

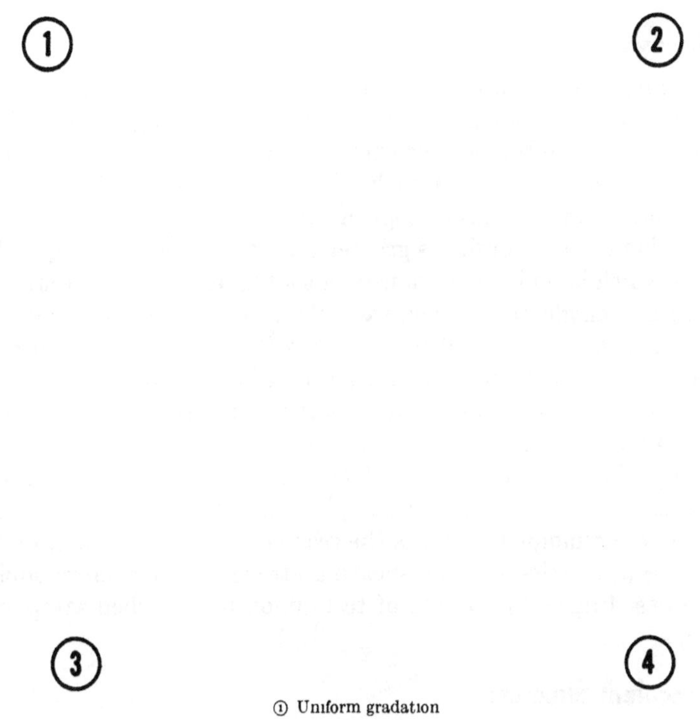

① Uniform gradation
② Uniform spheres, loose condition
③ Uniform spheres, dense condition
④ Well-graded soil

*Figure 13. Single-grain soil structures.*

ratio than do uniform soils. Typical values of the void ratio for various soils with single-grain structure are given in table II, below.

*b. Engineering Significance.* Loose, single-grain structures are inherently unstable, being particularly susceptible to shock or vibration. Soils which have dense, single-grain structures are basically stable and are generally able to withstand static load or vibration with relatively small decreases in void ratio. Coarse soils which have this structure in their natural state may, generally speaking, be disturbed and recreated in the laboratory with essentially the same properties as the natural soil.

**27**

| | Void ratio | |
|---|---|---|
| | Dense | Loose |
| Uniform sand_ _ _ _ _ _ _ _ _ _ _ _ _ _ _ _ _ _ _ _ _ _ _ _ _ _ _ _ _ _ _ _ _ _ _ _ _ _ _ _ _ | 0. 50 | 0. 85 |
| Well-graded sand_ _ _ _ _ _ _ _ _ _ _ _ _ _ _ _ _ _ _ _ _ _ _ _ _ _ _ _ _ _ _ _ _ _ _ _ _ _ | . 30 | 67 |
| Sandy, clayey gravel_ _ _ _ _ _ _ _ _ _ _ _ _ _ _ _ _ _ _ _ _ _ _ _ _ _ _ _ _ _ _ _ _ _ _ | . 25 | _ _ _ _ _ _ _ _ |
| Micaceous sand_ _ _ _ _ _ _ _ _ _ _ _ _ _ _ _ _ _ _ _ _ _ _ _ _ _ _ _ _ _ _ _ _ _ _ _ _ _ | . 80 | 1. 20 |

## 32. Honeycomb Structure

When fine, bulky grains of soil settle out in water they may build up the honeycomb structure which is illustrated in figure 14.   This structure is built up because, as each particle settles to the bottom, it tends to adhere to the grain with which it makes contact.   This is because, for fine grains in the range of silt sizes and below, the molecular attraction between grains is greater than the force of gravity.   The structure which is built up obviously has a higher void ratio than that occurring in a single-grain structure.   Typical values of the void ratio may range from perhaps 1.0 to 1.5.   A soil of this sort is also more compressible.   If the honeycomb structure is broken down the shearing resistance of the material is apt to be appreciably reduced; in extreme cases it may behave like a heavy liquid and present very difficult construction problems.   Many natural silt and clay soils are sensitive to disturbance of their natural structure, i. e. their properties are changed when they are remolded.   This is the reason why the determination of engineering properties such as shearing strength and compressibility must be based upon the results of tests upon undisturbed samples of these soils.

## 33. Flocculent Structure

The very fine particles which fall in the lower range of the clay sizes are colloidal in size and behave like colloids.   One of their characteristics is that they may remain in suspension indefinitely in certain circumstances; this is the so-called Brownian movement.   In this situation the particles carry like electrical charges and hence repel one another.   If conditions change and some of the particles then carry a neutral electrical charge, e. g. in the presence of sea water, which is an electrolyte, the particles will be attracted to one another and form *flocs*.   Each floc is made up of a group of very small particles and in itself has a very high void ratio.   When a floc becomes large enough it will settle out, along with comparatively large bulky grains which may still be in suspension.   The resultant structure may be highly complex, have a very high void ratio, and be highly compressible.   The general process of formation of a flocculent structure is illustrated in figure 14.   The engineering properties of soils which have flocculent structures vary somewhat, depending on the exact process of formation and the history of the deposit. All of them are compressible, and nearly all of them are sharply sensitive to remolding.   For example, a typical marine clay may be quite stiff and brittle in its natural state, and yet become very soft and sticky when its natural structure is disturbed.

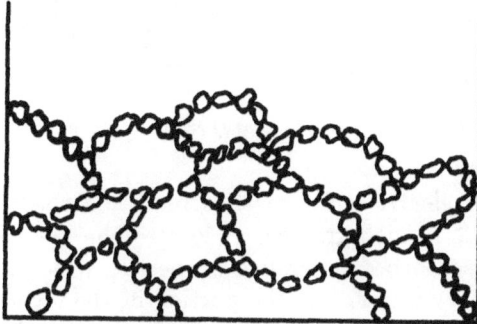

① Honeycomb structure.

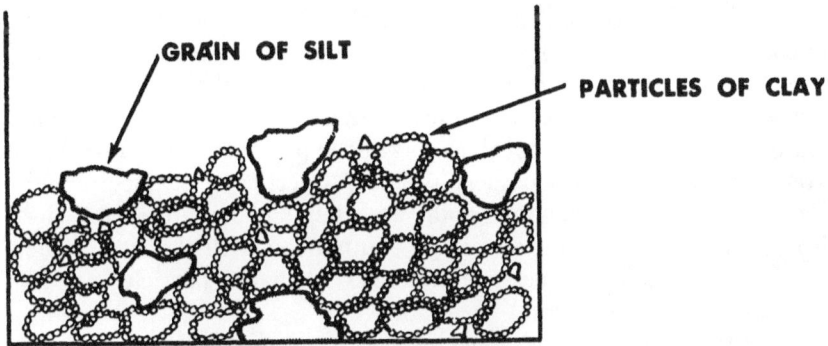

② Flocculent structure.

*Figure 14.   Honeycomb and flocculent soil structure.*

## 34. Adsorbed Water

*a.* In general terms, *adsorbed water*, is the water which may be present in the form of thin films which surround the separate soil particles.   When the soil is in an air-dry condition, the adsorbed water which is present is called hygroscopic moisture (par. 25).   The presence of adsorbed water is attributed to the fact that soil particles carry a negative electrical charge.   Water is dipolar and is attracted to the surface of the particle, and bound to it, as indicated in figure 15. The water films are affected by the chemical and physical structure of the soil particle and its relative surface area.   The relative surface area of a particle of fine-grained soil, particularly if it has a flaky or needle-like shape, is much greater than for coarse soils composed of

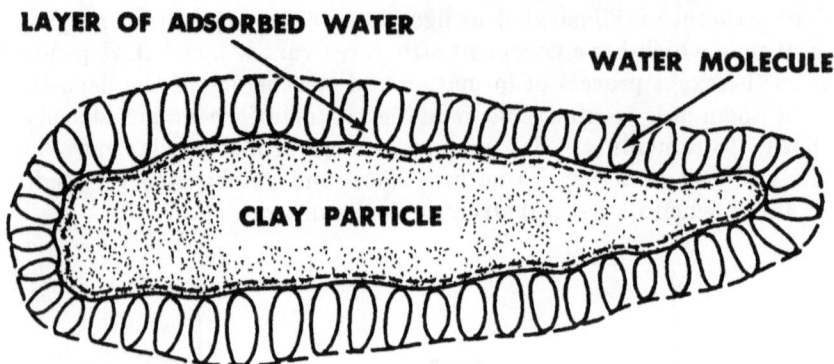

*Figure 15. Layer of adsorbed water surrounding a particle of soil.*

bulky grains; the electrical forces that bind adsorbed water to the soil particle also are much greater. Close to the particle, the water contained in the adsorbed layer has properties which are quite different from ordinary water. In the portion of the layer immediately adjacent to the particle the water may behave as a solid, while only slightly farther away it behaves as a viscous liquid. The total thickness of the adsorbed layer is very small, perhaps on the order of 0.00005 mm. for clay soils. In coarse soils the adsorbed layer is quite thin in comparison with the thickness of the soil particle. This, coupled with the fact that the contact area between adjacent grains is quite small, leads to the conclusion that the presence of the adsorbed water has little effect upon the physical properties of coarse-grained soils. By contrast, for finer soils and particularly in clays, the adsorbed water film is quite thick in comparison with the size of the particles. The effect is very pronounced when the particles are in the range of colloidal sizes.

*b.* Mention also may be made of the term *absorbed water.* This term is not widely used in soil engineering, but is used in concrete technology to describe the water which actually enters into the surface voids of aggregate particles.

## 35. Plasticity and Cohesion

Two very important aspects of the engineering behavior of fine-grained soils are directly associated with the existence of adsorbed water films. These are plasticity and cohesion.

*a.* By plasticity is meant the ability of a soil to deform without cracking or breaking. Soils in which the adsorbed films are comparatively thick, i. e. the clays, are plastic over a wide range of moisture contents. This presumably is due to the fact that the particles themselves are not in direct contact with one another. Plastic deformation can take place because of distortion or shearing of the outside layer of viscous liquid in the moisture films. Coarse soils, e. g. clean sands and gravels, are nonplastic. Silts also are essentially nonplastic materials, since they are usually composed predominantly of bulky grains; if flaky grains are present they may be feebly or slightly plastic.

*b.* Soils which are highly plastic are also cohesive. That is, they possess some *cohesion* or resistance to deformation because of the surface tension present in the water films. Thus, wet clays can be molded into various shapes without breaking and will retain these shapes, even though wet. Gravels, sands and most silts are not cohesive; they are called *cohesionless soils.* Soils of this general class can not be molded into permanent shapes and have little or no strength when dry and unconfined. Some of these soils may be slightly cohesive when damp. This is attributed to what is sometimes called *apparent cohesion*, which is also due to the surface tension in the water films between the grains.

## 36. Clay Minerals and Base Exchange

*a.* The very fine (colloidal) particles of clay soils consist of what are called *clay minerals*, which are crystalline in structure. They are complex compounds of hydrous aluminum silicate and are important because of the fact that the presence of even a small amount of these materials in a soil has a great influence upon its physical properties. By means of X-ray studies several different kinds of clay minerals have been identified which have somewhat different properties. Two of the extreme types are kaolinite and montmorillonite. Both of these have liminated crystalline structures, but they are quite different in behavior. One reason for this difference is that kaolinate has a very rigid crystalline structure while montmorillonites can swell by taking water directly into their lattice structure. Later, the flakes themselves may decrease in thickness as the water is squeezed out during drying; they are thus subject to detrimental shrinkage and expansion. An example of this type of material is seen in bentonite clay, which is largely made up of the montmorillonite type of clay mineral and which is widely used commercially because of its swelling characteristics. Most montmorillonites have much thicker films of adsorbed water than do kaolinites. Kaolinites hold water close to their particles and therefore tend to shrink and swell much less with changes in moisture content.

*b.* In addition to water, the adsorbed film around a soil particle may contain disassociated ions. For example, metallic cations, such as sodium, calcium, or magnesium may be present. The presence of these cations also has an effect on the physical behavior of the soil. For example, a montmorillonite clay in which calcium cations predominate in the adsorbed layer may have properties which are quite different from one of the same general type in which sodium cations predominate. The process of replacing cations of one type with cations of another type in the surface of the adsorbed layer is called *base exchange*. It is possible to effect this replacement and thereby alter the physical properties of a clay soil. For example the plasticity may be reduced or the permeability increased by this general process.

## 37. Capillary Phenomena

*a. General.* Capillary phenomena in soils are important in two general ways. First, the capillary movement of water into a soil from a free water surface. This aspect of capillarity is not discussed here, but is covered in chapter 4. Second, capillary phenomena are closely associated with the *shrinkage* and *expansion* (*swelling*) of soils. These items are discussed in the following paragraphs.

*b. Capillary Phenomena in Small Tubes.* The capillary rise of water in small tubes is a common phenomenon which is caused by surface tension. It is illustrated in figure 16. The water that rises upward in a small tube is in tension, hanging on the curved boundary between air and water (miniscus) as if from a suspending cable. It is important

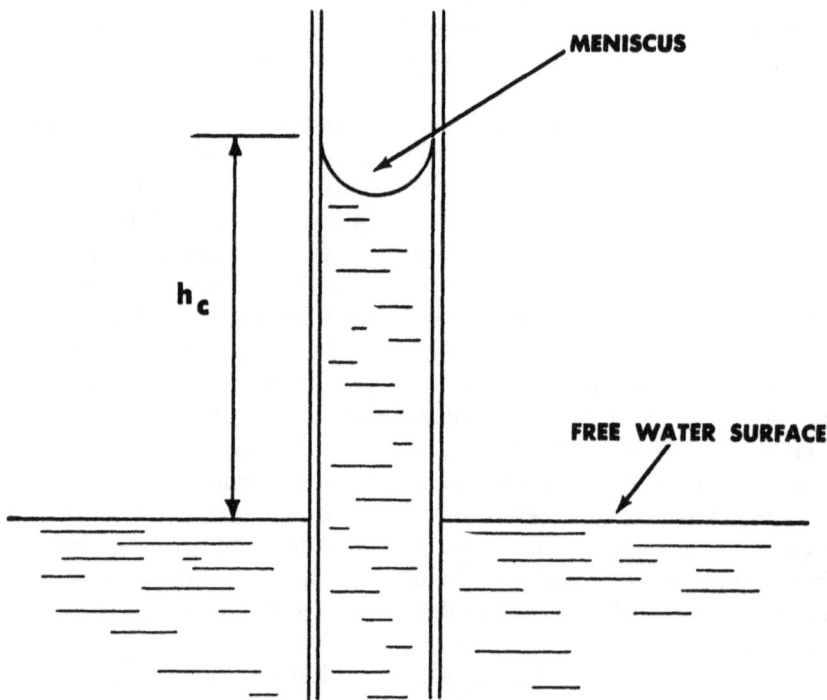

NOTE: FOR WATER AT STANDARD TEMPERATURE

$$h_c = \frac{0.306}{D} \text{ CM., WHERE D IS THE DIAMETER}$$

OF THE TUBE IN CM.

*Figure 16. Capillary phenomena associated with the rise of water in small tubes.*

to note that the tensile force in the miniscus is balanced by a compressive force in the walls of the tube. It is quite simple to analyze capillary phenomena in small tubes and to derive equations for the radius of the curved meniscus, the capillary stress (force per unit of area), and the height of capillary rise ($h_c$ in figure 16). Such expressions are of academic interest only in relation to soils, except as they contribute to general understanding of capillary phenomena. With relation to capillarity a soil mass may be regarded as being made up of a bundle of small tubes, the interconnected void spaces. These spaces form extremely irregular, tortuous paths for the capillary movement of water. An understanding of capillary action in soils is thus gained by analogy. Theoretical analyses indicate that the maximum possible compressive pressure which can be exerted by capillary forces is inversely proportional to the size of the capillary openings.

## 38. Shrinkage

Many soils undergo a very considerable reduction in volume when their moisture content is reduced. The effect is most pronounced when the moisture content is reduced from that corresponding to complete saturation to a very dry condition. This reduction in volume is called *shrinkage* and is greatest in clays. Some of these soils show a reduction in volume of 50 percent or more while passing from a saturated to an oven-dry condition. Sands and gravels, in general, show very little or no change in volume with change in moisture content. An exception to this is the *bulking* of sands, discussed in paragraph 40 below. The shrinkage of a clay mass may be attributed to the surface tension existing in the water films created during the drying process. When the soil is saturated a free water surface exists on the outside of the soil mass and the effects of surface tension are not important. As the soil dries out because of evaporation the surface water disappears and innumerable meniscuses are created in the voids which are adjacent to the surface of the soil mass. Tensile forces are created in each of these boundaries between water and air. They are accompanied by compressive forces which, in a soil mass, act upon the soil structure. In the case of the typical fairly dense structure of a sand or gravel the compressive forces are of little consequence and very little or no shrinkage results. In fine-grained soils, the soil structure is compressible and the mass shrinks. As drying continues the mass attains a certain limiting volume. At this point the soil is still saturated. The moisture content at this stage is called the *shrinkage limit*. Further drying will not cause a reduction in volume but may cause cracking as the meniscuses retreat into the voids. In clay soils the internal forces set up by drying may become very large. The existence of these forces also principally accounts for the rocklike strength of a dried clay mass. Both silt and clay soils may be subject to detrimental shrinkage with disastrous results in some practical situations. For example, the uneven shrinkage of a clay soil may deprive a concrete pavement of the uniform support for which it is designed. Severe cracking or failure may result when wheel loads are applied to the pavement.

## 39. Swelling and Slaking

If water is again made available to a soil mass which has undergone shrinkage but is still saturated, it will enter the voids of the soil mass from the outside and reduce or destroy the internal forces previously described. Thus, a clay mass will absorb water and expand or *swell*. The expansion force created as the water enters the soil may be very large. If the expansion is prevented, as by the weight of a concrete pavement, the expansion force may be sufficient to cause severe cracking of the pavement. If water is made available to the soil after the moisture content is below the shrinkage limit, the mass will generally simply disintegrate or *slake*. The phenomenon of slaking may be observed by putting a dry piece of clay into a glass of water. The mass will fall completely apart, usually in a matter of minutes. Construction problems associated with shrinkage and expansion are generally solved by removing the soils which are subject to these phenomena to a detrimental degree or by taking steps to prevent excessive changes in moisture content.

## 40. Bulking of Sands

*Bulking* means the increase in volume which may occur in a moist sand, as compared to dry sand, when the moist sand is disturbed and replaced in a loose condition. Meniscuses form at the points of contact of the grains and the forces of surface tension tend to hold the grains in a loose condition. The bulking of damp sands is the principal reason why sands to be used in concrete mixtures are measured by weight in modern construction practice rather than by volume. The meniscuses can be destroyed by submersion in water, which is the reason why sands used in a fill are sometimes flooded and vibrated to produce a dense structure before a pavement or a structure is placed upon them. The problem is of less practical consequence now than formerly, because of the improvements which have been made in compaction procedures and equipment. (ch. 9).

## 41. Consistency Limits of Fine-Grained Soils

*a. General.* As we have seen, a fine-grained soil can exist in any one of several different states, depending upon the amount of water which is present in the soil. The boundaries between the different states in which a soil may exist are moisture contents which are called *consistency limits*. They are also called *Atterberg limits* after the Swedish soil scientist who first defined them more than 40 years ago. Thus, the *liquid limit* is the boundary between the liquid and plastic states. Above the liquid limit the soil is presumed to behave as a liquid. The *plastic limit* is the boundary between the plastic and semi-solid states. The numerical difference between the liquid limit and the plastic limit is called the *plasticity index*. It is the range of moisture content over which the soil is in a plastic condition. The *shrinkage limit* is the boundary between the semi-solid and solid states. The Atterberg limits are important index properties of fine-grained soils. They are particularly important in classification and identification, as will be seen in the discussion of classification systems in the next chapter. They are also widely used in specifications to control the properties and behavior of soil mixtures.

*b. Test Procedures.* The limits are defined by more or less- arbitrary and standardized test procedures which are performed upon the portion of the soil which passes the No. 40 sieve. This portion of the soil is sometimes called the *soil binder*. Detailed test procedures to be used in determining the liquid limit and the plastic limit are presented in TB 5–253–1. The tests are performed with the soil in a disturbed condition.

## 42. Liquid Limit

In general terms the *liquid limit*, $w/_L$ or L. L., may be defined as the minimum moisture content at which the soil will flow upon the application of a very small shearing force. In the laboratory the liquid limit is usually determined by the use of the mechanical device which is shown in figure 17. A sample of the soil under examination

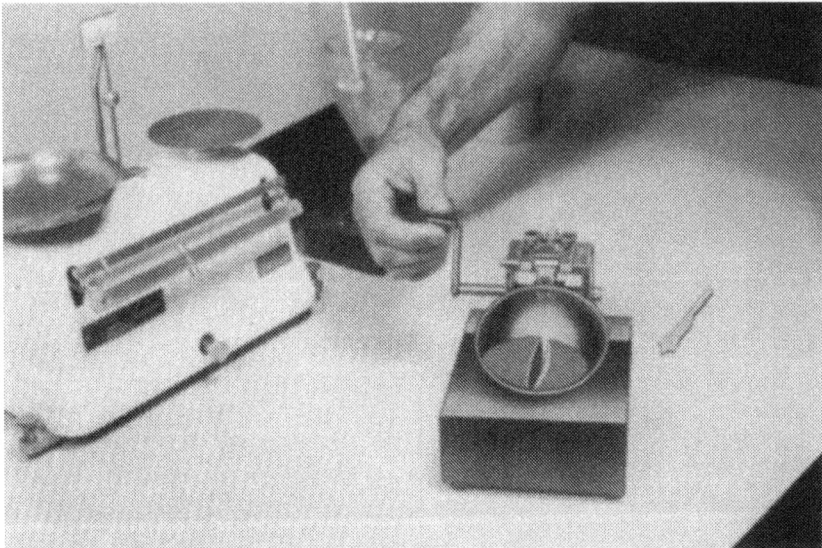

*Figure 17. Liquid limit test.*

is mixed with water, placed in a standard fashion in the brass cup and grooved with the special tool shown in the figure. The crank of the device is then turned at a standard rate. Each turn of the crank raises the cup from the hard rubber base, and then allows it to fall through a distance of one centimeter. The number of blows corresponding to closing of the groove over a distance of one-half inch is determined, and the moisture content measured. By definition, the liquid limit is the moisture content at which the groove will just close upon application of 25 blows. A number of trials are made with the soil at moisture contents both above and below the liquid limit, and a plot made to determine $w_L$. High values of the liquid limit indicate that the soil is highly compressible.

## 43. Plastic Limit

The *plastic limit*, $w_p$ or P. L., is rather arbitrarily defined and is the lowest moisture content at which the soil can be rolled into a thread ⅛ inch in diameter without crumbling or breaking. If a cohesive soil has a moisture content above the plastic limit, a thread may be rolled to less than ⅛-inch diameter without breaking. If the moisture content is below the plastic limit, the soil will crumble before this diameter can be reached   When the moisture content is just equal to the plastic limit, a thread can be rolled out by hand to ⅛-inch diameter and then will crumble or break into pieces. This procedure is indicated in figure 18. Some soils, e. g. clean sands, are non-plastic and the plastic limit can not be determined.

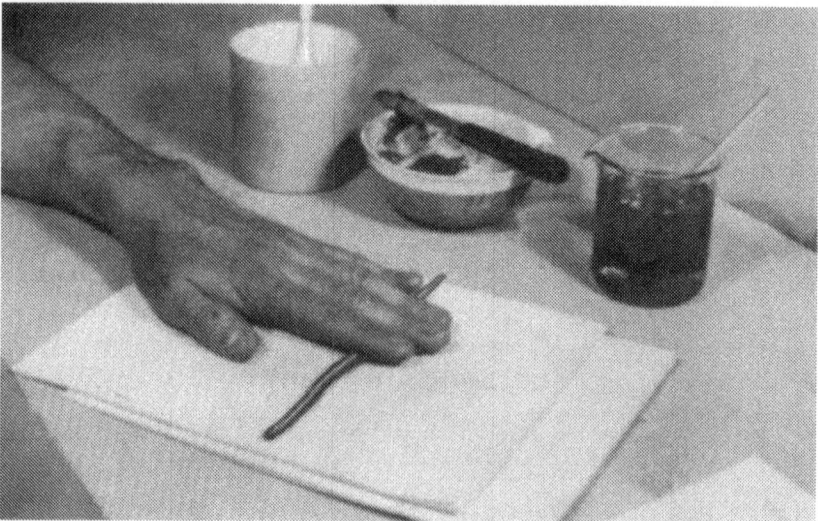

*Figure 18.   Plastic limit test.*

## 44. Plasticity Index

The *plasticity index* of a soil ($I_p$ or P. I.) is the numerical difference between the liquid and plastic limits. For example, if a soil has a liquid limit of 57.3 percent and a plastic limit of 21.3 percent, then the plasticity index, $I_p = w_L - w_p$, or $57.3 - 21.3 = 36$. Sandy soils and silts have characteristically low values of the plasticity index, while most clays have higher values. Soils which have high values of this index are highly plastic and, in general, are highly compressible; they also are highly cohesive. The plasticity index is inversely proportional to the permeability (par. 28) of a soil. Soils which do not have a plastic limit, such as clean sands, are reported as having a P. I. of zero.

## 45. Shrinkage Limit

The *shrinkage limit, $w_s$,* may be defined as the moisture content which corresponds to the attainment of minimum volume while the voids of a soil still are filled with water during the process of drying. This limit is not widely used among soil engineers at the present time, and for that reason will not be discussed further here.

## 46. Plasticity Chart

Relationships between the liquid limits and plasticity indexes of many soils were studied by Arthur Casagrande of Harvard University and lead to the development of the plasticity chart which is shown in table V, appendix II. The development of this chart and its use in classifying and identifying soils will be discussed in the next chapter.

# CHAPTER 3
# SOIL CLASSIFICATION

## Section I.  INTRODUCTION

### 47. Objectives of Soil Classification

The principal objective of any soil classification system is to be able to predict the engineering properties and behavior of a given soil upon the basis of a few simple tests which are performed in the laboratory or the field upon relatively small disturbed samples.   Using the results of the laboratory and/or field examination, the soil is identified and put into a group of similar soils, all of which have similar characteristics and properties.   It is doubtful if any existing classification system completely achieves the stated objective because of the large number of variables involved in soil behavior and the variety of soil problems encountered in engineering practice.   Considerable progress has been made toward this goal, particularly in relation to soil problems encountered in highway and airport engineering.   Soil classification should not be regarded as an end in itself, but rather as a tool to further the engineer's knowledge of soil behavior.   A detailed knowledge of existing classification systems is extremely useful in this respect.

### 48. Development of Classification Systems

Early attempts to classify soils were primarily based upon grain size.   These are the *textural classification systems*.   In 1908 a system which recognized other factors was developed by Atterberg in Sweden and primarily used for agricultural purposes.   Somewhat later a similar system was developed and used by the Swedish Geotechnical Commission.   In the United States the Bureau of Public Roads system was developed in the late 'twenties and was in widespread use by highway agencies by the middle 'thirties.   This system has been modified as time passed and, in a revised form, is widely used today.   The Airfield Classification System was developed by Professor Arthur Casagrande during World War II and is the basis for the Unified Soil Classification System, which is now used by the Corps of Engineers.   In recent years, as knowledge relative to soils has spread among engineers and soil problems have received increasing attention, a number of other systems have been developed and used by other agencies. Many times soils are classified informally by engineers who are concerned with a particular soil problem.

## 49. Scope of This Chapter

Attention will be focused primarily upon the Unified Soil Classification System in this chapter, as this is the system with which the military engineer should be most familiar. This system is particularly adapted to the field identification of soils, and this subject will be discussed thoroughly. Detailed attention will be given also to the revised Public Roads System, while the elements of the classification system used by the Civil Aeronautics Administration will also be covered. Brief mention will be given to agricultural and geological systems.

## Section II. UNIFIED SOIL CLASSIFICATION SYSTEM

### 50. Introduction

*a. Historical Development.* As has previously been stated, this system is based upon the Airfield Classification System developed by Professor A. Casagrande of Harvard during World War II. The system was modified slightly and became the Department of the Army Uniform Soil Classification System. In turn this system was slightly modified and adopted by the Corps of Engineers and the Bureau of Reclamation in January, 1952 under the name of the *Unified Soil Classification System.*

*b. General Nature.* As is the case with the majority of engineering soil classification systems, the Unified Soil Classification System recognizes the fact that natural soils seldom exist separately as sand, gravel, or any other single component, but are usually mixtures. Each component contributes its characteristics to the mixture. The Unified System is based primarily upon those characteristics of a soil which determine how it will behave as an engineering construction material.

*c. Master Chart.* Table V, appendix II, is a master chart which presents information which is applicable to the Unified Soil Classification System, and procedures which are to be followed in identifying and classifying soils under this system. Principal categories which are shown in the chart include soil groups, group symbols and typical soil names; laboratory classification criteria; field identification procedures; and information required for describing soils. Much of the information which is included in table V, appendix II, and some additional facts of importance are included in the discussions which follow.

## 51. Major Categories

In the Unified Soil Classification System all soils are divided into three major categories: coarse-grained soils, fine-grained soils, and highly organic soils. Coarse-grained soils are defined as those in which more than half the material is larger than (retained on) a No. 200 sieve. Fine-grained soils are those in which more than half the material is smaller than (passes) a No. 200 sieve. The third major category, highly organic soils, is not defined by numerical criteria: these soils are identified by visual and manual inspection.

## 52. Coarse-Grained Soils

*a. General.* The coarse-grained soils are divided into two major divisions: gravels and sands. A coarse-grained soil is classed as a gravel if more than half the coarse fraction (fraction retained on a No. 200 sieve) is retained on a No. 4 sieve. It is a sand if more than half the coarse fraction is smaller than a No. 4 sieve. The symbol G is used to denote a gravel and the symbol S to denote a sand. In general practice there is no clear-cut boundary between gravelly and sandy soils; as far as behavior is concerned, the exact point of division is relatively unimportant. Where a mixture occurs, the primary name is the predominant fraction and the minor fraction is used as an adjective. For example, a sandy gravel would be a mixture containing more gravel than sand. For the purpose of systematizing our discussion, it is desirable to further divide coarse-grained soils into three groups on the basis of the amount of fines (materials passing a No. 200 sieve) which they contain.

*b. Coarse-Grained Soils With Less Than 5 Percent Passing No. 200 Sieve.* These soils may fall into the groups GW, GP, SW, or SP, as follows. Reference is made to paragraph 19 for a discussion of the meaning of the terms well graded and poorly graded.

> (1) *GW and SW groups.* In the GW groups are well-graded gravels and gravel-sand mixtures which contain little or no non-plastic fines. The presence of the fines must not noticeably change the strength characteristics of the coarse-grained fraction, and must not interfere with its free-draining characteritics. The SW group contains well-graded sands and gravelly sands with little or no nonplastic fines. The grain-size distribution curves marked GW and SW in figure 9 are typical of soils included in these groups. Definite laboratory classification criteria have been established to judge if the soil is well graded. For the GW group $C_u$, the uniformity coefficient, must be greater than 6; for the SW group, greater than 4. For both groups $C_g$, the coefficient of gradation, must be between 1 and 3.

(2) *GP and SP groups.* The GP group includes poorly graded gravels and gravel-sand mixtures containing little or no non-plastic fines. In the SP group are contained poorly graded sands and gravelly sands with little or no non-plastic fines. These soils will not meet the gradation requirements established for the GW and SW groups. The grain-size distribution curve marked GP in figure 9 is typical of a poorly graded gravel-sand mixture while the curve marked SP is a poorly graded (uniform) sand.

  c. *Coarse-Grained Soils Containing More Than 12 Percent Passing No. 200 Sieve.* These soils may fall into the groups designated GM, GC, SM, and SC. The use of the symbols M and C is based upon the plasticity characteristics of the material passing the No. 40 sieve. The liquid limit and plasticity index are used in specifying the laboratory criteria for these groups. Reference is also made to the plasticity chart shown in table V, which is based upon established relationships between the liquid limit and plasticity index for many different fine-grained soils. The symbol M is used to indicate that the material passing the No. 40 sieve is silty in character. M is taken from the Swedish *Mo*, which usually designates a fine-grained soil of little or no plasticity. The symbol C is used to indicate that the binder soil is clayey in character.

(1) *GM and SM groups.* Typical of the soils included in the GM group are silty gravels and poorly graded gravel-sand-silt mixtures. Similarly, in the SM group are contained silty sands and poorly graded sand-silt mixtures. Gradation of these materials is not considered significant. For both of these groups, the Atterberg limits must plot below the A-line of the plasticity chart or the plasticity index must be less than 4.

(2) *GC and SC groups.* The GC group includes clayey gravels and poorly graded gravel-sand-clay mixtures. Similarly, SC includes clayey sands and poorly graded sand-clay mixtures. Gradation of these materials is not considered significant. For both of these groups, the Atterberg limits must plot above the A-line with a plasticity index of more than 7.

  d. *Borderline Soils.* Coarse-grained soils which contain between 5 and 12 percent of material passing the No. 200 sieve are classed as borderline and are given a dual symbol, e. g. GW–GM. Similarly, coarse-grained soils which contain more than 12 percent of material passing the No. 200 sieve, and for which the limits plot in the shaded portion of the plasticity chart (table V, app. II), are classed as borderline and require dual symbology, e. g. SM–SC. It is possible, in rare instances, for a soil to fall into more than one borderline zone and, if appropriate symbols were used for each possible classification, the result would be a multiple designation consisting of three or more symbols. This approach is unnecessarily complicated, and it is considered best to use only a double symbol in these cases, selecting the two that are believed to be most representative of the probable behavior of the soil. In cases of doubt, the symbols representing the poorer of the possible groupings should be used. For example, a well-graded sandy soil with 8 percent passing the No. 200 sieve, with LL 28 and a PI of 9, would be designated as SW–SC. If the Atterberg limits of this soil were such as to plot in the shaded portion of the plasticity chart (e. g. LL 20 and PI 5), the soil would be designated either SW–SC or SW–SM, depending on the judgment of the engineer.

## 53. Fine-Grained Soils

The fine-grained soils are not classified on the basis of grain size, but according to plasticity and compressibility. Laboratory classification criteria are based on the relationship between the liquid limit and plasticity index which is designated as the Plasticity Chart in table V, appendix II. This chart was established by the determination of limits for many soils, together with an analysis of the effect of limits upon physical charactersitics. Examination of the chart will show that there are two major groupings of fine-grained soils, These are the L groups, which have liquid limits less than 50, and the H groups, which have liquid limits in excess of 50. The symbols L and H have general meanings of low and high compressibility, respectively. Fine-grained soils are further divided with relation to their position above or below the A-line of the plasticity chart.

a. *ML and MH Groups.* Typical soils of the ML and MH groups are inorganic silts; those of low plasticity in the ML group, others in the MH. All of these soils plot below the A-line. In the ML group are included very fine sands, rock flours, and silty or clayey fine sands with slight plasticity. Loose-type soils usually fall into this group. Micaceous and diatomaceous soils generally fall into the MH group, but may extend into the ML group when their liquid limits are less than 50. The same statement is true of certain types of kaolin clays which have low plasticity. Elastic silts will fall into the MH group.

b. *CL and CH Groups* In these groups the symbol C stands for clay, with L and H denoting low or high liquid limits. These soils plot above the A-line and are principally inorganic clays. In the CL group are included gravelly clays, sandy clays, silty clays, and lean clays. In the CH group are inorganic clays of high plasticity, including fat clays, the gumbo clays of the southern United States, volcanic clays, and bentonite. The glacial clays of the northern United States cover a wide band in the CL and CH groups.

c. *OL and OH Groups.* The soils in these two groups are characterized by the presence of organic matter, hence the symbol, O. All of these soils plot below the A-line. Organic silts and organic silt-clays of low plasticity fall into the OL group, while organic clays plot in the OH zone of the plasticity chart. Many of the organic silts, silt-clays, and clays deposited by the rivers along the lower reaches of the Atlantic seaboard have liquid limits between 40 and more than 100, and plot below the A-line. Peaty soils may have liquid limits of several hundred percent, but will plot well below the A-line.

*d. Borderline Soils.* Fine-grained soils which have limits which plot in the shaded portion of the plasticity chart are borderline cases, and are given dual symbols, e. g. CL–ML. Several soil types exhibiting low plasticity plot in this general region on the chart, and no definite boundary between silty and clayey soils exists.

## 54. Highly Organic Soils, Pt Group

A special classification is reserved for the highly organic soils, such as peat, which have so many undesirable characteristics from the standpoint of their behavior as foundations and their use as construction materials. No laboratory criteria are established for these soils, as they generally can be readily identified in the field by their distinctive color and odor, spongy feel, and frequently fibrous texture. Particles of leaves, grass, branches, or other fibrous vegetable matter are common components of these soils.

## 55. Identification From Laboratory Test Results

In many field situations samples of the soils encountered in the field are obtained during the soil survey (ch. 13) and subjected to testing in the field laboratory. Principal tests are the mechanical analysis, liquid limit, and plastic limit. These tests are used on all soils except those in the Pt group, which are identified by visual inspection. With the percentages of gravel, sand, and fines available from the mechanical analysis, together with the results of the limit tests, where applicable, the soil may be classed into one of the groups of the Unified Soil Classification System by following the diagram of figure 19 from top to bottom. It is believed that the diagram, when viewed in conjunction with table V, appendix II, is self-explanatory. The diagram has been simplified in the portion which applies to the borderline cases in the coarse-grained soil groups, since these soils occur relatively infrequently. The portions of the diagrams which contain the words color and odor will be further discussed in the next paragraph. Although it is not specifically designed for the purpose, figure 19 can also serve as a guide in the planning of laboratory tests to be performed on any soil.

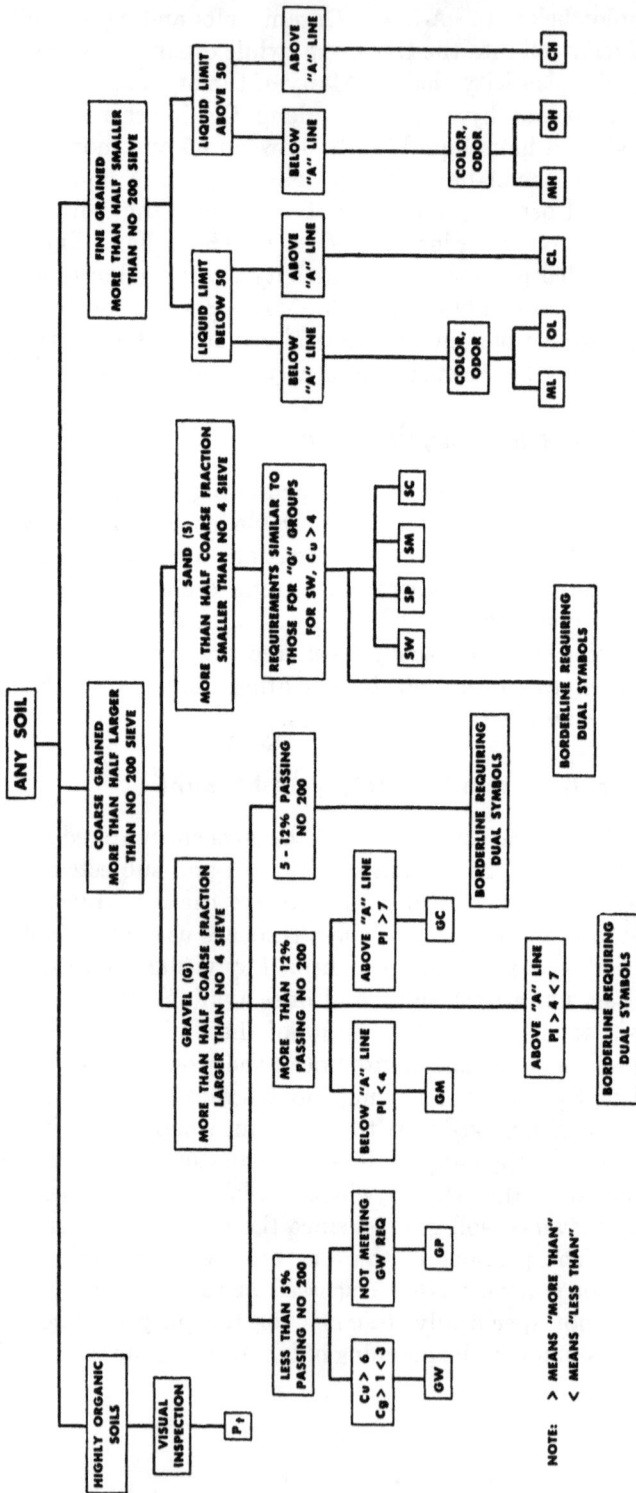

*Figure 19.  Identification procedure using laboratory test results.*

## 56. Field Identification Procedures

The Unified Soil Classification System is specifically designed to permit the identification of soils in the field. The easiest and best way to learn field soil identification procedures is under the guidance of an experienced soils engineer. Lacking this guidance, one should systematically compare the test results which are obtained for soils in each group with the observed behavior of the soil during the process of field identification. Specific guidance is given in table V, appendix II, for field identification procedures.

*a. Coarse-Grained Soils.* Principal items of importance in the field identification of coarse-grained soils are grain size, gradation, and the plasticity characteristics of the fraction passing the No. 40 sieve, if present. Gravel and sand particles can be readily identified with the naked eye. The size corresponding to the No. 200 sieve is about the smallest which can be detected in this fashion. A comparison of the size limits corresponding to the various gravel and sand fractions has previously been given (par. 16). Gradation and size can be judged by spreading out a representative sample on a flat surface. Well-graded soils show a wide range in grain sizes and substantial amounts of all intermediate particle sizes. Poorly graded soils show grains of essentially one size or a range of sizes, with some intermediate sizes missing. Although no mention of it is made in table V, some idea of the gradation of sands may be secured by shaking the sample in water in a glass jar and then allowing it to settle out. Approximate gradation is indicated by the separation of the particles in the jar, coarse to fine, from bottom to top. Silt sizes will remain in suspension for at least ½ minute while clay sizes will remain in suspension much longer. The plasticity characteristic of the fines which may be present in coarse-grained soils are judged by the procedures indicated below for fine-grained soils.

*b. Fine-Grained Soils.* The field identification procedures are to be performed upon the minus No. 40 sieve size particles, smaller than approximately ¹⁄₆₄ inch. For field classification purposes, screening of the sample is not intended; simply remove by hand the coarse particles that interfere with the tests. The three tests principally used in the field identification of fine-grained soils are the—(1) dilatancy or reaction-to-shaking test, (2) dry strength or crushing characteristics, and (3) toughness or consistency near plastic limit. These procedures are explained in detail in table V, together with the range of results of each test which are applicable to soils in each of the fine-grained soil groups. No further explanation of these tests or their use is believed necessary.

*c. Highly Organic Soils.* As previously indicated, these soils are readily identified by color, odor, spongy feel, and frequently by fibrous texture.

*d. Color.* In the field, color may be useful in distinguishing among different soil strata and, when coupled with local experience, may be valuable in identification. Organic matter is frequently indicated by the presence of olive greenish and light brown to black colors. Clean and bright-looking colors are characteristic of inorganic soils. Red to dark brown colors are frequently indicative of the presence of iron oxide, which may be a cementing agent. Soils which contain appreciable amounts of coral, kaolin, caliche, gypsum, talc, or diatomaceous earth may range from white or light gray to various shades of gray.

*e. Odor.* The existence of large amounts of organic matter may often be detected by a distinctive odor, that of decayed vegetation. The odor is particularly strong for fresh samples, but can be reintensified by heating a sample quickly.

## 57. Other Aids to Field Identification

Although they are not included in table V, certain other field tests may be of value in soil identification, as follows:

*a. Feel.* Sandy soils, even very fine sands, have a characteristic gritty feel when rubbed between the fingers. In contrast, dry silts, like rock flour, have a smooth, silky or floury feel. Some clays, when wet, feel slick, or even soapy. Unusually light or unusually heavy soils may be detected sometimes by holding a dry pat of soil in the hand and feeling its weight.

*b. Shine.* The shine test may be conducted by rubbing a dry or moist sample of soil with the finger nail or a knife blade. The surface of most inorganic clays will remain dull because of the high percentage of silt or sand sizes. In contrast, the surface or typical highly plastic clays will become shiny, because of the presence of colloidal, flaky grains.

*c. Acid Test.* Certain soils, such as coral, will appear to be cohesive because of the presence of calcium carbonate ($CaCO_3$), which acts as a cementing agent. If in doubt, this may be detected by dropping a little hydrochloric acid (HC1) on a piece of soil; if $CaCO_3$ is present, a fizzing reaction will be produced.

*d. Taste Test.* If a dry pat of extremely fine-grained soil is lightly touched to the tongue, it will tend to stick because of capillary forces.

## 58. Information Required for Describing Soils

Even though a classification system is used, the ability to write an accurate and comprehensive description of the soils which he encounters is important to the engineer. In table V, appendix II, are given specific suggestions for describing both coarse-grained and fine-grained soils, including information relative to the soil in an undisturbed condition, which is especially important when the soil is to serve as a foundation. Many of the items which are mentioned in this portion of the table have been discussed in previous paragraphs.

Some of them have not, however, and the following statements may serve to clarify the meaning.

*a. Coarse-Grained Soils (Disturbed Condition).* By angularity is meant the items which have been discussed under particle shape (par. 21). The coarse grains would be described as angular, rounded, etc. By surface condition is meant such descriptive terms as pitted, rough, smooth, etc. The hardness and durability of the coarse soil grains deserves careful examination in some cases. Grains derived from sound, igneous rock are easily identified and entirely desirable from the standpoint of hardness and durability. Weathered material may be recognized from its discolorations and the fact that the grains may be easily crushed. Coarse grains consisting of weathered granitic rocks and quartzite are not necessarily objectionable when used in foundations for pavements. However, soils which contain fragments of shaley rock may be dangerous because alternate wetting and drying may result in partial or complete disintegration. This tendency may be detected by a slaking test. Agricultural and geologic names are discussed later in this chapter.

*b. Coarse-Grained Soils (Undisturbed Condition).* By stratification is meant the occurrence of alternate layers of somewhat different soil types. The terms *stratified*, meaning alternate layers of different soils; *varved*, meaning repeated and alternate thin layers of different soils; *laminated*, when the repeating layers are very thin; and *banded*, meaning alternate layers in residual soils, are all in current use. The degree of compactness or density of coarse-grained soils in place is very important; the terms *loose, firm, dense* and *very dense* are frequently applied. Similar terms using the words *compact* or *compacted* are also used. The presence of cementation may be important because certain of the cementing agents, such as iron oxide and calcium carbonate, may later be decomposed, dissolved, and carried away, leaving the soil in a less stable condition. Additional explanation relative to the items, moisture content and drainage characteristics, is not believed to be necessary at this stage.

*c. Fine-Grained Soils.*

(1) Nearly all the items included in table V as information which is required in describing fine-grained soils have been discussed previously. Observations relative to structure and stratification are important. In addition to the terms mentioned as relating to stratification in *b* above, the term homogeneous is applied to soil structures which have uniform properties. Soil structure can frequently be detected easily, as in the case of loess, which has a loose, porous structure or glacial till, which has a dense, compact structure. It may be important to observe and record defects in the structure of fine-grained soils. These include fissures or

cracks resulting from frost action or shrinkage, rootholes, weathering, which is usually denoted by irregular discoloration, and slickensides, or former failure planes.

(2) The consistency of this type of soil is extremely important, both in the undisturbed and remolded condition. In the undisturbed condition terms ranging from very soft, firm, stiff, to hard, may be used. The consistency frequently is judged by squeezing (or attempting to squeeze) a small undisturbed sample between the fingers. The terms brittle, elastic, spongy, friable (crumbles easily), and sensitive (loses strength when remolded) are also used to describe consistency. It is particularly important to notice changes in consistency which occur when the undisturbed sample is remolded. The Terzaghi classification based upon undisturbed samples and the results of the unconfined compression test (par. 130) is as follows:

| Unconfined Compressive Strength, Tons/sq. ft. | Consistency |
|---|---|
| Less than 0.25 | Very soft. |
| 0.25–0.50 | Soft. |
| 0.50–1.00 | Medium. |
| 1.00–2.00 | Stiff. |
| 2.00–4.00 | Very stiff. |
| More than 4.00 | Hard. |

*d. Examples of Soil Description.* Two examples of adequate soil descriptions are given in table V, one for a coarse-grained soil and the other for a fine-grained soil.

## 59. Engineering Characteristics of Soil Groups

In table VI, appendix II, is given information relating the groups of the Unified Soil Classification System to soil characteristics which are pertinent to roads and airfields. Similarly, in table VII is given information concerning characteristics of soil groups pertinent to embankments and foundations. No attempt will be made at this stage to discuss the meaning of the information which is contained in these tables. Each item contained therein is discussed in the appropriate later section of this manual. In table III, below, is given an approximate correlation among the groups of the Unified System and those of the revised Public Roads and C. A. A. systems which are described in the next section.

## Section III.   RELATION OF UNIFIED SOIL CLASSIFICATION SYSTEM TO OTHER SYSTEMS

### 60. Purpose of Correlation

Information about soils is available from many sources, including publications, maps, and reports.   These sources may be of value to the military engineer in studying soils in a given area.   For this reason, it is important that the military engineer have some knowledge of other commonly used systems.   The majority of civil agencies concerned with highways in the United States classify soils by the revised Public Roads system; this includes the Bureau of Public Roads and most of the state highway departments.   The revised Public Roads system and the system used by the Civil Aeronautics Admin-

Table III.   Approximate Equivalent Groups of Unified Soil Classification System, Revised Public Roads System, and Civil Aeronautics Administration System

| Unified system | Public roads | Civil Aeronautics Administration | |
|---|---|---|---|
| GW | A–1–a | Gravelly soils not included directly.   Note upgrading permitted in table VII. | |
| GP | A–1–a | | |
| GM | A–1–a, A–2–4 or 5 | | |
| GC | A–2–6 or 7 | | |
| SW | A–1–b | E–1, 2 or 3 | E–4 or 5 (usually SM or SC) |
| SP | A–3 | | |
| SM | A–1–b, A–2–4 or 5 | | |
| SC | A–2–6 or 7 | | |
| ML | A–4 | E–6 | |
| CL | A–6, A–7–5 | | |
| OL | A–4, A–7–5 | E–6 | |
| MH | A–5 | E–10, 11 or 12 | E–8 (usually L group) E–9 (usually not CH) |
| CH | A–7 | | |
| OH | A–7 | | |
| Pt | | E–13 | |

Note: Groups are only approximately equivalent, since different limiting values are used in each system.

**49**

istration are discussed in some detail in this section. Some information is also given about agricultural and geologic classification systems, since this type of information may be of general value to the soils engineer.

## 61. Revised Public Roads System

The Public Roads system was originally presented in 1931. The portion of the original system which applied to uniform subgrade soils utilized a number of tables and charts based upon several routine soil tests to permit placing of a given soil into one of eight principal groups, designated A-1 through A-8. The system was put into use by many agencies. As time passed, it became apparent that certain of the groups were too broad in coverage, permitting the inclusion of somewhat different soils in the same group. Efforts were directed toward improving the system, and a number of the agencies using it modified it to suit their purposes. Principal modifications included the breaking down of certain of the broad groups into subgroups of more limited scope. Efforts directed toward improving the system were climaxed by a comprehensive committee report which appeared in the Proceedings of the 25th annual meeting of the Highway Research Board (1945). This same report contains detailed information relative to the Airfield Classification System and the C. A. A. system. For a brief time after the report became available the system was called the Highway Research Board system. At the present time and by common usage, it is known as the *revised Public Roads* system. It is primarily designed for the evaluation of subgrade soils, although it is useful for other purposes, also.

## 62. Basis for Classifying Soils (Revised Public Roads System)

*a.* In table VIII, appendix II, is shown the basis of the revised Public Roads classification system. It will be noticed that two very broad groups are utilized. One of these is designated *granular materials* i. e. soils which contain less than 35 percent of material passing a No. 200 sieve. The other is *silt-clay materials*, i. e. those which contain more than 35 percent passing a No. 200 sieve. There are seven major groups, numbered A-1 through A-7 inclusive, together with a number of suggested subgroups. The A-8 group of the original system, which contained the highly organic soils such as peat, is not included in the revised system. It was felt that no group was needed for these soils because of their ready identification by appearance and odor. Whether a soil is silty or clayey depends upon its plasticity index (P. I.). *Silty* is applied to material which has a P. I. of 10 or less, and *clayey* is applied to material which has a P. I. of 10 or more.

*b.* In figure 20 are shown the formula for group index and charts to facilitate its computation. The group index was devised to provide a basis for approximating within-group evaluations. Group indexes

## CHART A

### GRAIN SIZE AND P.I. RELATIONS

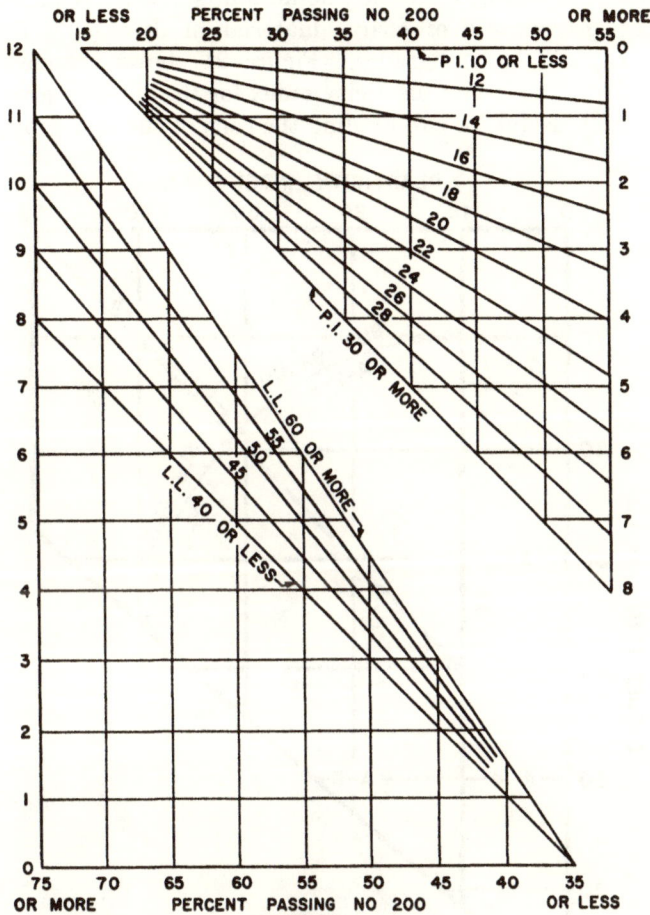

**CHART B**

**GRAIN SIZE AND L.L. RELATIONS**

## GROUP INDEX = SUM OF READINGS ON BOTH VERTICAL SCALES

The Group Index Formula is as follows:

Group Index $= 0.2a + 0.005ac + 0.01bd$ where

$a =$ that portion of percentage passing No. 200 sieve greater than 35 per cent and not exceeding 75 per cent, expressed as a positive whole number (1 to 40).

$b =$ that portion of percentage passing No. 200 sieve greater than 15 per cent and not exceeding 55 per cent, expressed as a positive whole number (1 to 40).

$c =$ that portion of the numerical liquid limit greater than 40 and not exceeding 60, expressed as a positive whole number (1 to 20).

$d =$ that portion of the numerical plasticity index greater than 10 and not exceeding 30, expressed as a positive whole number (1 to 20).

*Figure 20.  Group index formula and charts.  Revised Public Roads classification system.*

**51**

range from 0 for the best subgrade soils to 20 for the poorest. Increasing values of the group index within each basic soil group reflect the combined effects of increasing liquid limits, plasticity indexes, and decreasing percentages of coarse material in decreasing the load-carrying capacity of subgrades. Figure 21 shows graphically the ranges of liquid limit and plasticity index for the silt-clay groups. It is particularly useful for subdividing the soils of the A–7 group.

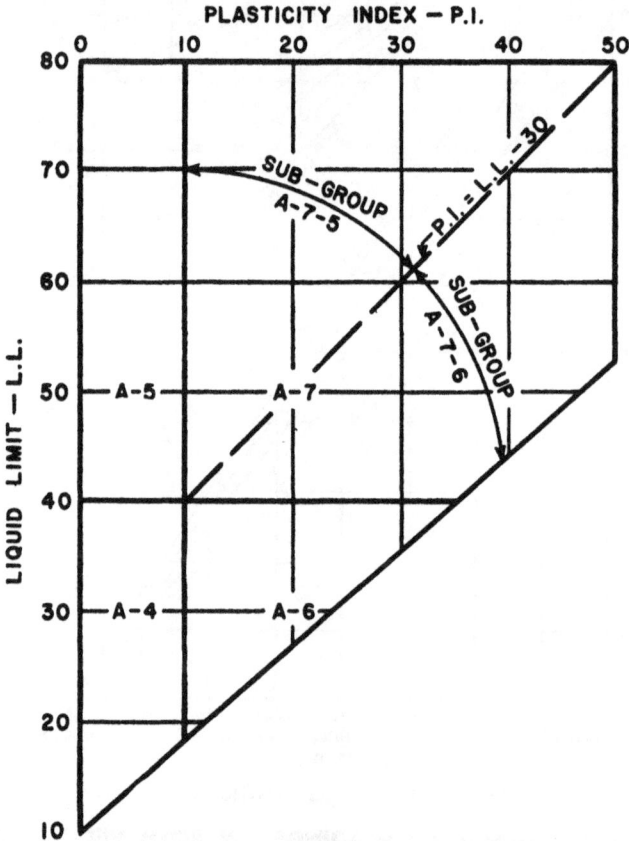

*Figure 21. Relationship between liquid limit and plasticity index for silt-clay groups. Revised Public Roads classification system.*

## 63. Classification Procedure (Revised Public Roads System)

Assuming that the results of the mechanical analysis and the limit tests are available, the classification procedure is quite simple, in fact almost automatic. Table VIII, appendix II, is used in a left-to-right elimination process and the given soil placed into the first group or subgroup into which it fits. In order to distinguish the revised from the old system, the group symbol is given, followed by the group index in parentheses, e. g. A–4(5). The fact that the A–3 group is placed ahead of the A–2 group does not imply that it is necessarily a better subgrade material. This arrangement is used to facilitate the elimination process. The classification of some borderline soils will require judgment and experience. The assignment of the group designation is frequently accompanied by the writing of a careful description, as in the Unified Soil Classification System. A number of detailed examples of the classification procedure are given in paragraphs 67 through 73.

## 64. Civil Aeronautics Administration Classification System

The Civil Aeronautics Administration uses a soil identification and classification procedure which is based upon two somewhat different items, as follows.

*a. Textural Classification.* The textural identification or classification of a soil is based upon a mechanical analysis of the material which passes a No. 10 sieve. With the percentages of sand, silt and clay sizes known, a triangular classification chart is used to assign a name to the soil. Agricultural terms are used. The triangular classification chart which is used by the CAA is very similar to that shown in figure 23, and discussed in the next section. For that reason, the CAA chart will not be presented here.

*b. Classification of Soils for Airport Construction.* The C. A. A. system of classifying soils for engineering purposes is based upon the mechanical analysis, liquid limit, and plasticity index. With these data available the appropriate soil group, ranging from E–1 to E–13, inclusive, is selected from table IX, appendix II. Two modifications of this procedure may be necessary. In such cases, the results of tests on fine-grained soils, groups E–6 to E–12 inclusive, may place the soil in more than one group. When this occurs, the chart of figure 22 may be used to make the final selection of a group designation. The other change which may be needed is made when the soil contains a considerable amount of material which is retained on a No. 10 sieve, since the classification is based on the material which is finer than this sieve size. As will be seen in table IX, upgrading of the material from one to two classes is permitted when the percentage of material retained on the No. 10 sieve exceeds 45 percent for soils on the E–1 to E–5 groups, and 55 percent for the remaining classes. In order for the material to be upgraded, the coarse fraction must consist of reasonably sound material and be reasonably well graded from the maximum size down to the No. 10 sieve. Stones or rock fragments scattered through a soil are not considered to be of sufficient benefit to warrant upgrading. It will be noticed that, as in the Unified Soil Classification System, a separate class, E–13, is reserved for highly organic soils.

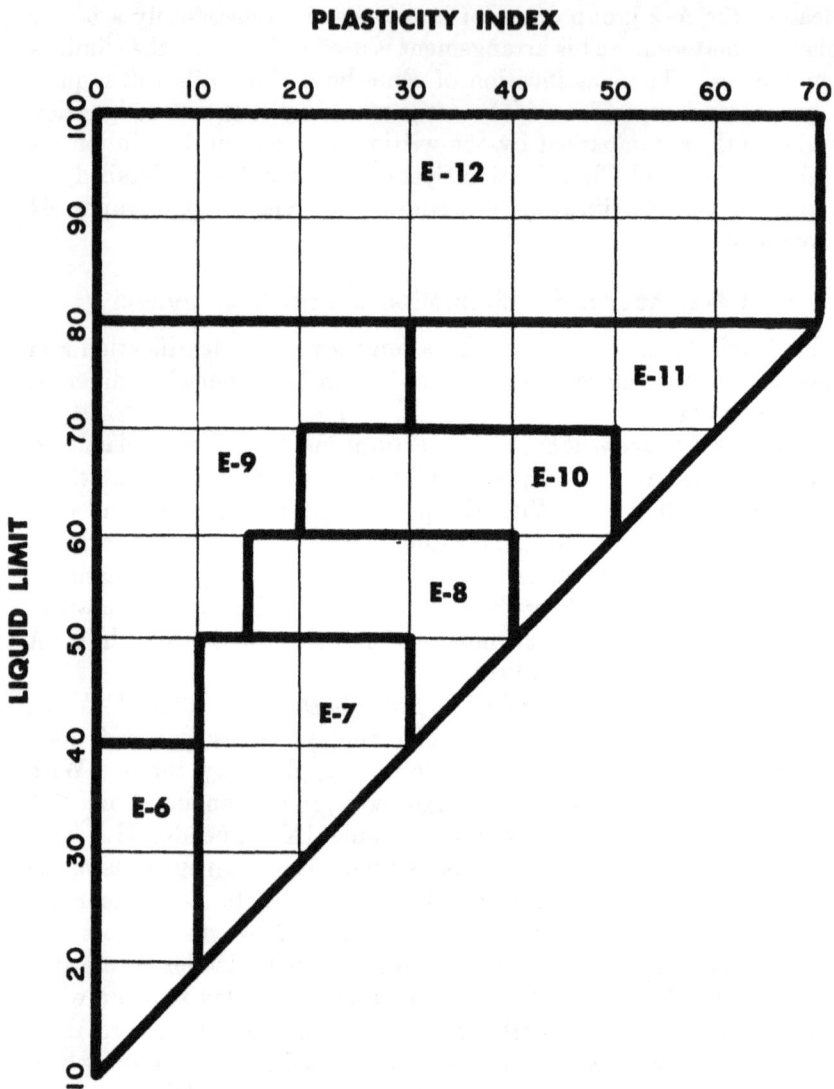

*Figure 22.  CAA classification chart for fine-grained soils.*

## 65. Agricultural Soil Classification System

Two phases of the soil classification systems used by agricultural soil scientists will be discussed. These are classification by grain size (textural classification) and the pedological classification system.

*a. Textural Classifications.* Information relative to two textural classification systems of the U. S. Bureau of Chemistry and Soils is contained in figure 23 and table X, appendix II. It is believed that the chart and table are largely self-explanatory. Examples of their use will be given in paragraphs 67 through 73. The grain size limits which are applicable to the categories shown are as follows:

Coarse Gravel: Retained on No. 4 Sieve
Fine Gravel: Passing No. 4, Retained on No. 10 Sieve
Coarse Sand: Passing No. 10, Retained on No. 60 Sieve
Fine Sand: Passing No. 60, Retained on No. 270 Sieve (0.05 mm.)
Silt: 0.05 mm. to 0.005 mm.
Clay: Below 0.005 mm.

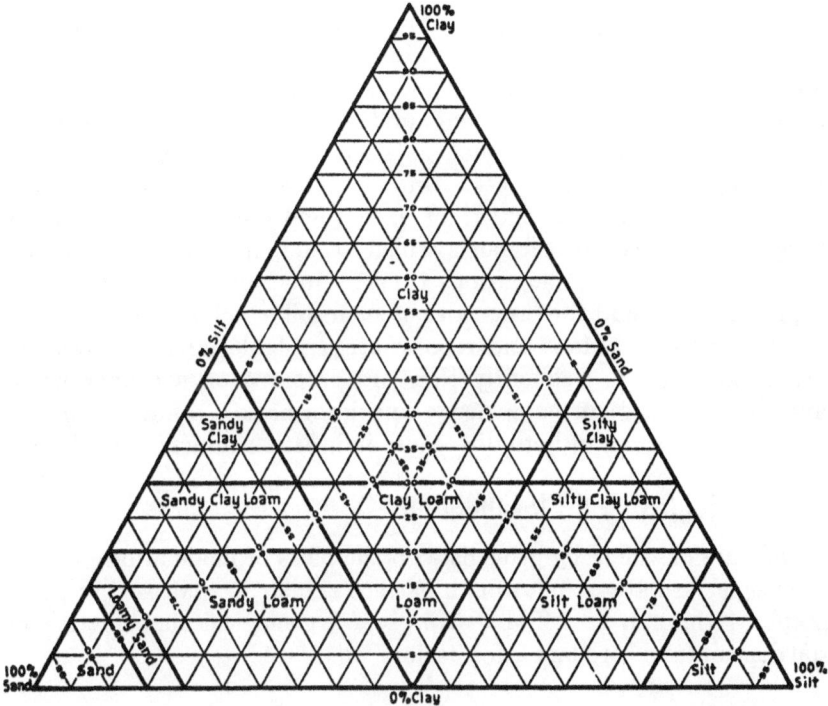

*Figure 23. Triangular textural classification chart, U. S. Bureau of Chemistry and Soils.*

*b. Pedological Classification.* Discussion has previously been given (par. 11) of the soil profile and pedology. Agricultural soil scientists have devised a very complete and quite complex system for describing and classifying surface soils wherever they are encountered on the surface of the earth. No attempt will be made to discuss this system in detail here. The portion of the system in which engineers are principally interested refers to the terms used in mapping limited areas. Mapping is based on—(1) series, (2) type, and (3) phase. The designation known as soil series is applied to soils having the same genetic horizons, which possess similar characteristics and similar profiles, and are derived from the same parent material. Series names follow no particular pattern, but are generally taken from some geographical place name near which they were first found. The soil type refers to the texture of the upper portion of the soil profile. Several types may, and usually do, exist within a soil series. By phase is meant some variation, usually of minor importance, in the soil type. A mapping unit may be something like Emmet loamy sand, gravelly phase, or any one of the large number of similar designations. Generally speaking, soils which are given the same designation have the same agricultural properties wherever they are encountered. Many of their engineering properties may also be the same. A given soil is placed in the proper mapping unit on the basis of the careful examination of the soil in an auger boring, highway cuts, natural slopes, and other places where the soil profile is exposed, Factors such as color, texture, organic content and consistency are particularly important. Other things, such as slope, drainage, vegetation, and land use, are of consequence. Agricultural soil maps prepared from field surveys show the extent of each important soil type and its geographical location. Reports which accompany the maps contain word descriptions of the various types, some laboratory test results, typical profiles, and the properties of each soil which are important to agricultural use. Maps and reports are available for many areas in the United States, frequently they are prepared on a county basis, and in many foreign countries. The information contained in the maps and reports is of both direct and indirect usefulness to engineers.

## 66. Geological Soil Classification

Soils are classified by geologists on the basis of mode of origin, i e. process of formation, following a pattern similar to that used in paragraphs 4 through 12. TM 5–545 gives a geological classification of soil deposits and corollary information.

## Section IV.  CLASSIFICATION OF TYPICAL SOILS

## 67. Introduction

The discussion in this section is concerned with the classification of four different inorganic soils upon the basis of available laboratory test data. Each soil will be classified under the Unified Soil Classification System, revised Public Roads system, CAA system, and the agricultural textural classification scheme.

## 68. Soil Test Data

The following information is available about each soil:

| Soil No. | 1 | 2 | 3 | 4 |
|---|---|---|---|---|

### Mechanical Analysis

*Percent passing, by weight*

| | 1 | 2 | 3 | 4 |
|---|---|---|---|---|
| 3-inch | ----- | ----- | ----- | 100. 0 |
| ¾- nch | ----- | ----- | ----- | 56. 0 |
| No. 4 | ----- | ----- | ----- | 30. 0 |
| No. 10 | 100. 0 | 100. 0 | 100. 0 | 16. 4 |
| No. 40 | 85. 2 | 97. 6 | 85. 0 | 7. 2 |
| No. 60 | ----- | ----- | 20. 0 | 5. 0 |
| No. 200 | 52. 1 | 69. 8 | 1. 2 | 3. 5 |
| No. 270 | 48. 2 | 65. 0 | ----- | ----- |
| $C_u$ | ----- | ----- | 2. 0 | 12. 5 |
| $C_z$ | ----- | ----- | ----- | 2. 2 |

### Plasticity Characteristics

*Percent, by weight*

| | 1 | 2 | 3 | 4 |
|---|---|---|---|---|
| Liquid limit | 29. 2 | 66. 7 | 21. 3 | ----- |
| Plasticity index | 5. 0 | 39. 0 | N. P. | ----- |

## 69. Unified Soil Classification System

(Table V, app. II)

*a. Soil No. 1.* Soil is fine grained, since more than half passes the No. 200 sieve. Liquid limit is less than 50, hence must be ML or CL, since it is inorganic. On plasticity chart it falls below the A-line, therefore *ML*.

*b. Soil No. 2.* Soil is fine grained, since more than half passes No. 200 sieve. Liquid limit is more than 50, hence must be MH or CH. On plasticity chart if falls above the A-line, hence *CH*.

*c. Soil No. 3.* Soil must be coarse grained, since very little passes No. 200 sieve. Must be a sand, since it all passes No. 10 sieve. Following the chart of table V, the soil contains less than 5 percent passing No. 200 sieve, therefore it must be either an SW or an SP. Value of $C_u = 2$ will not meet requirements for SW, hence *SP*.

*d. Soil No. 4.* Soil must be coarse grained, since again very little passes a No. 200 sieve. Must be a gravel, since more than half of the coarse fraction is larger than a No. 4 sieve. Again referring to table V, since the soil contains less than 5 percent passing the No. 200, it is either GW or GP. It meets the requirements relative to gradation, hence *GW*.

## 70. Revised Public Roads Classification System

(Table VIII, app. II, and figs. 20 and 21)

*a. Soil No. 1.* To calculate the group index refer to figure 20. From chart A, read 0; from chart B, read 3. Therefore, Group Index=0+3=3. Entering table VIII and using a left-to-right elimination process, soil can not be in one of the *granular materials* groups, since more than 35 percent passes a No. 200 sieve. It meets the requirements of the A-4 group. It is therefore A-4 (3).

*b. Soil No. 2.* As before, Group Index=8+10=18. From table VIII, this soil must fall into the A-7 group, since this is the only group which will permit a group index value as high as 18. Referring to the chart of figure 21, it falls in the A-7-6 subgroup; hence, A-7-6 (18).

*c. Soil No. 3.* As before, Group Index=0+0=0. This is one of the soils described as *granular materials*. It will not meet the requirements of an A-1 soil, since it contains practically no fines. It does meet the requirements of the A-3 group and is thus A-3 (0).

*d. Soil No. 4.* As before, Group Index=0+0=0. This is obviously a granular material and meets the requirements of the A-1-a subgroup, even though the plasticity index is not known. Therefore, A-1-a (0).

## 71. Civil Aeronautics Administration Classification System
(Table IX, app. II, and fig. 22)

*a. Soil No. 1.* Referring to table IX, soil contains more than 45 percent passing No. 270 sieve, hence it must be group E-6 or higher. It meets the requirements of the E-6 group, hence E-6.

*b. Soil No. 2.* As before, group must be E-6 or higher. Examination of table IX shows that it cannot be E-6 through E-9, inclusive. It can be either E-10 or E-11. Reference to figure 22 reveals that it is properly classed as E-10.

*c. Soil No. 3.* For this soil it is necessary to convert the values given to the basis used in figure 22. For this soil, percent passing No. 10 and retained on No. 60 (Coarse Sand)=100−20=80; percent passing No. 60 and retained on No. 270 (Fine Sand)=20−0=20. It will then be seen that this soil is an E-1.

*d. Soil No. 4.* The laboratory data are not given in such a way as to facilitate the use of figure 22, since the sieve analysis of the material passing a No. 10 sieve is not separately stated. It is probable that this material would also be classed as E-1.

## 72. Agricultural Textural Classification System
(Table X, app. II, and fig. 23)

*a. Introduction.* Although the values are not given in the previous tabulation, it will be assumed that 12 percent of Soil No. 1 and 35 percent of Soil No. 2 are in the range of clay sizes, i. e. below .005 mm. (par. 65).

*b. Soil No. 1.* This soil contains 100−48.2=51.8 percent sand by the definitions of paragraph 65, since the opening of a No. 270 sieve is 0.05 mm. The soil is then composed of 52 percent sand, 36 percent silt, and 12 percent clay. From figure 23, this soil is classed as a *Sandy Loam*.

*c. Soil No. 2.* This soil contains approximately 35 percent sand, 30 percent silt, and 35 percent clay. From figure 23, this would be classed as *Clay*.

*d. Soil No. 3.* This soil is 99 percent sand, and therefore can be classed only as *Sand*, figure 23.

*e. Soil No. 4.* This soil contains approximately 70 percent coarse gravel, 14 percent fine gravel, 13 percent sand, and 3 percent silt and clay combined. This cannot be classified from figure 23, because the chart does not cover gravels and gravelly soils. From table X, this material is *Gravel and Sand*.

## 73. Summary and Comparison

The following tabulation is a summary of the classification of the soils in question under the four different classification systems considered in paragraph 67 through 72.

| Soil No. | Unified classification system | Revised public roads system | CAA classification system | Agricultural textural classification |
|---|---|---|---|---|
| 1 | ML | A-4 (3) | E-6 | Sandy Loam. |
| 2 | CH | A-7-6 (18) | E-10 | Clay. |
| 3 | SP | A-3 (0) | E-1 | Sand. |
| 4 | GW | A-1-a (0) | E-1 | Gravel and Sand. |

# Section V.  DESCRIPTIVE SOIL NAMES

## 74. Introduction

Although the basic terms gravel, sand, silt, clay and combinations thereof constitute the basic terminology of the soil engineer, there are also many other terms which are used to describe soils of particular characteristics or in certain areas. Some of these terms are principally of local significance, while others are widely used. A number of these terms are defined in the next paragraph. Reference should be made to TM 5-545 for definitions of geologic terms, including rock types.

## 75. Descriptive Soil Names

*a. Adobe.* Calcareous silts and sandy-silty clays which are usually high in colloidal clay content, found in the semiarid regions of the southwestern United States and North Africa.

*b. Alluvium.* Deposits of mud, silt, and other material commonly found on the flat lands along the lower courses of streams.

*c. Argillaceous.* Soils which are predominantly clay or abounding in clays or clay-like materials.

*d. Bentonite.* A clay of high plasticity formed by the decomposition of volcanic ash. It has high swelling characteristics.

*e. Boulder Clay.* Another name, used widely in Canada and England, for glacial till (par. 8).

*f. Buckshot.* Clays of the southern and southwestern United States which, upon drying, crack into small, hard lumps of more or less uniform size.

*g. Bull's Liver.* This is a name used in some sections of the United States to describe an inorganic silt of slight plasticity. When saturated, it quakes like jelly from vibration or shock.

*h. Calcareous.* Soils which contain an appreciable amount of calcium carbonate, usually from limestone.

*i. Caliche.* This term is widely used in construction to describe certain soft limestone deposits which contain various amounts of silt and clay, as found in France, North Africa, Texas and other southwestern states.

*j. Coquina.* Consists essentially of marine shells which are held together by a small amount of calcium carbonate to form a fairly hard rock. *Coquina shells* (and oyster shells) are widely used for the granular stabilization of soils along the Gulf Coast of the United States.

*k. Coral.* Calcareous, rock-like material formed by secretions of corals and coralline algae (par. 9).

*l. Diatomaceous Earth.* Composed essentially of the siliceous skeletons of diatoms (extremely small unicelled organisms). It is composed principally of silica, is white or light gray in color and extremely porous (par. 9).

*m. Dirty Sand.* A slightly silty or clayey sand.

*n. Disintegrated Granite.* Granular soil derived from advanced weathering and disintegration of granite rock.

*o. Fat Clay.* Fine, colloidal clay of high plasticity.

*p. Fuller's Earth.* Unusually highly plastic clays of sedimentary origin, white to brown in color. Used commercially to absorb fats and dyes.

*q. Gumbo.* Peculiar, fine-grained, highly plastic silt-clay soils which become impervious and soapy, or waxy and sticky, when saturated.

*r. Hardpan.* A general term used to describe a hard, cemented soil layer which does not soften when wet.

*s. Lateritic Soils.* Residual soils which are found in tropic regions. Many different soils are included in this category and they occur in many sections of the world. They are frequently red in color, and in their natural state have a granular structure with low plasticity and good drainability. When they are remolded in the presence of water they often become plastic and clayey to the depth disturbed.

*t. Lean Clay.* Silty clays and clayey silts, generally of low to medium plasticity.

*u. Limerock.* A soft, friable, compact, cream-white, high-calcium limestone found in the southeastern United States which consists of coral and other marine remains which have been disintegrated by weathering.

*v. Loam.* A general agricultural term which is applied most frequently to sandy-silty topsoils which contain a trace of clay, are easily worked, and are productive of plant life.

*w. Loess.* Silty soil of aeolian origin characterized by a loose, porous structure, and a natural vertical slope. It covers extensive areas in North America (especially in the Mississippi Basin), Europe and Asia (especially North Central Europe, Russia and China). Loess is further described in paragraph 10.

*x. Marl.* A soft, calcareous deposit mixed with clays, silts, and sands, often containing shells or organic remains. It is common in the Gulf Coast area of the United States.

*y. Micaceous Soil.* Soil which contains a sufficient amount of mica to give it distinctive appearance and characteristics.

*z. Muck (mud).* The very soft, slimy silt or organic silt which is frequently found on lake or river bottoms.

*aa. Muskeg.* Peat deposits found in northwestern Canada and Alaska.

*ab. Peat.* A term which is frequently applied to fibrous, partially decayed organic matter or a soil which contains a large proportion of such materials. Large and small deposits of peat occur in many areas and present many construction difficulties. Peat is extremely loose and compressible.

*ac. Red Dog.* The residue from burned coal dumps.

*ad. Rockflour.* A fine-grained soil, usually sedimentary, of low plasticity and cohesion. Particles are usually in the lower range of silt sizes. At high moisture contents it may become "quick" under the action of traffic.

*ae. Talus.* A fan-shaped accumulation of mixed fragments of rock that have fallen, because of weathering, at or near the base of a cliff or steep mountainside.

*af. Topsoil.* A general term applied to the top few inches of soil deposits. Topsoils usually contain considerable organic matter and are productive of plant life.

*ag. Tufa.* A loose, porous deposit of calcium carbonate which usually contains organic remains.

*ah. Tuff.* A term applied to compacted deposits of the fine materials ejected from volcanoes, such as more or less cemented dust and cinders. Tuffs are more or less stratified and in various states of consolidation. They are prevalent in the Mediterranean area.

*ai. Varved Clay.* A sedimentary deposit which consists of alternate thin layers of silt and clay.

*aj. Volcanic Ash.* Uncemented volcanic debris, usually made up of particles less than 4 mm. in diameter. Upon weathering, a volcanic clay of high compressibility is frequently formed (par. 10). Some volcanic clays present unusually difficult construction problems, as do those in the area of Mexico City and along the eastern shores of the island of Hawaii.

# CHAPTER 4

# MOVEMENT OF WATER THROUGH SOILS

## 76. Introduction

The movement of water into or through a soil mass is a phemonenon of great practical importance in engineering design and construction. For example, water may be drawn by capillarity from a free water surface into the subgrade beneath an airport runway. Water then accumulated may greatly reduce the shearing resistance of the subgrade soil, with the consequence that the pavement fails under wheel loads. Seepage flow may be responsible for the erosion or failure of an open cut slope, the "blow-up" of the bottom of a cofferdam, or the failure of an earth dam. This chapter is concerned with fundamental facts relative to the movement of water into and through soils and, to some extent, with practical measures which are undertaken to control this movement.

## 77. Terminology

*a. Water.* The liquid which is present in varying amounts in all natural soils is almost invariably referred to simply as water or moisture, although it may contain varying amounts of other substances in solution. In certain types of work the presence and nature of these dissolved substances is of considerable importance, as when underground utilities or foundation elements must be protected against corrosion or disintegration. Ordinarily, however, the chemical composition of water in soils is of less importance than the way in which it is held or migrates through the soil.

*b. Capillary Moisture.* As has been mentioned previously in paragraph 37, because of the effects of surface tension, water may be drawn into the voids of a soil just as water may be drawn into a capillary tube. This movement may take place in any direction in soil, but most commonly is associated with an upward movement of water from the water table. The moisture which is present at any point in a soil because of such movement is called *capillary moisture.* Capillary phenomena are discussed in more detail in the paragraphs which follow.

*c. Gravitational or Free Water.* Water which is free to move through the voids of a soil mass under the force of gravity or some other applied force, such as the pressures created by the application of loads, is called *gravitational water* or *free water*. Below the water table it completely fills the voids of the soil, or nearly so. The *water table* or *free water surface* is the upper boundary of the zone of free water. Water in this zone, generally speaking, moves through the soil in response to the same laws of hydraulics which govern the flow of water in conduits. The level of the water table may be established by observing the level to which water will rise in a test hole, although special techniques may be required in fine-grained soils. The movement of free water through a soil mass frequently is termed *seepage*.

## 78. Height of Capillary Rise

As was mentioned in paragraph 37, for the purpose of visualizing the upward movement of moisture by capillarity from a free water surface it is convenient to visualize the voids of a soil mass as a bundle of interconnected small tubes which provide extremely irregular, tortuous paths for the flow of capillary water. In figure 16 it is indicated that, other things being the same, the height of capillary rise in a small tube is inversely proportional to the diameter. That is, the smaller the tube, the greater the height of capillary rise. By analogy, although the comparison is not strictly correct since the voids in a soil have variable diameters, the height of capillary rise in a soil depends upon the size of the individual void spaces. Going a step further, it seems logical that the individual void spaces would be small when the soil particles are small. Hence, the height of capillary rise would be expected to be greatest in fine-grained soils. This reasoning agrees well with practical experience. From the standpoint of susceptibility to capillary action and the detrimental effects which sometimes accompany the capillary flow of water in soils, the most critical condition is believed to occur in a silt or a very fine sand. The eventual height of capillary rise is greater in clay soils than in silts, but the upward flow of water in clay is much slower. Thus, the time required for a clay to become saturated (or to attain a high degree of saturation, since some air may be trapped in the voids) may be very long. The eventual height of capillary rise in a fine-grained soil may be as much as 50 feet in extreme cases, and frequently is of the order of 9 or 10 feet. Coarse sands and gravels are not normally susceptible to capillary action because of the comparatively large size of the individual void spaces. The height of capillary rise may vary from practically zero to a few inches in these soils. If the height of capillary rise is more than the distance from the ground surface to the water table, there will be movement of capillary moisture upward to the surface as it is lost by evaporation.

## 79. Zones of Capillary Moisture

Terzaghi has suggested that capillary moisture in soils located above the water table may be conveniently visualized as occurring in three zones; the zones of capillary saturation, partial capillary saturation, and contact moisture. In the zone of capillary saturation the soil is essentially saturated. The height of this zone depends not only on the soil but also on the history of the water table, since the height will be greater if the soil mass has been saturated previously. The height of the zone of partial capillary saturation is likely to be considerably greater than that of the zone of capillary saturation; it also depends upon the water table history. Its existence may be explained as being due to the fact that a few large voids may serve effectively to stop capillary rise in some parts of the soil mass. Capillary water in this zone is still interconnected or "continuous", while the air voids may not be. Above these two zones water which percolates downward from the surface may be held in the soil by surface tension. It may fill the smaller voids or be present in the form of water films between the points of contact of the soil grains. Water may also be brought into this zone from the water table by evaporation and condensation. This moisture is termed *contact moisture*.

## 80. Effects of Contact Moisture

Two of the effects of contact moisture are the following. One effect is that of *apparent cohesion*. One evidence of the existence of apparent cohesion is presented by the behavior of sand on certain beaches. On these beaches the dry sand located back from the edge of the water and above the height of capillary rise is generally dry, very loose, and has little supporting power when unconfined. Closer to the water's edge, and particularly during periods of low tide, the sand is very firm and capable of supporting stationary or moving automobiles and other vehicles. This apparent strength is due primarily to the existence of contact moisture left in the voids of the soil as the tide went out. Very close to the edge of the water the surface soil may be within the zone of partial or complete capillary saturation. Somewhat similarly, capillary forces sometimes may be used to consolidate loose cohesionless deposits of very fine sands or silts in which the water table is at or near the ground surface. This is accomplished by lowering the water table by means of drains or well points. If the operation is properly carried out within the limits of the height of capillary rise, the soil above the lowered water table remains saturated by capillary moisture. The effect is to place the soil structure under capillary forces (tension in the water, compression on the soil structure) which compress it. The soil may be compressed as effectively as though an equivalent external load had been placed upon the surface of the soil mass.

## 81. Practical Methods of Controlling Effects of Capillarity

Brief mention will be made here of methods which are commonly used to control the detrimental effects of capillarity, particularly with relation to roads and airport pavements. Additional attention will be given to this subject in the next chapter, which is devoted to the closely allied subject of frost action. As has been noted, if the water table is closer to the surface than the height of capillary rise, water will be brought up to the surface to replace that which is removed by evaporation. If evaporation is wholly or partially prevented, as by the construction of an impervious pavement, water will accumulate and may cause a reduction in shearing strength or swelling of the soil, particularly when a fine-grained soil or a coarse soil which contains a detrimental amount of plastic fines is involved. One obvious solution is to excavate the material which is subject to capillary action and replace it with a granular material. This is frequently quite expensive and usually may be justified only in areas where frost action is a factor. Another approach is to include in the pavement structure a layer which is unaffected by capillary action. This is one of the functions of the base which is invariably used in flexible pavements (ch. 10). The base serves to interrupt the flow of capillary moisture, in addition to its other functions. Under certain circumstances the base itself may have to be drained to insure the removal of capillary water, as indicated in figure 24. This also is usually not justified unless other circumstances, such as frost action, are of importance. Still another approach is to lower the water table. This may sometimes be accomplished by the use of side ditches. Subdrains may be installed for the same purpose, as indicated in figure 25. This approach is particularly effective in soils which are relatively pervious or free-draining. Some difficulty may be experienced in lowering the water table by this method in flat country, because of the difficulty of finding an outlet for the drains. An alternative used in many areas where the permanent water table is at or near the ground surface is simply to build the highway or runway on a fill. Material which is not subject to detrimental capillarity is used to form a low fill. The bottom of the base is normally kept a minimum of 3 or 4 feet above the natural ground surface, depending on the soil used in the fill and other factors.

## 82. Permeability

*Permeability* is that property of a soil which permits water to flow through it. Water may move through the continuous voids of a soil in much the same way as it moves through pipes and other conduits. As has been indicated, this movement of water through soils is frequently termed seepage and may also be called *percolation*. Soils vary greatly in their resistance to the flow of water through them. Rela-

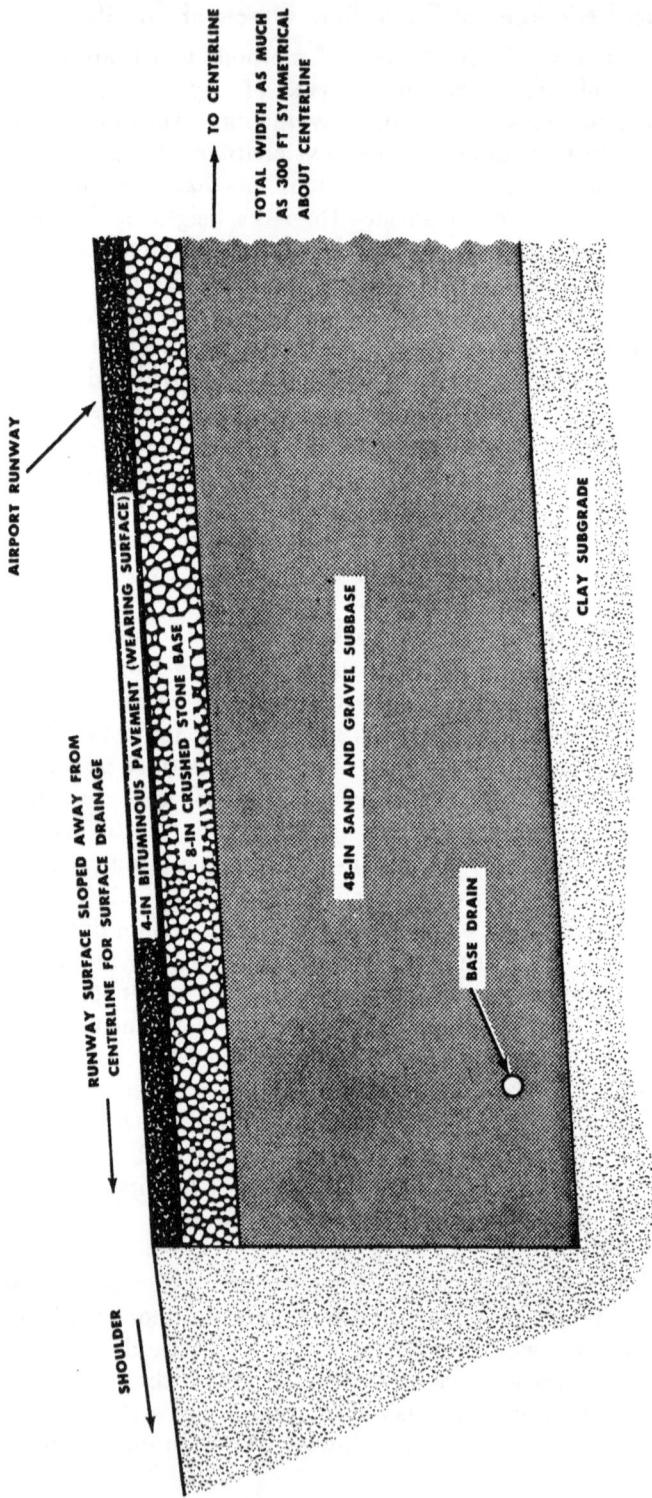

*Figure 24.   Base drains in an airport pavement.*

TO CENTERLINE

TOTAL WIDTH AS MUCH
AS 300 FT SYMMETRICAL
ABOUT CENTERLINE

AIRPORT RUNWAY

RUNWAY SURFACE SLOPED AWAY FROM
CENTERLINE FOR SURFACE DRAINAGE

4-IN BITUMINOUS PAVEMENT (WEARING SURFACE)

8-IN CRUSHED STONE BASE

48-IN SAND AND GRAVEL SUBBASE

BASE DRAIN

CLAY SUBGRADE

SHOULDER

66

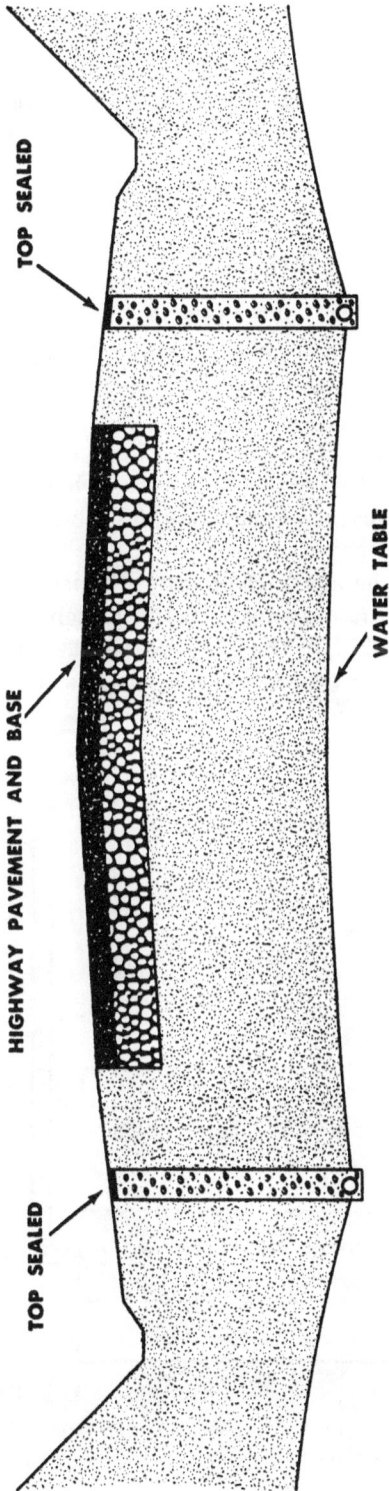

TOP SEALED

HIGHWAY PAVEMENT AND BASE

WATER TABLE

TOP SEALED

NOTE: REQUIRED DEPTH OF SUBDRAIN DEPENDS ON CAPILLARITY OF SUBGRADE SOIL

Figure 25.   Subdrains used to lower the water table.

tively coarse soils, like clean sands and gravels, offer comparatively little resistance to the flow of water; these are said to be *permeable* or *pervious soils*. Fine-grained soils, particularly clays, offer great resistance to the movement of water through them and are said to be *relatively impervious* or *impermeable*. Some water does move through these soils, however. The permeability of a soil reflects the ease with which it can be drained. Soils are thus sometimes classed as *well-draining*, *poorly draining* or *impervious soils*. Permeability is closely related to frost action and to settlement of soils under load.

## 83. Head, Hydraulic Gradient, and Pressure

In order to understand the flow of water through soil masses and the forces which govern that flow, it is necessary to briefly review some of the principles of hydraulics. In figure 26 is shown a portion of a soil mass through which the flow of water is occurring. The points 1 and 2 represent the ends of one path along which flow occurs. At these points standpipes or *piezometer tubes* have been inserted to observe the level to which the water will rise. The distance which the water will rise in the tube above the point at which it is inserted is the *piezometric head*; in the drawing $h_1$ is the piezometric head at point 1, and $h_2$ that at point 2. In the drawing, points 1 and 2 are at the same elevation; the *position head* between the two points is zero.

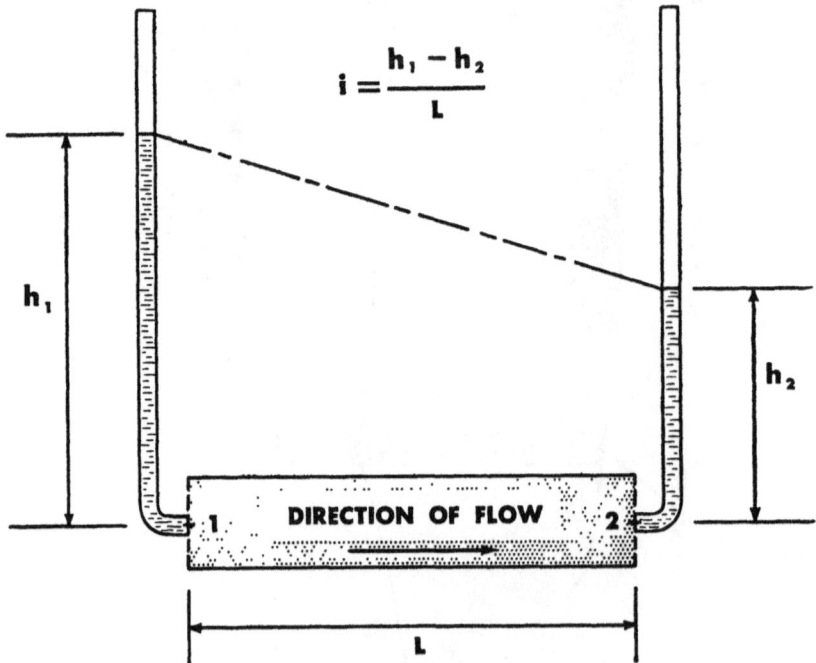

$$i = \frac{h_1 - h_2}{L}$$

DIRECTION OF FLOW

*Figure 26. Diagram indicating meaning of piezometric head, hydraulic head, and hydraulic gradient.*

In other situations, point 1 might be either above or below point 2 If the level of water in the two piezometer tubes is the same, the system is in a state of equilibrium and no flow can occur between points 1 and 2, regardless of the magnitude of the position head. Flow occurs only if there is a difference in the level of water in the piezometer tubes at the two points. This difference in elevation is called the *hydraulic head, h*; in figure 26, $h = h_1 - h_2$. The ratio between the hydraulic head and the length of the path along which this drop in head occurs is called the hydraulic gradient, $i = h/L$. There is also a difference in pressure at the two points. The pressure at point 1 is equal to the height of water above that point divided by the unit weight of water; similarly for point 2 for the conditions shown in the drawing. The difference in pressure between the two points is called *excess hydrostatic pressure;* it is equal to $h\gamma_w$ and is given the symbol $u$. The hydrostatic excess is what causes the water to move through the soil from point 1 to point 2. Since the velocity of the water is very small the velocity head, $v^2/2g$, is neglected in this and subsequent cases.

## 84. Types of Flow of Water in Soils

In the large majority of cases the flow of water through soils is *laminar flow*. By laminar flow is meant a general condition which is typical of the flow of water at low velocities. In this type of flow the particles of water move along smooth, orderly paths in the direction of flow and losses of energy or head, in a general sense, are directly proportional to the velocity. *Turbulent* flow, which is marked by irregular, helter-skelter movement of the water particles and much higher losses of head, may occasionally occur in coarse gravels.

## 85. Darcy's Equation and the Coefficient of Permeability

In the usual case of the flow of water in soils, i. e. laminar flow and relatively small values of the hydraulic gradient, use is made of Darcy's equation, which has been found to be applicable to the flow of water through soil both by experiment and experience. One form of Darcy's equation is $v = ki$. In this equation $v$ is the *discharge velocity*, being defined as the quantity of water which moves through a soil which has a unit cross-sectional area (perpendicular to the direction of flow) in a unit of time. In the English system the velocity is expressed in terms of feet per second or per day or per year. In the metric system, $v$ is usually expressed as centimeters per second. The term $i$ is a dimensionless number and is the hydraulic gradient defined in paragraph 83. The term $k$ is called the *coefficient of permeability;* it has the units of velocity and may be regarded as the discharge velocity under a unit hydraulic gradient. The coefficient of permeability is dependent upon the properties of the fluid involved and upon the soil. Since water is the fluid normally involved in soil problems and since its properties do not vary enough to affect most practical problems, the coefficient of permeability is regarded as a property of the soil. Principal factors which determine the coefficient of permeability for a given soil include grain size, void ratio, and structure. The relationships among these different variables for typical soils are quite complex, and preclude the development of formulas for the coefficient of permeability, except for the simplest cases. For the usual soil, $k$ is determined experimentally, either in the laboratory or the field. These methods are discussed briefly in the next paragraph. Typical values of the coefficient of permeability for the soil groups of the Unified Soil Classification System are given in column 8 of table VII, appendix II.

## 86. Methods of Determining the Coefficient of Permeability

Several different methods may be employed to determine the coefficient of permeability of a given soil, as follows:

*a. Calculation From Grain Size.* An approximate value of the coefficient of permeability for coarse, uniform sands with rounded grains was developed experimentally by A. Hazen. This expression is $k = C(D_{10})^2$, where $k$ is in centimeters per second and $D_{10}$, Hazen's effective size (par. 19), is expressed in centimeters. The value of $C$ varies from 100 to 150. The application of this expression is quite limited, and it gives only approximate results, even for other coarse-grained soils. It reflects the general concept that permeability increases with increasing grain size, since the size of the individual void spaces also increases. Shape of the individual grains is also important, however. In general terms again, it has been found that the coefficient of permeability increases approximately as the square of the void ratio for coarse-grained soils.

*b. Laboratory Determinations.* The coefficient of permeability may be determined experimentally in the laboratory, generally by tests upon undisturbed samples. The tests require careful technique and equipment which is not normally available to the military engineer in the field. Applicable test methods are described in most standard soil mechanics textbooks.

*c. Pumping Tests.* Large scale field pumping tests are sometimes conducted to determine the permeability of coarse-grained soils located below the water table. This procedure is usually employed only on large jobs. The test is conducted by drilling one well from which water is pumped and a number of others in which the level of the water table is observed. Two straight lines of observation wells are generally used, one approximately in the direction of the existing movement of ground water, and the other perpendicular to it. Pumping is begun and continued until a steady state of flow is established. The quantity of water pumped from the well per unit of time is measured and the level of the water table in the other wells observed simultaneously. Theoretical equations have been developed from which the coefficient of permeability may be estimated from the results of a field pumping test.

## 87. Percolation Test

Although they do not provide a means of determining the coefficient of permeability, field *percolation tests* have been used by civil engineers for various purposes, particularly for determining the rate at which domestic sewage may be applied to a tile absorption field. One method of conducting such a test is to dig a hole about one foot square to the depth at which it is proposed to lay the drain tile. The hole is then filled with water, and the water allowed to seep away. When the water falls to within 6 or 8 inches of the bottom of the hole, the rate at which the level of water has fallen is observed. This process is continued until the soil is saturated, and the water seeps away at a constant rate. After this has been achieved, the time $(t_p)$ required for the water to drop one inch is determined. Kiker has suggested the use of a percolation coefficient $C_p$, based on empirical observations and given by the following equation: $C_p = (t_p + 6.24)/29$, in which $t_p$ is expressed in minutes per inch. The *percolation coefficient* of an absorption bed or tile field is the reciprocal of the rate, in gallons per day per square foot of trench bottom, at which sewage may be applied safely to a field that has been properly constructed and designed with a life expectancy of 20 years. It also represents the number of square feet of bottom trench area required for each gallon of sewage applied daily to a tile field. The test is subject to some practical difficulties, including the time required for equilibrium conditions to be established in some soils, the fact that tests conducted in pervious soils which are underlain by impervious ones do not mean much, and the fact that the test is not effective when the water table is high or in frozen soils. However, the percolation test does give valuable practical information when properly conducted.

## 88. Effects of Soil Structure on the Coefficient of Permeability

Water will move more easily through a soil which has a loose single-grain structure than it will through one with a dense structure. As was indicated in paragraphs 4 through 12, many soils, particularly those deposited in relatively still water, are formed of alternate layers or *strata*. In such soils, the coefficient of permeability is frequently much greater in a direction parallel to the planes of bedding than it is at right angles to them. Thus, the value of $k$ may be 10 or 20 times as much in a horizontal direction as in the vertical. If a permeability test is conducted upon these soils the sample, in addition to being undisturbed, must be properly oriented or an erroneous result will be obtained. Advantage may sometimes be taken of this variation in permeability with direction of flow in order to speed up drainage processes in the field Weathering due to frost action or alternate wetting and drying may change the structure of a natural soil or result in the formation of cracks and fissures, with resultant changes in the permeability. Also, the root holes formed by grass and other forms of vegetation may affect the permeability.

## 89. Intergranular and Neutral Pressures

It is convenient to delineate here two types of pressure or stress (force per unit of area) which may exist in a saturated soil mass. These are the *intergranular pressure* or *effective stress*, which is the pressure transmitted from grain to grain, and the *neutral pressure* or *neutral stress*, which is the stress which acts within the water which fills the voids. The distinction between these two types of stresses is important, since only the intergranular pressure is effective in causing compression in a soil mass and in producing frictional resistance between the soil grains. These concepts may be explained by imagining a horizontal plane located $z$ feet below the water table in a soil mass. For convenience, also assume that the water table is located at the surface of the ground. At depth $z$, the total pressure (vertical), $p$, on a unit area located in a horizontal plane is equal to $\gamma z$, where $\gamma$ is the saturated unit weight of the soil. The total pressure is made up of the intergranular pressure, $\bar{p}$, and the neutral pressure, $u_w$. Thus, $p = \bar{p} + u_w$. Since the water in the voids of the soil is continuous, the neutral pressure must be equal to $\gamma_w z$, where $\gamma_w$ is the unit weight of water. Hence, $\gamma z = \bar{p} + \gamma_w z$ or $\bar{p} = \gamma z - \gamma_w z = z(\gamma - \gamma_w)$. The term, $(\gamma - \gamma_w)$, is frequently designated as $\gamma'$, the submerged unit weight. This concept is valid only if no flow is occurring through the soil mass.

## 90. Seepage Pressures

When seepage occurs through a soil mass the intergranular pressure may be either increased or decreased, as compared with that existing when there is no flow. This phenomenon is attributed to the frictional drag of the flowing water on the soil particles. The existence of seepage pressures may be of great practical importance in engineering design and construction. For example, the seepage pressures which accompany the flow of water through an earth dam may result in a decrease in intergranular pressure and hence the shearing resistance (ch. 7) of the soil on the downstream face of the dam; the slope may then be eroded and the soil washed away. Seepage pressures may also be of concern in the analysis of the uplift pressures on the bottom of a concrete dam. When water is pumped from a well or a sump, there may be a sudden drawdown of the water table. The seepage forces which accompany the flow of water toward the well may greatly increase the intergranular pressure in the soils adjacent to the well or sump. This increase in pressure may cause a sizeable amount of consolidation (ch. 6) in soils which have a loose structure. In clays the effect of seepage forces may be less apparent because of the cohesion possessed by these soils.

## 91. Critical Hydraulic Gradient

The upward flow of water through a soil mass and the seepage forces which accompany it will cause a reduction in intergranular pressures. Such a phenomenon is of greatest concern in granular soils and is illustrated in figure 27, which illustrates an experiment which may be carried out in the laboratory. Across the bottom of the soil sample shown, the total downward force is equal to $\gamma LA$; the total upward force of the water is $(h+L)\gamma_w A$. It is apparent that, if the value of $h$ is increased sufficiently, a condition will be reached when the total downward force due to the weight of the soil and the total upward force due to the water will just equal one another. The value of the hydraulic gradient corresponding to this condition is called *critical hydraulic gradient;* it is approximately equal to one for many coarse-grained soils. When this value is achieved, the intergranular pressure is equal to zero. This condition is known as *quicksand*. If the gradient exceeds the critical value the phenomenon known as *boiling* may occur, in which the soil grains at the surface become visibly agitated and may be washed away. This condition may occur,

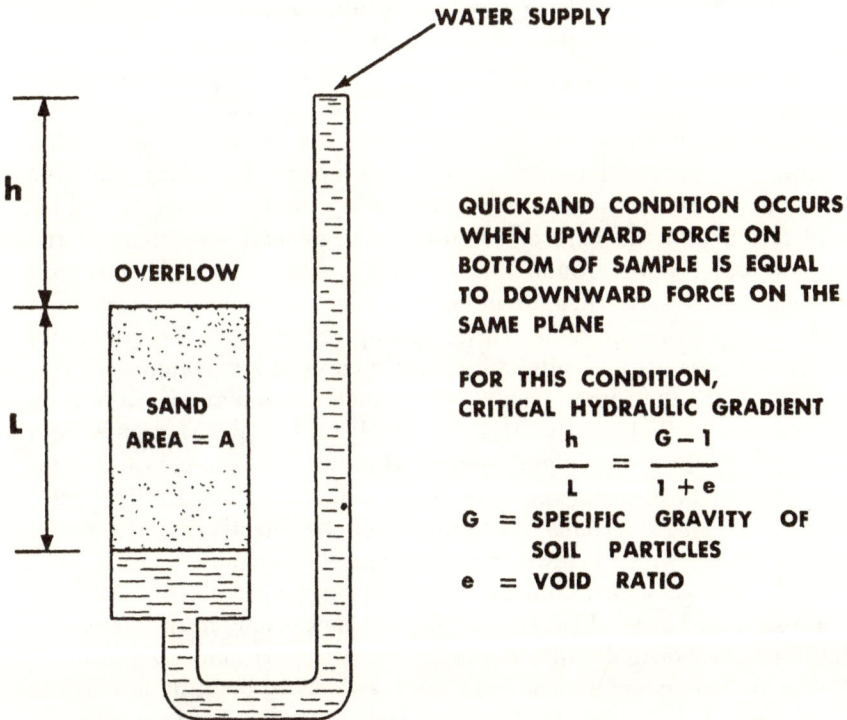

WATER SUPPLY

h

OVERFLOW

SAND
AREA = A

L

QUICKSAND CONDITION OCCURS WHEN UPWARD FORCE ON BOTTOM OF SAMPLE IS EQUAL TO DOWNWARD FORCE ON THE SAME PLANE

FOR THIS CONDITION,
CRITICAL HYDRAULIC GRADIENT

$$\frac{h}{L} = \frac{G-1}{1+e}$$

G = SPECIFIC GRAVITY OF SOIL PARTICLES

e = VOID RATIO

*Figure 27. Sketch illustrating meaning of term critical hydraulic gradient.*

for example, on the inside and at the bottom of cofferdams which are being used to excavate below the water table in cohesionless soils. A similar phenomenon sometimes occurs beyond the toe of masonry dams, earth dams, or levees when the water level becomes sufficiently high behind the structure. Under some circumstances boiling may begin and, as more and more material is washed away, a *pipe* formed beneath the structure. *Piping* can cause the failure of structures of this type. Boiling and piping may be prevented in several ways. One of these is to increase the weight of material in the zone where seepage velocities and seepage forces are high; a rock toe may be used for this purpose in an earth dam. Another way is to increase the distance the water must travel, thus reducing the hydraulic gradient. This might be done, for example, by driving steel sheet piling beneath a dam founded on pervious soils. Measures for the control of seepage are discussed in detail in paragraphs 263 through 272.

## 92. Drainage Characteristics of Soil Groups of the Unified Soil Classification System

The general drainage characteristics of soils classified under the Unified Soil Classification System are given in column (11) of table VI, appendix II. Soils may be divided into three general groups on the basis of their drainage characteristics, as follows:

*a. Well-Draining Soils.* Clean sands and gravels, such as those which are included in the GW, GP, SW, or SP groups fall into this classification. These soils may be drained readily by gravity systems. In road and airport construction, for example, open ditches sometimes may be used in these soils to intercept and carry away water which comes in from surrounding areas. This approach is very effective when used in combination with the sealing of the surface to reduce infiltration into the base or subgrade. In general statement, if the ground water table around the site of a construction project is controlled in these soils, then it will be controlled under the site, also.

*b. Poorly Draining Soils.* In this group are included inorganic and organic fine sands and silts, and organic clays of low compressibility, together with the coarse-grained soils which contain an excess of non-plastic fines. Soils in the ML, OL, MH, GM, and SM groups, and many of the Pt group might generally fall into this category. Drainage by gravity alone is apt to be quite difficult for these soils. Subsurface drainage systems may sometimes be effective, if the water table can be lowered below the effective height of capillary rise.

*c. Impervious Soils.* Fine-grained, homogeneous, plastic soils and coarse-grained soils which contain plastic fines belong in this category. This would normally include GC, SC, CL, CH, and OH groups. Subsurface drainage is so slow in these soils as to be of little value in improving their condition. Any drainage process is apt to be difficult and expensive.

## 93. Filters for Subdrains

The material which is used to backfill around the pipe in a subsurface drain (fig. 25), and for similar purposes, is generally known as a *filter* and must be very carefully selected if the drain is to fulfill its intended function, particularly where the drain is located in cohesionless or slightly plastic fine sand and silt. The reason for this is that, as the water moves toward the drain, seepage pressures are produced which tend to move the soil grains toward the pipe. The finer cohesionless materials may be washed out, which may lead to the settlement of the soil surrounding the pipe, the formation of erosion channels in the drained soil, or clogging of the drain. Criteria which have been developed to insure against these occurrences are as follows:

a. *Well-Graded Soils.*

$$\frac{D_{15} \text{ (filter)}}{D_{85} \text{ (protected soil)}} \quad \text{should be less than 5.}$$

$$\frac{D_{15} \text{ (filter)}}{D_{15} \text{ (protected soil)}} \quad \text{should be greater than 25.}$$

*Note.* By $D_{15}$ is meant the grain diameter which is larger than 15% of the soil grains, i. e. the 15% size on the grain-size distribution curve. Similarly for $D_{85}$.

b. *Uniform Soils.*

$$\frac{D_{15} \text{ (filter)}}{D_{85} \text{ (protected soil)}} \quad \text{should be less than 4.}$$

$$\frac{D_{15} \text{ (filter)}}{D_{15} \text{ (protected soil)}} \quad \text{should be greater than 5.}$$

c. *Relationship Between Size of Filter Material and Size of Perforations in Pipe.*

$$\frac{D_{85} \text{ (filter)}}{\text{Size of Pipe Perforations}} \quad \text{should be greater than 2.}$$

## 94. Other Drainage Processes

Drainage processes which do not utilize gravity alone are frequently used in construction operations. Most important of these is pumping which is most frequently done from sumps or well points. Additional attention will be given to pumping in chapter 12 of this manual. Fine silts and clay soils are very difficult to drain either by gravity or by pumping. They may be drained by desiccation (evaporation), consolidation, or sometimes by electrical methods. None of these processes is particularly important in ordinary construction operations.

# CHAPTER 5

# FROST ACTION

## 95. General

One of the most important soil problems with which the military engineer must deal in many parts of the world is frost action. *Frost action* is defined as the accumulation of ice lenses in soil under natural freezing conditions. Frost action is of particular concern to engineers who must design and build highways and airports, since these structures are usually built directly upon the ground surface. It is also of consequence in the design of foundations for structures, retaining walls and, to a lesser extent, embankment slopes. Two principal effects are associated with frost action, both of which can be very serious from an engineering point of view. One of these effects is frost heave. *Frost heave* is the raising of the ground surface or, more directly, the highway or airport pavement because of the growth of ice lenses in the upper soil layers, as indicated in figure 28. The amount of

*Figure 28. Typical ice lenses formed in a fine-grained soil as a result of frost action*

heave reflects directly the total thickness of ice lenses accumulated. in some cases, uniform heaving may occur; this usually causes little damage. More often, non-uniform or differential heaving occurs, which can be very damaging to pavements. The other general effect which is associated with frost action accompanies thawing of the soil usually in the spring of the year. This effect is sometimes categorically termed the *spring break-up*. During this period an excess of water may be accumulated in the subgrade soil, and failures may result because of the decreased supporting power of the soil. Damage which accompanies thawing may frequently be much more severe than that which is due to heaving.

## 96. Frost Heave

From a theoretical standpoint, frost heave is a very complex process which seems to defy purely analytical solution. However, the general nature of the process is quite well understood. When the air temperature drops below 32° F. and remains there for some length of time, the temperature in the upper soil layers also drops below 32° F. The plane or line above which the soil temperature is equal to or below 32° F. is called the *plane of freezing temperature* or the *frost line*. The depth to the frost line is frequently termed the *depth of frost penetration*. It is dependent upon a number of things, including the length and intensity of the cold weather and the density of the soil. Since the temperature of the soil is below 32° F. in the zone of frost penetration, it would be expected that the water in the larger voids would freeze. If the soil is saturated, it would also be expected that the soil would increase in volume, since water expands as it freezes. However, when water changes from the liquid to the solid state, the increase in volume is limited to about 9 percent. If the porosity of the soil is 50 percent and the voids are completely filled with water, then theoretically the increase in volume would be 0.09 (0.50)=0.045, or about 5 percent. Thus, if the saturated, frozen layer were 3 feet thick, then the increase in thickness would be on the order of 3 (12) .05=1.8 inches, which would generally not be enough to cause damage However, under certain circumstances, this soil in the field might suffer a heave of 6 inches or more, an increase in thickness which certainly can not be attributed to simple freezing of the water in the voids of the frozen layer. This fact, coupled with the observation that the moisture content of a frozen soil is frequently much greater than that existing before the soil is frozen, has lead to the understanding of the formation of ice lenses in soil as the basic cause of frost heave.

## 97. Formation of Ice Lenses

The general process of the formation of ice lenses in a soil mass is illustrated and explained in figure 29. The formation of ice lenses at the plane of freezing temperature is based upon two facts which have

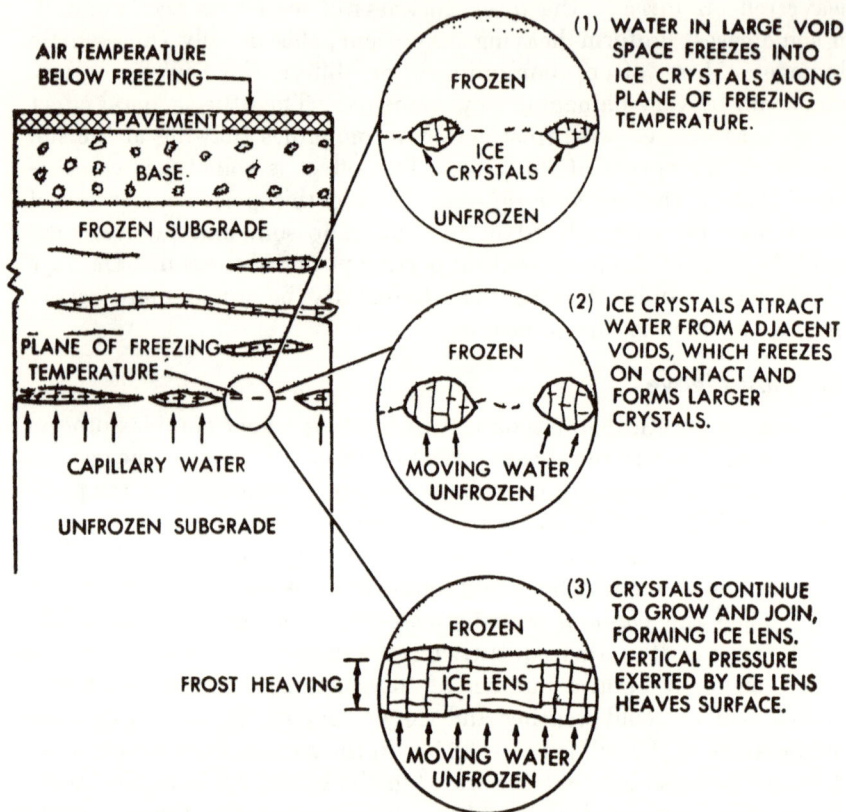

AIR TEMPERATURE BELOW FREEZING

PAVEMENT

BASE

FROZEN SUBGRADE

PLANE OF FREEZING TEMPERATURE

CAPILLARY WATER

UNFROZEN SUBGRADE

(1) WATER IN LARGE VOID SPACE FREEZES INTO ICE CRYSTALS ALONG PLANE OF FREEZING TEMPERATURE.

FROZEN

ICE CRYSTALS

UNFROZEN

(2) ICE CRYSTALS ATTRACT WATER FROM ADJACENT VOIDS, WHICH FREEZES ON CONTACT AND FORMS LARGER CRYSTALS.

FROZEN

MOVING WATER UNFROZEN

(3) CRYSTALS CONTINUE TO GROW AND JOIN, FORMING ICE LENS. VERTICAL PRESSURE EXERTED BY ICE LENS HEAVES SURFACE.

FROST HEAVING

FROZEN

ICE LENS

MOVING WATER UNFROZEN

*Figure 29. Sketch showing formation of ice lenses at plane of freezing temperature.*

been well established by years of experience and investigation. First is the fact that, while water in the larger voids will freeze at 32° F., the water in the smaller void spaces will not; lower temperatures are required. Thus, if an ice crystal is formed in a large void space or in a crack, water in the smaller voids is not frozen and is free to move through the soil. This unfrozen water is then attracted to the ice crystal, freezes upon contact, and the ice crystal increases in size. The increase in volume of the soil may still be quite small at this stage, *unless* there is a continuing supply of water which can be attracted to the ice crystals. A second fact of importance is that when water freezes it exerts a force which is similar to surface tension, the crystallization force. This force can pull water from the water table and even from saturated or partially saturated soils above the water table Thus, as in figure 29, capillary water from the unfrozen soil beneath the plane of freezing temperature is attracted to the ice crystal, frozen, and accumulates to form a layer or lens of ice. As the frost line penetrates deeper into the ground lenses are formed successively at different depths. The vertical pressure exerted by the ice lenses causes heave of the surface. In general statement, the direction of heat transfer parallels the direction of pressure exerted; this is normally upward. However, in the case of retaining walls and similar structures, a lateral pressure may be exerted as a result of the freezing process.

## 98. Thawing

*a. Favorable Situation.* If the air temperature remains slightly below freezing for a long period of time the ice accumulated in the upper soil layers may gradually thaw from the bottom upward because of the outward conduction of heat from the interior of the earth. This process may be aided by an insulating blanket of snow. In such a situation the excess water may simply drain downward through the soil; little damage to a road or airport pavement results.

*b. Frost Boils.* If the thawing process described above takes place rapidly the soil may not be able to dissipate the water from the melting ice lenses. This is particularly likely to happen in plastic soils. The upper layer of frozen soil may then become so thin that it is not able to support the stresses caused by wheel loads. It then breaks in localized areas, thus permitting the entrapped, supersaturated soil to extrude, forming what are commonly called *frost boils.*

*c. Unfavorable Situation.* If the air temperature suddenly rises from below freezing to well above that point, and if it remains there, as is usual in many areas during the spring thaw, a very unfavorable situation is created. In this case practically all of the thawing will be from the surface downward, leaving the accumulated water trapped above the deeper layers which are still frozen. The water then cannot escape except by surface runoff. Additional water may be present because of melting snow. In such a case the soil upon which the support of the pavement or structure depends may be converted to a liquid. Great damage can result.

## 99. Conditions Which Are Necessary for Detrimental Frost Action

As has been indicated, in order for detrimental frost heave to occur, three principal conditions must exist. These are—(1) the soil must be susceptible to frost action, (2) the temperature gradient in the soil must be favorable, and (3) there must be a supply of water available.

## 100. Frost-Susceptible Soils

*a. General.* In general statement, the most severe frost heaves are likely to occur in silt soils. The coarse-grained soils which contain little or no fines are not subject to severe heaving, since their void spaces are comparatively large, and the water in them freezes as does ordinary water. As previously stated, these are well-draining soils which have low capillarity. Clays are susceptible to frost heave, but they are so impervious that the amount of water which can be brought up to form ice lenses is quite limited. Severe local damage may sometimes be caused by the accumulation of ice in cracks or fissures in a clay soil. Column 9 of table VI, appendix II, gives information relative to potential frost action in soils classified under the Unified Soil Classification System.

*b. Criteria Based on Grain Size.* Soils which contain less than 3 percent by weight of material finer than 0.02 mm. in diameter are not frost-susceptible. Soils which contain more than 3 percent of material of this size generally are susceptible to frost action, except that uniform fine sands which contain up to 10 percent of material finer than 0.02 mm. generally are not susceptible to frost action.

## 101. Favorable Temperature Gradient

As has been noted, when the air temperature drops below 32° F. frost penetrates the ground. The variation of temperature with depth is frequently assumed to be linear and is the *temperature gradient.* The greater the drop in air temperature and the greater the conductivity of the soil, the steeper is the temperature gradient and the faster the rate of frost penetration. The rate of penetration is also dependent upon the conductivity of any cover materials which may be present, including snow, and in the case of subgrades, the surface and base courses. After frost has once penetrated the soil, any further drop in air temperature causes the depth of frost penetration to increase. The rate at which the frost line advances downward in the soil has a great effect upon the number and thickness of the ice lenses which are formed and the amount of heave which will occur. If the air temperature drops very sharply and remains well below freezing, it is not likely that severe frost heave will take place. This is because the zone of soil beneath the surface in which the temperature is below freezing and yet the water in the voids is not frozen, is quite thin. In such cases the ice lenses do not have time to form and the water content of the soil is not appreciably increased. The most severe frost heaves are likely to occur when the air temperature drops slightly below freezing and remains there over a long period of time. There will then be a thick zone in which the temperature is below 32° F., but the water in the smaller voids remains unfrozen. In this situation there will be ample time for the ice lenses to form and for the underlying, unfrozen soil to bring up additional water by capillary action to develop thick, closely spaced ice lenses.

## 102. Available Water

In order for severe frost action to occur there must be an ample supply of water available. This generally means that the soil must be saturated, and that the freezing zone must penetrate enough to be close to the water table. If the water table is at or near the surface of the ground, severe frost heave will occur very rapidly, assuming that the other requisites for frost action are present. In some locations difficulty may be experienced because of a perched water table, i. e. water which is trapped above a layer of impervious soil in a limited area. Severe frost heave may also occur when the water table is somewhat deeper but within the effective height of capillary rise, so that water may be drawn from the underlying soil into the freezing zone. In general terms, if the water table is 10 feet or more below the top of a frost-susceptible soil, severe frost heave will not occur. If the water table is 5 feet below or less, frost action will occur if other necessary conditions are present. Local heave may sometimes occur because of the accumulation of water near the surface because of poor drainage or the leakage of water through a pavement structure. A limited amount of heave may occur in saturated, fine-grained soils located above the water table. Clays which have a natural water content equal to or near the plastic limit are apt to cause trouble regardless of depth of the water table. A detrimental amount of heave is not likely if the soil is partially saturated.

## 103. Effects of Frost Action on Pavements

*a. Uniform Heaving.* Heaving of highway and airport pavements will occur when the combined thickness of pavement and base is not great enough to prevent freezing of a frost-susceptible soil. Under some conditions, frost heave may be quite uniform. This may occur when the rate of freezing, water supply, surface covering, and the soil are uniform over an area. This condition may exist, for example, at the site of an airfield on a level plain where ground-water and soil conditions are uniform. Little damage generally results from uniform heaving.

*b. Differential (Nonuniform) Heaving.* When ground-water and soil conditions are not uniform, differential or nonuniform frost heave will occur. This condition is likely to occur in locations where the subgrade soils vary from clean sands to silty soils and the water table is close to the surface of the ground. Differential heave may result in severe damage to either rigid or flexible pavements.

    (1) *Rigid pavements.* In this manual and in accordance with prevailing practice in the United States, the term *rigid pavement* is applied to pavements constructed of Portland cement concrete. Other pavement types are, with very few exceptions, called *flexible.* Rigid pavements may freeze so tightly to the subgrade soil that they are cracked or broken by differential heave. A failure of this sort is illustrated in figure 30.

    (2) *Flexible pavements.* Flexible pavements are somewhat better at withstanding the effects of differential heaving than are

Figure 30. Cracking of a rigid pavement caused by differential frost heave.

rigid pavements. However, even if failure of the surface does not take place, the bumps and waves caused by differential frost heave may present a serious hazard to traffic, since it is not unusual for small sections to heave as much as one foot. Surfaces which have become badly pitted or worn are more susceptible to destruction by differential heaving. Severe damage is also caused by alternate freezing and thawing of the soil beneath flexible pavements. Differential heave of a flexible (bituminous) road surface is illustrated in figure 31.

*Figure 31. Differential heaving of a flexible pavement.*

## 104. Effects of Thawing on Pavements

As has been previously indicated, pavements are frequently damaged severely as a result of the thawing of frozen soils. Damage is principally attributed to the accumulation of excess moisture in the subgrade.

*a. Rigid Pavements.* The weakening of the subgrade soil beneath a concrete pavement may not be of consequence, in that the pavement itself possesses considerable beam strength which may permit it to bridge over weak spots in the subgrade without breaking. If the subgrade is in a soupy condition traffic may have to be limited to light loads. One very severe problem is associated with the thawing of frozen soils beneath concrete pavements, however. This problem is *pumping*, which may be defined as the extrusion of a portion of the subgrade material at joints, cracks, and along the edges of the pavement. Pumping through joints will only occur if heavy wheel loads occur frequently, there is surplus water in the subgrade soil, and the soil is susceptible to pumping. Soils which are susceptible to pumping generally contain 45 percent or more of silt and/or clay. The amount of soil removed by pumping may be sufficient to cause a sizeable reduction in subgrade support for the slab, and may result in eventual damage to the pavement. Pumping through a transverse joint in a concrete pavement is shown in figure 32.

*b. Flexible Pavements.* The most common failures which occur in flexible pavements from frost action are those associated with the weakening of the subgrade soil because of the accumulation of an

*Figure 32. Pumping through a transverse joint in a concrete pavement.*

excess amount of water during the thawing period. Since the pavement has little or no beam strength, the presence of a weakened subgrade may result in rutting, cracking, and disintegration, if traffic is allowed on the road during this period. Wearing courses made from gravel, crushed rock, or similar materials may be lost entirely after one year of alternate freezing and thawing, since the material may sink into and mix with the soft subgrade soil under the action of traffic. Damage may be particularly severe when frost boils are

*Figure 33   A frost boil at the shoulder of a flexible surface.*

formed, since continued traffic may set up a pumping action which results in complete failure of the pavement in the neighborhood of the frost boil. This type of damage is shown in figure 33, which shows mud breaking through a flexible surface at the shoulder.

## 105. Effects of Frost Action on Other Structures

*a. Foundations for Structures.* Severe frost heaving, particularly if it is nonuniform, can cause damage to small buildings, small bridges, culverts, walls, and so on, if they have shallow foundations located above the depth of frost penetration on soils which are susceptible to frost action. This type of damage is illustrated in figure 34. During the thawing period such structures may fail because of the reduction in shearing strength of the soils upon which they are supported.

*Figure 34.  Cracked headwall caused by frost heaving.*

*b. Retaining Walls.* When poorly draining and frost-susceptible soils must be used for the backfill behind a retaining wall, frost action may greatly increase the lateral (horizontal) thrust exerted against the wall. Since a retaining wall is not normally designed to withstand this pressure, it may be forced out of line, cracked, or toppled over.

*c. Other.* Utilities, such as pipe lines which carry water or sewage, and other conduits may be damaged, either by frost heaving or by the lack of support during thawing. During thawing periods, also, exposed slopes of cuts and fills have a tendency to slump or slough because of the excess water present.

## 106. Preventive Measures

Preventive measures which may be used to eliminate or reduce the effects of frost action are generally based upon eliminating or controlling the principal contributing conditions previously described. Design and construction measures which are necessary to attain this end may be based upon careful study of the results of the soil survey (ch. 13). Particular attention should be given to the location and extent of frost-susceptible soils, water table locations, and the possibilities of lowering or eliminating ground water by drainage. Preventive measures which are detailed below may then be used singly or in combination with one another.

## 107. Removal of Frost-Susceptible Soil

One of the most effective ways of eliminating detrimental frost action is simply to remove frost-susceptible soils to the depth of frost penetration and replace them with clean granular soils. This is the approach most frequently used in cases where clean sands and gravels are readily available. Even though most of the soils encountered in the location of a road or airport are not susceptible to frost action, there almost invariably will be localized areas of undesirable soils which must be detected, removed, and replaced with selected granular material. Unless this is carefully done in such cases, differential heave will usually result.

## 108. Control of Ground Water

Because of frost action, as much attention must be given to the control of ground water by the use of subdrains as is given to the control of the flow of surface water. In some situations it may be possible to lower the water table, as indicated in figure 25. It must be emphasized that for this treatment to be effective, the water table must be lowered enough to be below the effective height of capillary rise for the soil concerned. Occasionally it may be possible to intercept the seepage flow which would contribute to detrimental frost action; one such case is illustrated in figure 35. Where a perched water table presents difficulties, it is sometimes possible to remove the water by punching through the impervious layer, thus permitting the water to drain into the more pervious underlying soil.

## 109. Prevention of Capillary Rise

In many cases it may be neither feasible nor practical to lower the water table sufficiently to prevent the capillary rise of moisture. This may be the case, for example, in swampy areas where it is not feasible to find an outlet for water that might be collected in sub-

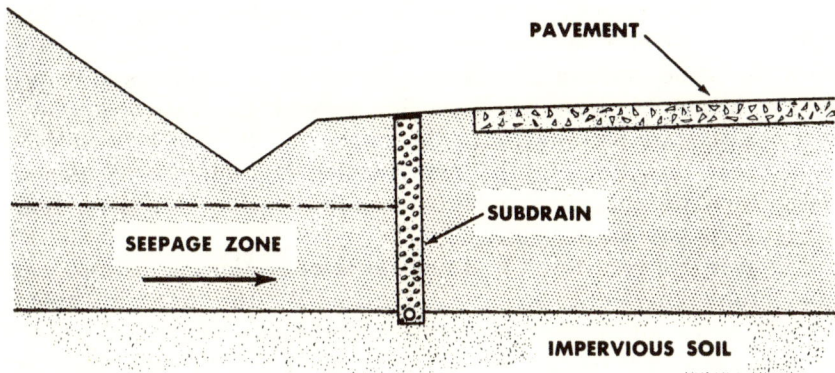

*Figure 35. Use of an intercepting subdrain to prevent detrimental frost action.*

drains. It may then be necessary to use some treatment which will successfully intercept the rise of capillary water. This may be done by placing a 6-inch layer of clean and pervious granular material 2 or 3 feet beneath the surface. If the depth of frost penetration is not too great it may be cheaper simply to backfill completely with granular material. Soil admixtures, such as those used in bituminous soil stabilization and soil-cement (ch. 10), used to form a six-inch layer immediately below the frost line will generally be effective in reducing the movement of capillary water. Another method which has been used effectively, although it is relatively expensive, is to excavate to the frost line, lay prefabricated bituminous surfacing, and backfill with granular material.

## 110. Insulation

One obvious approach to the problem of frost action is to provide some form of insulation to reduce the depth of frost penetration.

*a. Additional Thickness of Pavement and Base.* As far as pavements are concerned, probably the most generally used method of protecting them against the effects of detrimental frost action is simply to increase the combined thickness of pavement and base (nonfrost-susceptible material) enough to prevent freezing of the subgrade. Although the same material which is used in the base may sometimes be used when the thickness is increased, this additional thickness is frequently termed a *subbase.* Clean granular materials, like gravel, crushed gravel, or crushed rock, are commonly used in the subbase. Detailed design methods used in determining the required thickness of pavement and base (and subbase) for given traffic and soil conditions where frost action is a factor are beyond the scope of this manual. Base drains (par. 81) sometimes have been successful in controlling frost action.

*b. Shoulders.* During freezing weather, if the wearing surface of a road or airport runway or taxiway is kept free from snow, then the shoulders must also be kept free of snow. This is necessary because snow has some insulating qualities. Thus, if the shoulders are covered with snow, freezing will begin first beneath the wearing surface. Water then may be drawn into this zone from the soil beneath the shoulders. However, if the snow is removed from the shoulders, freezing will begin in the shoulder area, and the effects of frost action beneath the pavement reduced.

*c. Slopes.* Slopes in cut and embankment sections sometimes require protection from frost action, particularly if they are fairly steep (e. g. 1 to 1), and if they are subject to the flow of ground water. In such cases, a first step usually is to use an intercepting ditch at the top of the slope to divert the surface water (fig. 74). Drains may be used to intercept seepage. The slope may then be insulated by applying a blanket of granular material, or straw, hay, mulch, or similar material. More permanent protection may be found in seeding or sodding the slope, where this is feasible.

## 111. Other Methods

*a. Foundations for Structures.* The problem of frost action in relation to structures which have shallow foundations is most frequently solved by the standard practice of putting the bottom of the foundation below the anticipated frost line.

*b. Retaining Walls.* Retaining walls are ideally designed to support clean, granular fills and with drainage systems which are adequate to provide for the quick removal of surface and ground water. If this is done properly, frost action is not a problem. Treatments which can be used when this ideal solution is not feasible are discussed in detail in chapter 12.

*c. Utilities.* Damage which results to water and sewer pipes and other conduits from frost heave is usually prevented simply by burying them below the frost line. Because of the high conductivity of iron, iron pipelines should be at least 25 percent deeper than the thickness of the frost zone. If pipelines can not be placed below the surface of the ground they must be effectively insulated.

# CHAPTER 6
# COMPRESSIBILITY

## 112. General

As used here, the definition of the term *compressibility* is somewhat restricted in scope. It is taken to mean that property of a soil which permits it to deform under the action of an external compressive load. Loads with which we are concerned in this chapter are primarily static loads which act, or may be assumed to act, vertically downward. Brief mention will be made of the effects of vibration in causing compression. Thus, we are principally concerned here with the property of a soil which permits a reduction in thickness (volume) under a load like that of the weight of a bridge pier, a building, or a highway embankment. The compression of the underlying soil may lead to the settlement of such a structure. Settlement is discussed in more detail in chapter 8. Deformations which are caused by shear, in which one portion of the soil mass moves (or slides) with respect to another, are not considered in this chapter.

## 113. Soil Behavior Under Compressive Load

In a general sense, all soils are compressible. That is, they undergo a greater or lesser reduction in volume under compressive static loads. This reduction in volume is attributed to a reduction in volume of the void spaces in the soil rather than to any reduction in size of the individual soil particles, or of the air and/or water which is contained in the voids. If the soil is saturated before the load is applied, it is apparent that some of the water must be forced from the voids before compression can take place. Under our assumptions, if the water can not escape from the voids there will be no compression. This is not strictly true for some partially saturated soils, since there may be some compression of the gas in the voids even if the water can not escape. Fortunately, this latter case is of little practical consequence. The rate of compression is a function of how quickly the water can escape, i. e. the permeability. The magnitude of the compression which will occur in a given soil under a given set of conditions depends upon a number of different factors, including density or void ratio, grain size and shape, structure, past history of the soil deposit, the magnitude and method of application of the load, and the degree of confinement of the soil mass. In most of the discussion which follows it is assumed that the soil mass undergoing compression is completely confined, generally by the soil which surrounds it.

## 114. Cohesionless Soils

*a. General.* From a practical standpoint the compression of confined coarse-grained cohesionless soils, such as sand and gravel, is rarely a matter of concern. This is true because the amount of compression is likely to be quite small in a typical case and the compression will occur very rapidly after the load is applied, provided only that the water can escape. Thus, generally speaking, all the compression which is going to take place will take place during the period of load application (construction). Deformations which are thus produced in sands and gravels are essentially permanent in character. There is little tendency for the soil to return to its original dimensions or *rebound* when the load is removed. A sand mass in a compact condition may eventually attain some degree of elasticity upon repeated applications of load.

*b. Factors Affecting Compressibility.* The primary factor which determines the compressibility of a sand or gravel is relative density (par. 29). The compressibility of a loose sand deposit is much greater than that of the same sand in a relatively dense condition. It is infrequently necessary to determine the relative density quantitatively, and this property generally may be judged by visual inspection or simple penetration tests conducted in the field during soil exploration (ch. 13). In the field it is generally undesirable to found structures upon very loose sand deposits; they should be avoided if possible or compacted to a greater density before the load is applied. Some essentially cohesionless soils, including certain very fine sands and silts, have loose structures with medium compressibility. In a general way, both gradation and grain shape influence compressibility of a cohesionless soil. Gradation is of indirect importance in that a well-graded soil will generally have a greater natural density than one of uniform gradation. Soils which contain flaky particles are more compressible than those composed entirely of bulky grains. A fine sand or silt which contains mica flakes may be quite compressible.

*c. Effects of Vibration.* Although this chapter is concerned primarily with soils under static loads, passing mention must be made of the effects of vibration. Vibration may greatly increase the relative density of cohesionless soils. A loose sand deposit which is subjected to vibration during construction, as by pile driving or blasting, may change to a dense condition. This change in density may have disastrous effects upon the structures involved. Advantage frequently is taken of this fact to compact or "densify" cohesionless soils as a planned part of construction operations. Cohesive soils are relatively insensitive to the effects of vibrations.

## 115. Consolidation of Cohesive Soils (Clays)

The compression of cohesive, fine-grained soils, particularly clays, is quite different from the compression of cohesionless soils. Under comparable static loads, the compression of a clay may be much greater and the time required for the compression to occur may be very long. In many cases of practical importance the settlement of a structure is due to the compression of a saturated clay stratum which is located between layers of sand or stiffer clay, or bordered on the lower side by rock. Such a clay layer is almost completely confined, because of adhesion between the clay and the material which lies above and below it. A specific name is given to the process of compression of a confined, saturated clay soil under the action of static loads. This process is called *consolidation*. The consolidation of thick, compressible clay layers is a serious matter and a source of annoyance, or the cause of structural damage or failure, in many instances. Uniform settlement, in which the various portions of a structure settle approximately equal amounts, may be only annoying, as in the case of a building which settles several inches below the surrounding ground, or relatively serious as in the case of a building which settles enough to break sewer and water lines. If the settlements of the different parts of a structure due to consolidation are not uniform, *differential settlement*, there may be serious, structural consequences. A building may suffer damage to brick or plastered walls, or actual structural failure, when different portions of the building settle different amounts. Similarly, a highway pavement may be damaged by the nonuniform settlement of an embankment founded on a compressible soil.

## 116. Mechanics of Consolidation

The process of consolidation involves three closely related phenomena. These are the gradual transfer of pressure caused by the superimposed load from the water contained in the voids of the soil to the solid phase, the gradual escape of water from the voids because of the hydraulic gradient created by the application of load, and the gradual compression or reduction in void ratio of the soil mass. This concept can be further clarified by considering the changes in stresses which occur during consolidation. As has been previously explained (par. 89), the stress conditions which exist at a point in a soil mass may be expressed by the equation, $p = \bar{p} + u_w$, in which $\bar{p}$ is the total pressure, $\bar{p}$ the intergranular pressure, and $u_w$ the neutral pressure. For the purposes of this discussion, assume that this equation defines the vertical stresses which exist at a point in a saturated soil mass located below the water table. There is no flow through the soil and the soil is completely consolidated under the weight of soil above this point. Now imagine that a load is applied to the surface, as by the weight of a building  Because of this load the total pressure at the point with which we are concerned increases by an amount $\Delta p$.  The instant after the load is applied and because of the low permeability of a clay soil, none of the water has escaped, no compression has taken place, and the added stress is carried entirely by the water phase.  The applicable equation is $p + \Delta p = \overline{p} + (u_w + \Delta p)$.  The increase in neutral pressure creates a hydraulic gradient, and the water begins to flow from the stressed point toward points which are under less stress, or where less resistance is offered to flow.  As the water flows away from the point which we are considering, the neutral stress decreases and the intergranular pressure increases.  At some later time, the applicable equation might be $p + \Delta p = (\overline{p} + 0.5\Delta p) + (u_w + 0.5\Delta p)$, indicating that half the increase in pressure has been transferred to the solid phase.  As we have seen, many clays have highly compressible structures.  Under the increased intergranular pressure the soil particles are forced closer together and the void ratio decreases.  At some later time, the added stress is completely transferred to the solid phase and consolidation is 100 percent complete.  The applicable equation then is $p + \Delta p = (\overline{p} + \Delta p) + u_w$.  As previously mentioned, the complete consolidation of a thick clay layer may take a very long time, many years, in fact.  The explanation which has been presented here is somewhat simplified from a theoretical standpoint.  It is useful, however, in gaining a general understanding of the consolidation process.

## 117. Consolidation Test

A determination of the consolidation characteristics of a compressible soil is necessary to the rational design of many large structures which are founded on or above soils of this type.  Consolidation characteristics generally are determined by laboratory *consolidation tests* performed upon undisturbed samples; samples in which the natural structure, void ratio, and moisture content are preserved as carefully as possible.  This test is definitely not one which is conducted in the normal field laboratory.  A brief explanation of the test is presented here for the purpose of further clarifying understanding of consolidation and as a background for an explanation of the consolidation characteristics of clay soils.  As will be seen later in this chapter, approximate determinations of compressibility may be made from water content and limit tests.  A simplified sketch of a laboratory consolidometer is shown in figure 36.  In conducting a consolidation test an undisturbed sample is trimmed to fit the exact size of the consolidometer ring; a typical size is 2½ inches in diameter and 1 inch thick.  Tests may occasionally be performed upon disturbed or remolded samples, although the consolidation characteristics of most clays are sharply altered by remolding.  The sample is completely confined and drainage generally is permitted through porous stones

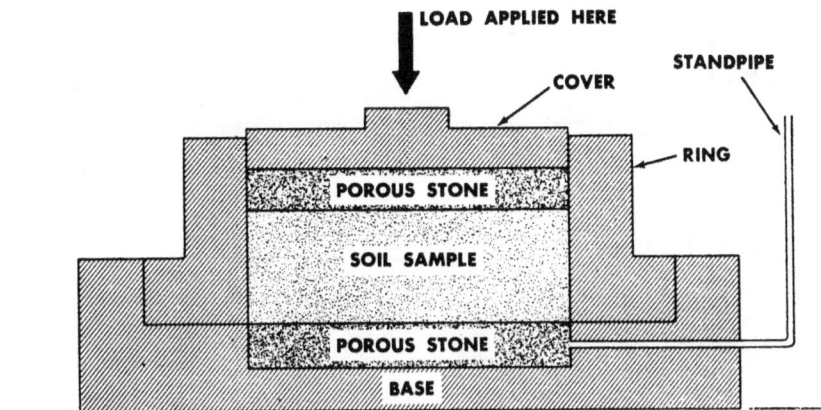

*Figure 36. Simplified sketch of a laboratory consolidometer.*

placed at the top and bottom of the sample. The equipment is then assembled, arrangements made for reading the change in thickness of the sample under load, and the first load applied. Dial readings indicating the reduction in thickness are taken at selected time intervals. The lead is allowed to remain in place until virtually all movement ceases. The load on the sample is then increased and the process repeated. Typical increments of load are ¼, ½, 1, 2, 4, 8 and 12 tons per square foot. After the sample has consolidated under the last load increment, the load may be decreased gradually if information is desired about the "rebound" characteristics of the soil. A typical test would require about a week to perform.

## 118. Results of Consolidation Test

Details of the calculations of the results of a consolidation test are beyond the scope of the manual, as are some of the details of interpretation. It is important to note, however, that two principal sets of data are obtained. One is the relationship between void ratio and pressure for the test as a whole, the *e-p curve*. The other consists of the time-consolidation curves for each increment of load. In figure 37 are shown two plots of the void ratio-pressure relationship obtained from a consolidation test upon a normally loaded clay of low sensitivity. In one case arithmetic scales are used for plotting both *e* and *p*. In the other the vertical scale (*e* values) is arithmetic while the horizontal scale is logarithmic (*p* values). The semilogarithmic plot is the one generally used in the presentation of consolidation data.

## 119. Characteristics of Normally Loaded Clays

A clay deposit in nature is said to be *normally loaded* if it has never been subjected to vertical pressures which are greater than those which act on it at the present time. As noted, figure 37 relates to a normally loaded clay of low sensitivity. By *sensitivity* is meant, in

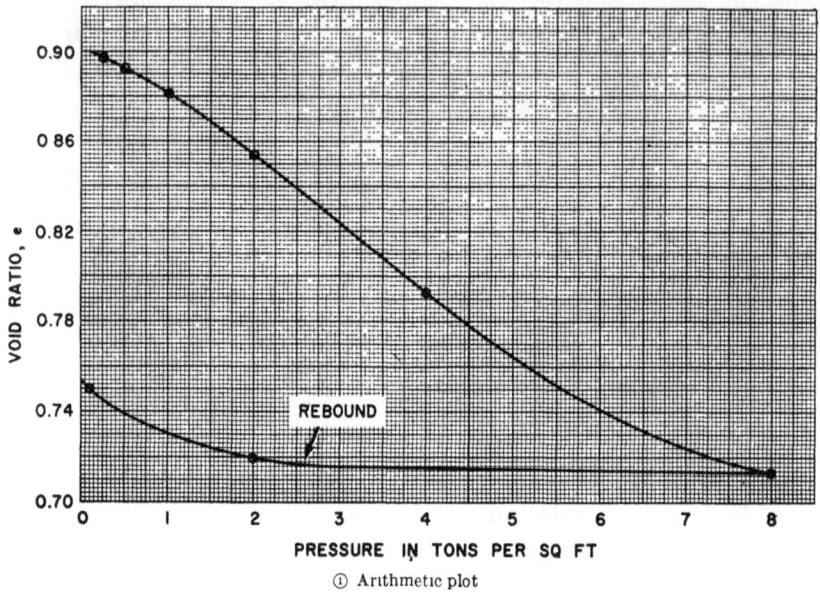

① Arithmetic plot

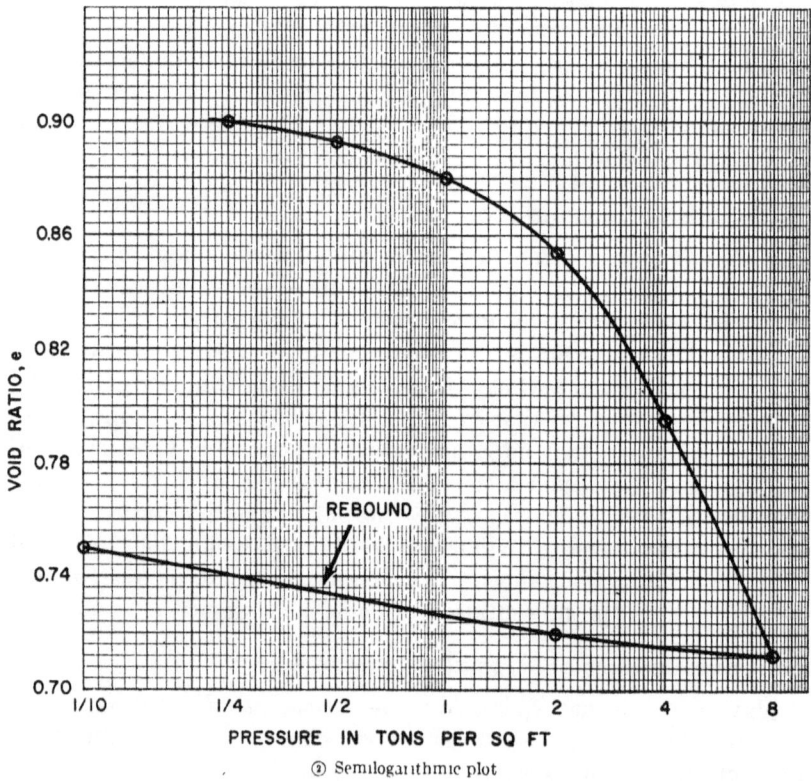

② Semilogarithmic plot

*Figure 37.   Void ratio-pressure curves for a normally loaded clay of low sensitivity*

general terms, the susceptibility of the clay to loss in strength when remolded. Specific criteria defining sensitivity in terms of the ratio between the unconfined compressive strength of a clay in an undisturbed condition and that of the same soil in a remolded condition are presented in paragraph 138. The semilogarithmic plot of figure 37 has two typical portions; the relatively flat curving section in the range of lower pressures and the straight-line portion in the range of higher pressures. The consolidation of a clay soil is not an elastic, reversible process. If the pressure causing consolidation is removed, the soil generally will increase in volume, but the amount of expansion generally will be much less than the original compression. If the load is applied again to the same soil, a recompression curve similar to the original will be obtained. At higher pressures the straight-line relationship will be developed as before. The straight line represents, at least approximately, the behavior of a field layer of normally loaded clay. The slope of this portion of the curve can be extrapolated to permit the calculation of the settlement to be expected of a structure, as indicated in paragraph 163. The slope of the straight-line portion of the curve is defined as the *compression index*, $C_c$. On the semilogarithmic plot, it is given as $C_c = (e_1 - e_2)/\log_{10} (P_2/P_1)$, where $e_1$ and $e_2$ are the void ratios corresponding to $P_1$ and $P_2$. In other words, $e_1 - e_2$ represents the decrease in void ratio which results from an increase in pressure from $p_1$ to $p_2$. To determine $C_c$ numerically, it is convenient to select an arbitrary value of $p_1$ and a value of $p_2$ equal to $10p_1$. Then $C_c = e_1 - e_2$, since $\log_{10} (10/1) = 1$. For the clay of figure 37, $C_c$ is 0.35. $C_c$ is indicative of the compressibility of the soil. The higher the value of the compression index, the greater is the compressibility of the soil, provided that compression is within the straight-line portion of the curve.

## 120. Relationship Between Compression Index and Liquid Limit for Normally Loaded Clay Soils

An approximate value of $C_c$ for normally loaded clays of low to moderate sensitivity may be obtained from the relationship $C_c = 0.009$ $(w_L - 10)$, where $w_L$ is the liquid limit, expressed in percent. A rough idea of the compressibility of a clay of this type may be obtained from this relationship, which is shown graphically in figure 38. This approach should be used with some caution, however, since the experience of some investigators has indicated that the relationship is of little value for other than glacial clays.

## 121. Compressibility of Highly Sensitive Clays

Values of the compression index derived from the equation of the preceding paragraph are likely to be too small for highly sensitive clays. Included in this group are the volcanic clays found in Mexico City, some highly organic clays, and certain marine clays. Extremely

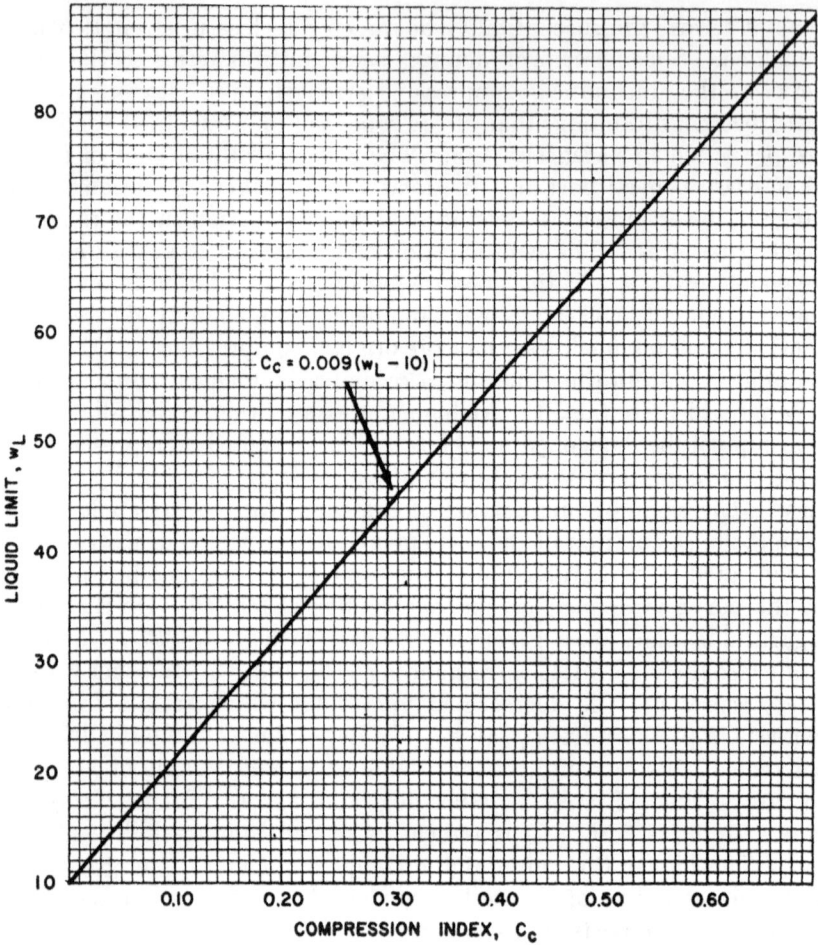

Figure 38. *Approximate relationship between compression index and liquid limit for normally loaded clays of low to moderate sensitivity.*

high compressibility is indicated if a clay contains a high percentage of organic material, has a liquid limit of 100 or more, or has a natural water content greater than the liquid limit at a considerable depth below the surface. If such conditions are encountered at the site of a major structure the prediction of settlements should be based upon the results of consolidation tests upon undisturbed samples.

## 122. Precompressed Clays

Quite frequently clay strata which are encountered in the field have been *precompressed* or *preconsolidated*. In general terms, this means that in the past the clay layer has been subjected to greater pressures than those which act on it at the present time. The greater pressures of the past may have been caused by the weight of glacial ice, the existence of a greater depth of overburden which has since been removed by erosion, or by drying (desiccation). Clays which have been heavily precompressed can sustain very heavy loads without appreciable settlements. The value of the compression index computed from the equation of paragraph 120 will be several times too large for such a soil. Many hard and dense clays are of this type. Recognition of a precompressed clay and settlement predictions for an important structure should again be based upon the careful laboratory testing of undisturbed samples. Recognition of a clay of this type may be based upon a knowledge of the geological history of the area and the fact that, in many cases, the natural moisture content of a precompressed clay is nearer to the plastic limit than to the liquid limit.

## 123. Time Rate of Consolidation

As has been indicated, the consolidation of a clay soil does not take place instantaneously with the application of load. Time is required for the consolidation of a laboratory sample or a clay stratum in the field, due primarily to the low permeability of these soils and the slow escape of water from the voids. Thus, the settlement of a structure due to the consolidation of a thick clay layer may take many years to complete. This action is further indicated by the plot of figure 39, which is the time-consolidation curve for one increment of load in a laboratory consolidation test. As indicated, the decrease in thickness

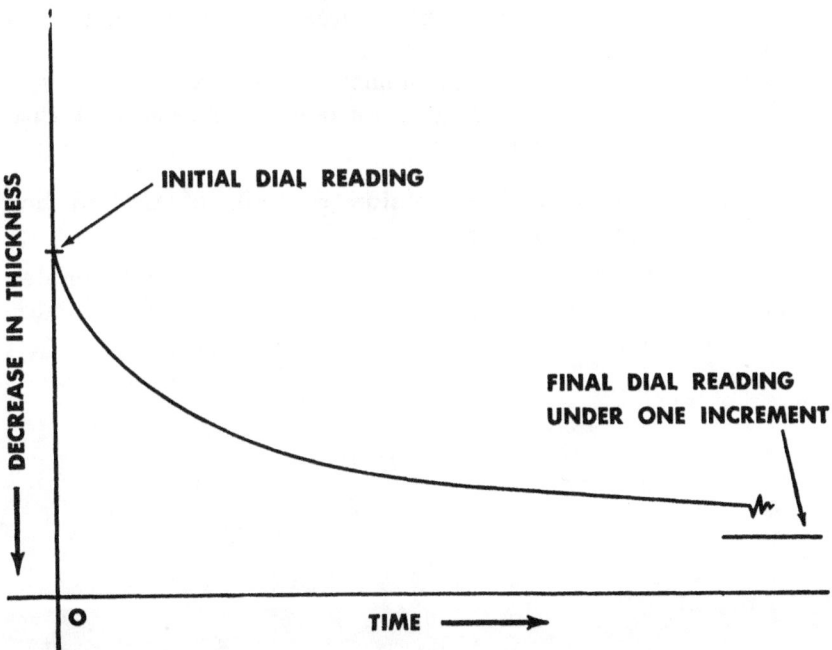

Figure 39. Time-consolidation curve for one increment of load in a laboratory consolidation test.

of a laboratory sample, and of a typical layer in the field, occurs quite rapidly at first and then more slowly until a fairly constant rate is achieved until the consolidation is 100 percent complete under the given load. Prediction of the rate of settlement of a structure can be made by use of the laboratory test data. Methods of making time-rate predictions are not discussed in this manual because of their complex theoretical nature and the fact that they are seldom necessary, except for very large and important structures. In addition to the consolidation characteristics of the clay and the loads involved, the time rate of consolidation in the field is primarily dependent upon the thickness of the layer and the drainage conditions, i. e. whether the water can escape only at the top or bottom of the layer, or at both top and bottom.

## 124. Expansion

Soils which have been compressed generally do not return to their original dimensions if the load causing compression is removed. Some soils do expand enough to be damaging under certain circumstances. Such soils are sometimes termed *elastic soils* and are most likely to present problems when they are used as subgrades for roads and air-port runways. Problems associated with elasticity are characteristic of certain fine-grained soils. For example, silts which contain large amounts of mica, some diatomaceous earths, and fine-grained soils which contain large amounts of organic colloids may possess enough elasticity to be troublesome in certain situations. Fortunately, soils of this general character do not occur very frequently. Damage due to excessive expansion has been minimized in recent years by recognition of these soils and avoiding or removing them whenever their presence would be harmful.

## 125. Compressibility and Expansion of Soils of Unified Soil Classification System

Column 10 of table VI, appendix II, gives the general compressibility and expansion characteristics of the soil groups of the Unified Soil Classification System.

# CHAPTER 7

# SHEARING RESISTANCE

## 126. Introduction

From an engineering viewpoint, one of the most important properties which a soil possesses is *shearing resistance* or *shear strength*. The shearing resistance which a soil may possess under given conditions is related to its ability to withstand load. It is especially important in its relation to the supporting power, or stability, of a soil which is used as a base or subgrade beneath a road surface or an airport runway, and the pressure which can be used in the design of the foundation of a building or a bridge pier. It is also of consequence in determining the stability of the slopes used in a highway cut, an embankment or an earth dam, and in estimating the pressures which will be exerted against an earth-retaining structure, such as a retaining wall.

## 127. Stresses

In our explanation of shearing resistance, we are greatly concerned with the stresses which occur in a mass of soil which is subjected to load. Here the term *stress* is taken to mean *unit stress*, i. e. force per unit of area. In soil problems two basic types of stresses are of principal concern. These are illustrated in figure 40, which represents the stress conditions existing on some plane through a point within a stressed body. *Shear stress* (*s*) is the force per unit of area acting on the plane in a direction which is parallel to the plane under consideration. *Normal stress* (*p*) is the force per unit of area acting on the plane in a direction which is normal or perpendicular to the plane. The normal stress which is shown in the figure acts toward the plane and is a *compressive stress*. If it were acting away from the plane it would be a *tensile stress*. If the normal stress acts on a plane upon which there is no shearing stress (*s*=0), then it is termed a *principal stress*. It is possible to completely define the stress conditions at a point within a stressed body by determining three independent, mutually perpendicular principal stresses which act on three independent, mutually perpendicular principal planes. Stress conditions at a point may be defined in other ways, also. The principal stress which has the largest value is called the *major principal stress*; the smallest, the *minor principal stress*, and the one of intermediate value, the *intermediate principal* stress. In soil problems we primarily are concerned with normal stresses which are compressive. This somewhat simplifies any analysis of stresses which may be necessary. Our approach is further simplified by limiting ourselves to a two-dimensional stress picture. Stresses which act in a direction perpendicular to, or at some other angle to, the plane of the paper are not considered. Figure 40 is a two-dimensional stress picture.

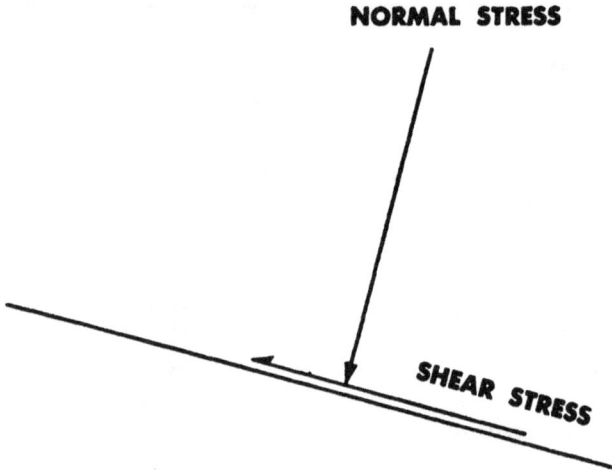

*Figure 40. Shear stress and normal stress.*

## 128. Mohr's Theory of Failure

Failures that occur in soil masses are frequently explained in terms of *Mohr's theory of failure*, which has been found to be particularly applicable to this type of material and to concrete. The general notion involved in Mohr's theory of failure is that failure at a point within a stressed body does not occur because the normal stress on some plane through that point reaches a limiting value, nor because the shear stress on some plane through the point reaches a critical value, but rather that failure will occur when there is a critical combination of shear stress and normal stress on a plane through the point. The occurrence of stresses which correspond to failure at one point in a soil mass may not necessarily mean that the entire mass will fail, but when failure conditions are reached along a surface or in a zone of considerable extent the mass will fail. Failure of a soil generally involves movement or sliding of one portion of the mass with respect to another. In a failure of this sort it is frequently convenient to visualize what happens by saying that the shearing forces exceed the available shearing resistance along the plane of failure. Failures of this type are frequently, although somewhat loosely, termed *shear failures*. A shear failure in a soil mass may cause the failure of a structure founded upon it. A structure may fail in other ways, also, as by excessive differential settlement caused by consolidation.

## 129. Mohr's Envelope of Rupture

Assume that a set of coordinate axes are established as indicated in figure 41. On the horizontal axis will be plotted values of normal stress (compression) with values increasing from zero at the origin toward the right. On the vertical axis will be plotted values of shear stress increasing from zero at the origin toward the top. The coordinates of a point such as A in the figure $(p, s)$ represent the normal stress and the shear stress on some plane through a point in a stressed body. A combination of $p$ and $s$ which corresponds to failure may plot at point B. If a plot is made of the values of $p$ and $s$ (obtained by testing) which correspond to failure for a given soil mass under given conditions a line will be formed which is called *Mohr's envelope of rupture*. Stress combinations which plot below this line will not cause failure. Stress combinations which plot on the envelope are, of course, critical ones. Stress combinations which plot above the envelope are theoretically impossible, since failure would have already occurred. Mohr envelopes for typical soils under carefully prescribed conditions are well established, both by laboratory experiment and field experience. Certain of these envelopes are discussed in detail in the paragraphs which follow.

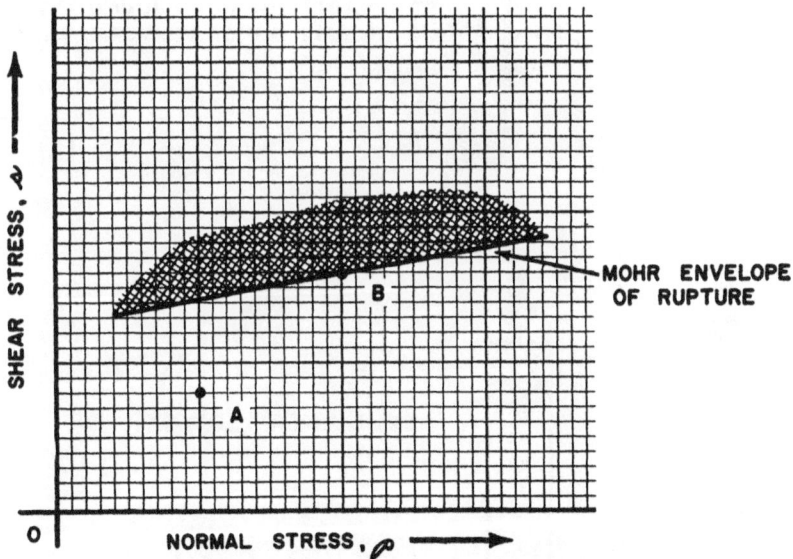

*Figure 41. Mohr's envelope of rupture.*

## 130. Laboratory Determinations of Shear Strength

Three test procedures are commonly employed in soil mechanics laboratories to determine the shear strength of a soil. These are the direct shear, triaxial compression, and unconfined compression tests. The basic principles involved in each of these tests are illustrated in the simplified drawings of figure 42.

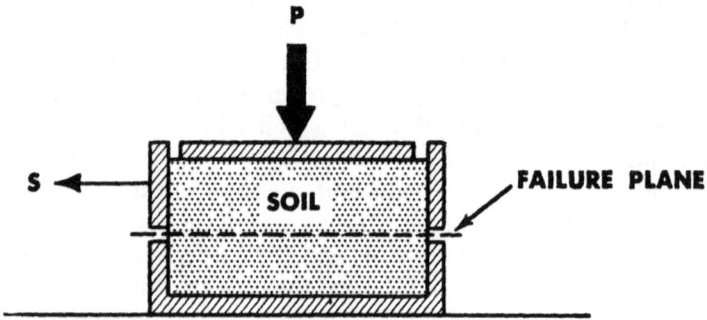

① Direct shear test.

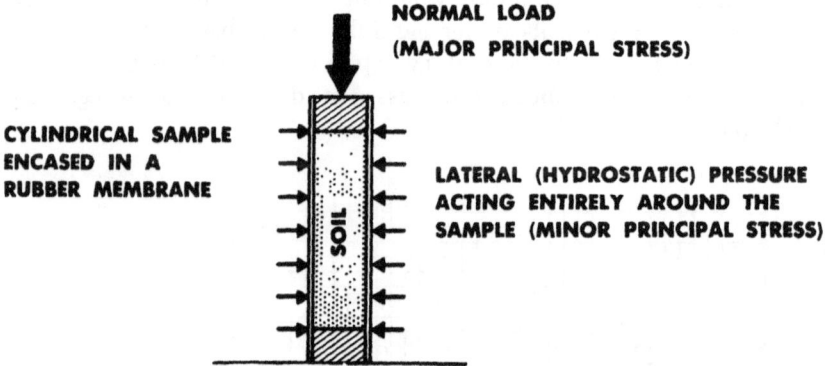

② Triaxial compression test.

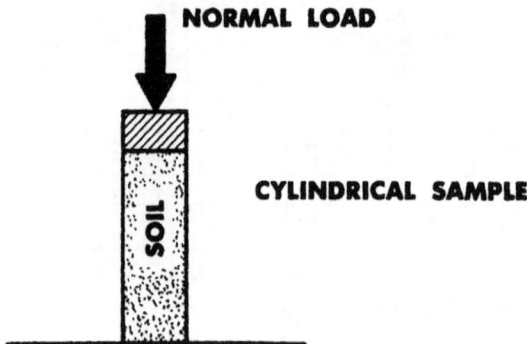

③ Unconfined compression test.

Figure 42. Laboratory shear tests.

*a. Direct Shear Test.* As indicated in figure 42 ①, in the direct shear test a normal load is applied to the top surface of a confined soil sample; a horizontal (shearing) load is then applied and this load increased until a shear failure occurs. The normal load $(p)$ is held constant throughout the test and, since the location of the failure plane is known, the shearing stress on this plane at the moment of failure $(s)$ can easily be determined. The Mohr envelope may be defined by performing two or more tests upon the given soil in a given condition, using a different normal load for each test.

*b. Triaxial Compression Test* (fig. 42 ②). In this test a cylindrical soil sample is encased in a thin rubber membrane. The rather elaborate equipment is assembled and lateral (hydrostatic) pressure applied to the exterior of the sample. The lateral pressure is held constant throughout the test and the vertical load increased until failure occurs. In this particular situation, the lateral pressure represents the minor principal stress (and the intermediate principal stress, since the lateral pressure acts completely around the outside of the sample). The vertical pressure is the major principal stress. At failure, the values of the major principal stress and the minor principal stress are determined. Both the location of the failure plane and the stresses on the plane of failure may be calculated by methods which are beyond the scope of this manual. The Mohr envelope may be defined by performing two or more tests at different lateral pressures. This test is used in many soil mechanics laboratories in preference to the direct shear test, principally because of the fact that more uniform stress conditions are created in the triaxial test, and it is easier to control the drainage of the sample during the conduct of the test.

*c. Unconfined Compression Test.* As indicated in figure 42 ③, the unconfined compression test is a simple compression test which usually is performed upon a small (typically 1 or 1½ inches in diameter and 2 or 3 inches in height) cylindrical sample. It resembles the usual compression test of a concrete cylinder, except that the sample and the loads are much smaller. A procedure for performing the unconfined compression test is described in TB 5–253–1. The principal information which is gained from the test is the value of the compressive unit stress which corresponds to failure. This value frequently is designated as $q_u$ and commonly is expressed in pounds, or tons, per square foot. The relationship between stress and strain (deformation or, in this case, shortening) during the progress of the test also may be determined, if desired. The unconfined compression test can not be performed upon cohesionless soils, since these soils generally will not stand unsupported. It is used very widely to estimate the shearing strength of clay soils. For these soils the following approximate relationship is used: $c = \frac{1}{2}q_u$, where $c$ is the cohesion (par. 137) and $q_u$ is the unconfined compressive strength. The test normally is performed upon relatively undisturbed samples. Because of the complete lack of confinement during the test and the fact that the test is run quickly, the shearing strength obtained is regarded generally as the minimum strength of a cohesive soil at a given water content. Its use in design is therefore conservative in many instances.

## 131. Shearing Resistance of Dry Sands and Gravels

The Mohr envelope for a dry sand (and/or gravel) is shown in figure 43. As shown, the envelope is a straight line which passes through

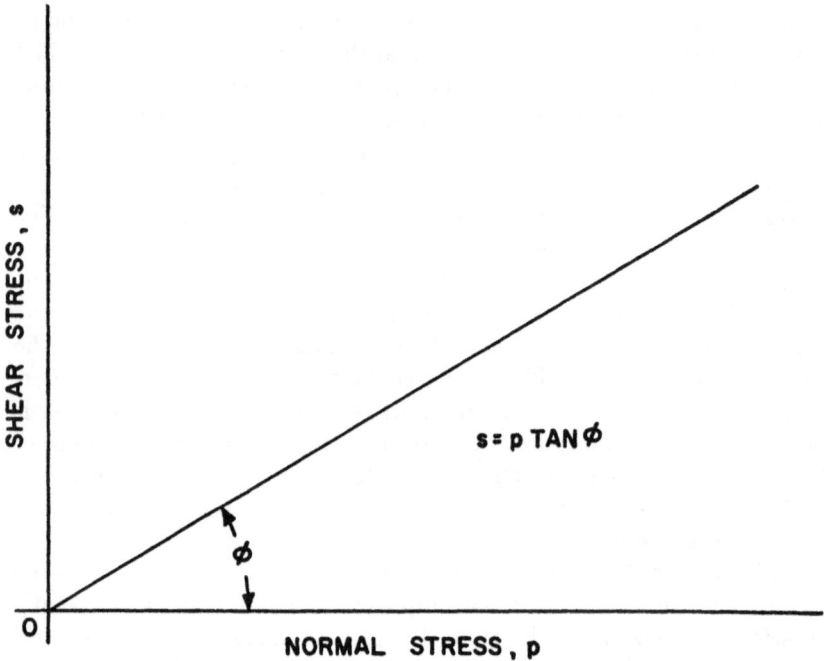

Figure 43. Mohr envelope for dry sands and gravels.

the origin. The equation for shearing resistance is $s = \bar{p} \tan \phi$, where $s$ is the shear strength, $\bar{p}$ is the effective (intergranular) pressure on the plane of failure, and $\phi$ is the *angle of internal friction*. For a dry sand or gravel the value of the neutral stress $(u_w)$ is zero, since there is no water present and $\bar{p} = p$, the total pressure on the plane of failure. The designation of $\phi$ as the angle of internal friction stems from the fact that the relationship indicated in figure 43 is the same as that which is obtained when a sliding block is subjected to a normal force, $P$. The shear force, $S$, which is required to maintain equilibrium is given by the equation, $S = P \tan \phi$, where in the case of a sliding block, $\tan \phi$ is the coefficient of friction. The shearing resistance of a dry sand or gravel is thus a frictional phenomenon.

## 132. Factors Which Affect Value of Angle of Internal Friction of Dry Sand or Gravel

The most important factor which influences the value of the angle of internal friction of a dry sand or gravel is relative density (par. 29). The greater the relative density, the greater will be the angle of internal friction for the same granular soil. Two other important factors are grain shape and gradation. In general, the value of $\phi$ is higher for a soil composed of angular grains than for one composed of rounded grains. Similarly, the value of $\phi$ for well-graded material would be higher than for one of uniform gradation. It is not possible to give precise values of the angle of internal friction which correspond to general values of density, gradation, and grain shape. However, it is possible to indicate the range of values to be expected. Typical values which may be of considerable usefulness in the solution of practical problems are given in table IV below. Very high values of $\phi$ are frequently exhibited by dense, angular, well-graded gravels and gravelly sands, because of the additional resistance which is provided by the interlocking of the coarse particles. The presence of water has very little effect upon the value of the angle of internal friction of a sand or gravel; $\phi$ is perhaps 1° less in a saturated condition than in a dry state. However, as is emphasized in paragraph 134, the shearing resistance of a saturated sand may be much less because of a reduction in the intergranular pressure.

*Table IV. Typical Values of the Angle of Internal Friction of Granular Soils*

|  | Angle of internal friction, degrees | |
|---|---|---|
|  | Loose condition | Dense condition |
| Angular, well graded_____ | 38 | 45 |
| Angular, uniform_____ | 34 | 42 |
| Rounded, well graded_____ | 32 | 40 |
| Rounded, uniform_____ | 29 | 36 |

## 133. Volume Changes Accompanying Shear

Volume changes take place in a dry cohesionless soil which is subjected to shear failure. If the sand is in a loose condition before shearing, the increase in shear stress causes a decrease in volume. This is caused by the rearrangement of the soil particles into a more dense mass, which has a lower void ratio. On the other hand, if the soil is in a dense condition before shearing begins the increase in shear stress causes an increase in volume. In order for shearing failure to occur, the particles must be forced over one another; the void ratio thus increases, and a dense soil swells during shear. As shear continues beyond the point of failure, both loose and dense sands tend to approach the same void ratio or density. This condition is sometimes termed the *critical density* or *critical void ratio*. If the void ratio of a natural sand deposit is higher than the critical, it is in a very loose condition and tends to be unstable.

## 134. Shearing Resistance of Saturated Sands and Gravels

*a. Equation for Shearing Resistance.* The equation for shearing resistance given in paragraph 131 is applicable to saturated sands and gravels. However, since the value of the neutral stress $(u_w)$ may be of extreme importance in these soils, it is desirable to write it in the form $s = (p - u_w) \tan \phi$, where the symbols have meanings previously given. Four principal effects are of importance in considering the shearing strength of saturated sands and gravels. These are the effects of submergence, seepage flow, volume changes, and capillary forces.

*b. Effects of Submergence.* If a cohesionless soil is located a distance, $z$, below a free water surface, e. g. below the water table in a natural soil deposit, the intergranular pressure is reduced by an amount equal to the head of water above the plane which is being considered. This effect is discussed in paragraph 89. If the value of the neutral stress becomes sufficiently large, the shearing resistance may become quite low; under applied loads a failure may then result. This sometimes is the cause of landslides, particularly in cases where an inclined stratum of cohesionless soil located adjacent to a hillside or a cut slope is confined between two strata of impervious soil, so that drainage is prevented.

*c. Effects of Seepage Flow.* When seepage flow occurs through a cohesionless soil, the shearing strength may be reduced very considerably. The effects of seepage flow are discussed in paragraph 90. No additional discussion of these phenomena is necessary here, although a general understanding of the effects of seepage on cohesionless soils is enhanced by an understanding of the shearing resistance of such soils.

*d. Effects of Volume Change.* As has been noted previously, volume changes accompany shear. If the soil is saturated, these volume changes must be accompanied by changes in stress in the water contained in the soil, unless the water can escape. If a dense sand is subjected to quick shear, i. e. shear which occurs in a very short period of time, the soil expands, and the water is placed in tension. This can result in a large negative increase in neutral pressure, and a correspondingly large increase in intergranular pressure and shearing resistance. On the other hand, a loose sand decreases in volume, the water is placed in compression, the neutral pressure increased, and the intergranular pressure and shearing resistance decreased.

Such behavior is not of great concern with the course-grained soils, since their permeability is high, and the water may readily escape. Nor is it of concern if the shear occurs slowly, as the water may escape. However, it is of great concern in some situations where fine cohesionless soils of low permeability, e. g. very find sands and silts, which have a loose structure, are encountered and are subject to quick shear. In these soils, vibration or even seepage stresses, may cause the soil to break down and flow like water; this is termed a *temporary quick condition*. The situation is particularly acute in loose silts with honeycomb structure. The temporary liquefaction of such soils presents a great hazard to construction and extreme care must be exercised in handling such soils in the field.

*e. Effects of Capillarity.* The effects of capillary tension upon cohesionless soils are discussed in paragraph 80.

## 135. Shearing Resistance of Silts and Silty Sands

The shearing resistance of dry silts and silty sands may be computed by means of the equation previously given for dry sands and gravels. However, the values of the angle of internal friction are typically somewhat less for these soils than for comparable sands. In a loose condition, $\phi$ for a dry soil of this type may be in the approximate range of 26° to 30°. For dense dry soils in this category $\phi$ may be as high as 35°. As indicated in the preceding discussion, saturated soils of this general character frequently experience a reduction in shearing strength because of the fact that the water cannot readily escape. This phenomenon leads to apparent values of the angle of internal friction which are substantially lower than those given above for dry silts and silty sands.

## 136. Shearing Resistance of Cohesive Soils

The shearing resistance of a clay soil is a much more complex phenomenon than is that of a cohesionless soil. Clays are much different from sands in permeability, compressibility, the effects of adsorbed water and other things. They are also markedly different from sands in their resistance to shear stresses. Many variables influence the shearing resistance of a clay soil under field loading conditions, including such things as void ratio, structure, moisture content, past history of the deposit, the speed at which shear occurs, and many other things. Many saturated clays which are loaded so rapidly that the water cannot escape behave as though the angle of internal friction is equal to zero. In other words, they are loaded so quickly that the increase in pressure does not contribute to shearing resistance. Since, in general statement, the shearing resistance which is exhibited by a saturated clay in quick shear (this condition of loading and failure is also termed *undrained shear*) is a conservative value, it is widely used to estimate the shearing strength of a soil of this type.

## 137. Saturated Clays in Quick Shear

The Mohr envelope of a saturated clay in quick shear is shown in figure 44. As can be seen, the envelope is a straight line which is parallel to the horizontal axis. The intercept with the vertical axis, $c$, is the *cohesion*. As indicated previously, $c$ is most frequently determined by the use of an unconfined compression test (par. 130), and $c = \frac{1}{2}q_u$, where $q_u =$ unconfined compressive strength. It may also be determined by triaxial compression or direct shear tests, although less conveniently. So many variables affect the shearing strength of clays that it is impossible, as well as inadvisable, to tabulate typical values of the cohesion which would be of general usefulness. The numerical value of $c$ may vary from practically zero for very soft clay soils to 4000 pounds per square foot or more for very hard clays. In general, the shearing strength of a given clay soil increases with a decrease in moisture content.

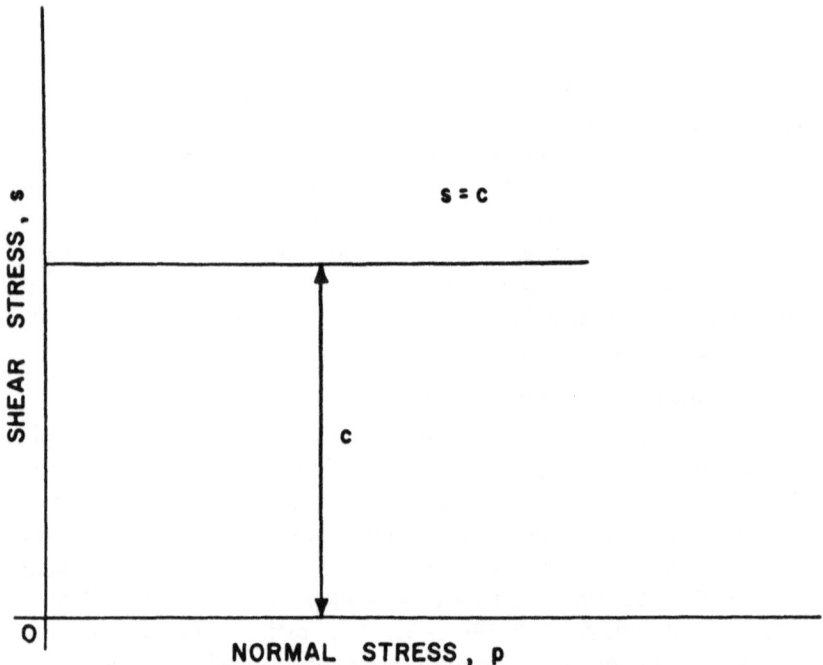

*Figure 44. Mohr envelope for a saturated clay soil in quick shear.*

## 138. Effects of Remolding and Sensitivity

Natural clay soils generally show a decrease in shearing resistance when they are remolded, i. e. when their natural structure is disturbed, even though the water content remains unchanged. This loss in strength is attributed to a breakdown of the adsorbed water films and the soil's natural structure. For this reason, shear tests are normally performed upon relatively undisturbed samples. Some clays, particularly those which have void ratios of one or less, suffer little loss of strength on complete remolding. Others show much greater decreases in strength. The ratio between the unconfined compressive strength of the undisturbed clay and that of the remolded clay is termed the *sensitivity*, $S_t$. If the sensitivity of a clay is less than 4, it is termed an *ordinary* or *relatively insensitive clay*. Clays which have values of $S_t$ between 4 and 8 are termed *sensitive clays*. If $S_t$ is more than 8, they are called *highly sensitive* or *extrasensitive clays*. Some clays particularly stiff or brittle clays, contain structural defects, such as hair cracks or slickensides. Such structural defects tend to lower the shear strength of a clay deposit. The strength probably will be between the remolded strength and the undisturbed strength of a sample of the same deposit where the defects are not present.

## 139. Effect of Consolidation

If a load is applied slowly to a clay soil, the shearing strength generally will increase with normal load. In other words, the soil will behave as if it possessed frictional resistance. This increased strength might be determined in the laboratory, for example, by consolidating a sample and then shearing it. The procedures involved generally require considerable time and this fact, coupled with the fact that it is difficult to duplicate field loading conditions in the laboratory, has limited attempts to accurately determine the increase in strength which accompanies consolidation, except on very large and important projects. The approach most frequently used, as indicated previously, is to rely upon the results of unconfined compression tests, since strengths thus obtained are generally regarded as *minimum* values.

## 140. Shearing Resistance of Partially Saturated Clays

Partially saturated clays behave as though they possessed some frictional resistance. A typical Mohr envelope for such a soil is shown in figure 45. The equation for shearing resistance is $s = c + p \tan \phi$, where the symbols have meanings previously given. The value of $\phi$ seems to depend largely upon the air content; it may vary from $0°$ for saturated clays to perhaps $30°$ for some dry clays. This equation for shearing resistance is frequently applied to soil mixtures. For example, it is convenient to visualize the shearing resistance of a sand-clay mixture as being made up in part of the cohesion contributed by the clay, and the internal friction which is supplied by the sand. This concept may also be applied in a general way to cemented soils, such as the loess of the Mississippi Valley and the lime soils of the southwest, in which the soil grains are held together by a cementing material. The presence of the cementing agent gives the soils an initial strength under no load. After the strength of the cementing material has been exceeded the shearing resistance is largely frictional in nature. In a general way, also, the concept of figure 45 is useful in visualizing how the strength of a cohesionless soil may be increased by adding a cementing agent, such as a bituminous material. The bituminous material, or other stabilizing agent, will supply cohesion in order that the soil will have initial strength under no load or low pressures, and will also contribute to an increased shearing resistance under greater pressures.

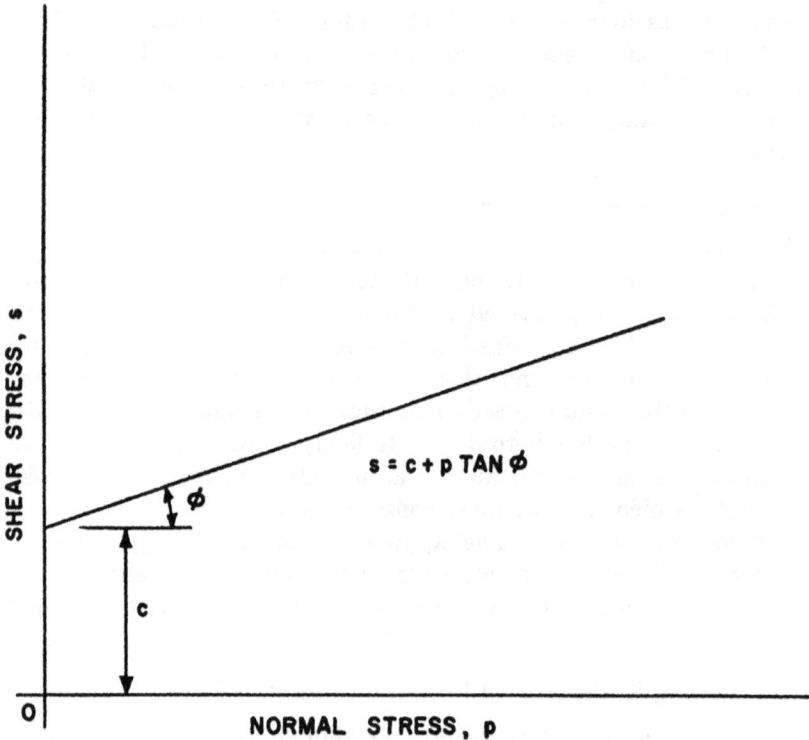

Figure 45. Mohr envelope for a partially saturated clay.

## 141. California Bearing Ratio

*a. General.* The California Bearing Ratio is a measure of the shearing resistance of a soil under carefully controlled conditions of density and moisture. The CBR is determined by a penetration shear test and is used in conjunction with empirical curves for designing flexible pavements; recommended design procedures are discussed in paragraph 260. The CBR test procedure for use in design consists of two principal steps; (1) the preparation of soil test specimens, and (2) the penetration test performed upon the prepared soil samples. Although a fixed, standardized procedure has been established for the penetration portion of the test, it is not possible to establish one fixed procedure for the preparation of test specimens, since soil conditions and construction methods vary widely. The soil test specimen is prepared so that it duplicates the soil conditions existing in the field as nearly as possible. The method of preparing the test specimens and the number of test specimens depend upon whether a road or airfield is to be constructed, the type of airfield, and the soils encountered at the site. Although penetration tests are most frequently performed upon laboratory-compacted test specimens, they may also be performed upon undisturbed soil samples, or in the field upon the soil in place. Detailed procedures for determining the CBR are contained in TB 5–253–1.

*b. Penetration Test Procedure.* Elements of the penetration portion of the CBR test are shown in figure 46. The piston is forced into the soil sample at a specified rate and values of the load corresponding to various deflections noted. A plot is made of the relationship between the load, in pounds per square inch, and deflection. The curve is corrected, if necessary, and the values of the unit load corresponding

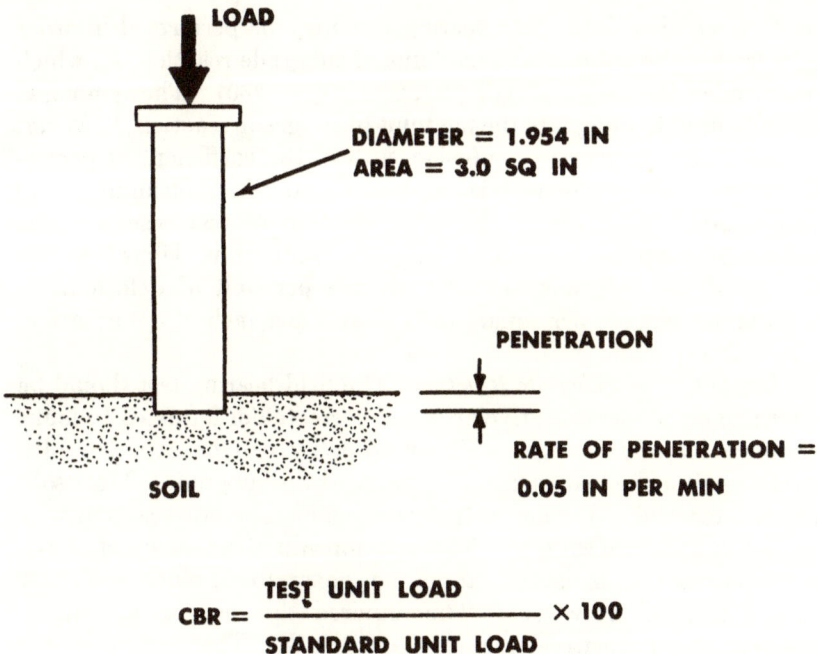

$$CBR = \frac{\text{TEST UNIT LOAD}}{\text{STANDARD UNIT LOAD}} \times 100$$

NOTE: TEST MAY BE PERFORMED ON LABORATORY-COMPACTED SPECIMEN, UNDISTURBED SAMPLE, OR IN PLACE IN THE FIELD. SURCHARGE WEIGHT USED TO SIMULATE CONFINING EFFECT OF PAVEMENT NOT SHOWN.

*Figure 46. Elements of penetration portion of California Bearing Ratio test procedure.*

to 0.1- and 0.2-inch penetration determined. CBR values are obtained by dividing the loads, in pounds per square inch, at 0.1 and 0.2 inch by the standard loads of 1,000 and 1,500 pounds per square inch, respectively. The standard loads are the unit loads required to effect the stated penetrations into a standard sample of crushed rock. Each ratio is multiplied by 100 to obtain the CBR in percent. The CBR corresponding to 0.1-inch penetration is the one usually selected for design. If the CBR at 0.2 inch is greater, the test should be redone; if check tests give similar results, the CBR at 0.2-inch penetration should be used.

c. *Values of CBR for Typical Soils.* In column (14) of table VI, appendix II, are shown typical ranges in value of the field CBR for soils of the Unified Soil Classification System. As will be noted, values of the field CBR may range from as low as 3 for highly plastic, inorganic clays (CH), and some organic clays and silts (OH), to as high as 80 for well-graded gravels and gravel-sand mixtures.

## 142. Field Plate Bearing Test

a. *General.* The field plate bearing test may be performed in order to determine the value of the modulus of subgrade reaction, $k_s$, which is required in the design of rigid pavements (par. 260). The symbol, $k$, is usually used to designate the modulus of subgrade reaction; however, $k_s$ is used here to prevent confusion with $k$, the coefficient of permeability (par. 85). The test is expensive and time consuming, and requires good judgment in the interpretation of test results. The *modulus of subgrade reaction* or *subgrade modulus* is defined as the reaction of the subgrade per unit of area per unit of deformation. Its units are pounds per square inch of area per inch of deformation, or pounds per cubic inch.

b. *Evaluation of Subgrade Reaction.* The field bearing test should be performed on a representative area and corrected or adjusted to make allowance for any changes which reasonably can be expected ultimately to alter the load-supporting value of the subgrade. Most soils exhibit a marked reduction in the modulus of subgrade reaction with increase in moisture content. This relationship of reduction of modulus to increase in moisture cannot be generalized, since each soil constitutes a separate case. The required thickness of a concrete pavement is not sensitive to small changes in $k_s$. Therefore, it is only necessary to fix the bracket of $k_s$ values, rather than to fix its absolute value. Conditions of moisture, density, and type of material all enter into the interpretation of test results in order to give a design value which will represent conditions that ultimately will obtain.

c. *Preparation of Test Area.* A small area of the subgrade should be stripped to the proposed elevation of the subgrade surface. The selected location for test should be uncovered for a sufficient area to eliminate surcharge or confining effects. If the subgrade is to be composed of fill material, a test embankment about 30 inches in height should be constructed after necessary stripping. The subgrade should be compacted at optimum moisture to the density specified. If ordinary compaction equipment is not available, approximate compaction may be accomplished by hand tamping in thin layers. The field bearing test should be made with the soil at or close to its optimum moisture content. Insofar as practical, it is advisable also to conduct comparative tests on subgrades representative of the conditions after the pavement has been in place for some time. Methods of correction are suggested (*i* below) for use when it is impractical to simulate true conditions.

*d. Loading System.* Loads are applied by means of a hydraulic jack working against a jacking frame, and through a circular bearing plate 30 inches in diameter. The loading plate must be sufficiently rigid to insure maximum vertical strain in the subgrade. This is usually accomplished by stacking plates 30 inches, 24 inches, 18 inches, and 12 inches in diameter. A cushion layer of fine, dry sand or a plaster of Paris mixture may be used between the bearing plate and the subgrade to insure uniform contact.

*e. Loading Reaction.* The reaction for jacking can be a truck trailer, anchored frame, or any other device such that its reactions are at least 8 feet from the bearing plate. The test load should be applied with a ball joint between the test load and the jack, or between the jack and the bearing plate, to avoid eccentricity in loading.

*f. Measuring Deformation.* The movements resulting from applied loads should be measured by at least three dial gages, located at the third points of the circumference of the plate. The gages should be supported by an independent framework such that their positions are unaffected by loading operations.

*g. Loading Procedure.* The loading system and bearing plate should be seated by applying a load of 1 pound per square inch in 30 seconds and releasing immediately, after which the zero readings of the dial gages are recorded. This procedure usually eliminates the steep slope of the load-deformation curve at the origin. A load of 7,070 pounds to obtain a unit pressure of 10 pounds per square inch should be applied in 10 seconds, and held until practically complete deformation has taken place. After the movement of the dials is recorded, the load is released. For some clay soils, it may be necessary to plot' a time-settlement curve to aid in determining when practically complete deformation has been obtained. In general, the load should be held until the deflection of the plate is less than 0.002 inch per minute, which rate of settlement generally indicates that the major portion of the settlement has taken place.

*h. Evaluation of Test Results.* The modulus of subgrade reaction with the subgrade at or near optimum moisture ($k_u$) shall be determined by the equation, $k_u = 10/d$, in which $d$ represents the total

**113**

movement which took place during the application of the load. The most usable method of correcting $k_u$ for saturation is by the use of the consolidation test (par. 117).

*i. Correction for Saturation.* Consolidation tests should be performed upon representative undisturbed or remolded samples from the location of the field bearing test. One specimen should be tested at its field moisture content, the other after saturation in the consolidation device. Results of the two tests are plotted, and the deformation corresponding to a unit load of 10 pounds per square inch in each test determined. The following expression then may be used: $k_s = (d/d_s)k_u$, in which

$\quad$ $k_s$=subgrade modulus, corrected,

$\quad$ $k_u$=subgrade modulus, uncorrected,

$\quad$ $d_s$=deformation of a saturated specimen under a unit load of 10 psi, and,

$\quad$ $d$=deformation of a specimen at field conditions under a unit load of 10 psi.

*j. Interpretation of Corrections.* Rarely is the value of $k_u$ used directly in design without correction. In some instances, the correction does not fulfill the intended purpose and is not satisfactory. Conditions of this sort are encountered in testing cohesionless, or nearly cohesionless, silts and fine sands that "pump", and uniformly, poorly graded, not well-compacted sands (like Florida sands) that tend to shift at pavement joints. For these conditions, a value of $k_s$ should be assumed for design. Coarse, granular soils may be used without correction, providing compaction requirements are fulfilled. A saturation correction for cohesive soils in arid regions of low water table may be smaller than that obtained from the consolidation tests. If examination of pavements in the area indicates no loss of the subgrade supporting value at joints due to infiltration or leakage, and the moisture content is lower than the optimum for compaction, field bearing tests on the subgrade in place may be acceptable without correction.

*k. Values of $k_s$ for Typical Soils.* In table VI, appendix II, the range of values of $k_s$ for the soil groups of the Unified Soil Classification System is shown in column (15). It will be noted that the value may range from a high of 300 or more for the GW and GP groups to a low of 50 for the CH and OH groups. For many practical purposes, the value of $k_s$ selected from table VI will be satisfactory, since a variation in this factor has relatively little effect upon the required thickness of a rigid pavement.

# CHAPTER 8
# BEARING CAPACITY OF SOILS

## Section I.  INTRODUCTION

### 143. General

In general terms, the *bearing capacity* of a soil is its ability to support loads which may be applied to it by an engineering structure. The load may be that of a building, a bridge, a highway pavement, or an airport runway, and the moving loads which may be carried thereon, an embankment, or any one of a number of other things.  A soil which does not have sufficient bearing capacity to support the loads which are applied to it may simply fail by shear, allowing the structure to move or sink into the ground, or it may fail because it undergoes excessive deformation, with consequent damage to the structure.  The ability of a soil to support load is also sometimes simply termed its *stability*.  Bearing capacity is directly related to the allowable load which may be safely placed upon a soil; this allowable load is sometimes termed the *allowable soil pressure*.

### 144. Scope of Chapter

This chapter is devoted to a discussion of the bearing capacity of soils beneath structural foundations.  The material is presented in two sections, one pertaining to shallow foundations and the other to deep foundations.  Most of the latter section is concerned with pile foundations.  Material relating to the bearing capacity of soils which support highway and airport pavements is presented later (ch. 10), and the stability of foundations of earth structures is discussed in chapter 11.  This chapter is primarily devoted to the effects of static loads, including the usual dead and live loads used in structural design. The effect of vibrations upon bearing capacity is briefly discussed.

## Section II.  SHALLOW FOUNDATIONS

### 145. Foundation Types

A shallow foundation is one which is located at, or slightly below, the surface of the ground.  A typical foundation of this type is seen in the *shallow footings*, either of plain or reinforced concrete, which may support a building.  Footings are generally square or rectangular in shape. The width of a square or rectangular footing is given the symbol $b$, while the longer dimension of a rectangular footing is designated as $L$. Long continuous or *strip footings* are also used, particularly beneath basement or retaining walls. Another type of shallow foundation is the *raft* or *mat*, which may cover a large area, perhaps the entire area occupied by a structure.

## 146. Criteria for Successful Action of Foundation

The principal function of a foundation is to transmit the weight of the structure and the loads which it carries to the underlying soil or rock. In order for the foundation to function satisfactorily, two principal, specific criteria must be satisfied; these criteria are discussed below, along with other requirements of a more general nature.

*a.* A foundation must be designed so as to be safe against a shear failure in the underlying soil. This means that the load which is placed upon the soil must not exceed its ultimate bearing capacity. Types of failure which may take place when the ultimate bearing capacity is exceeded are illustrated in figure 47. Such a failure may involve tipping of the structure, with a bulge at the ground surface on one side of the structure. Failure may also take place on a number of surfaces within the soil, usually accompanied by bulging of the ground around the foundation. The ultimate bearing capacity is not only a function of the nature and condition of the soil involved, but upon the method of application of the load.

*b.* A foundation must also be designed so that the structure will not be subjected to detrimental settlement. Uniform settlement of a building may not be particularly harmful, unless it is so great as to break utility lines, water pipes, sewer lines, etc. With some bridges, any settlement of a pier or abutment may be serious in that stresses are created beyond those used in the design. Differential settlement of adjacent parts of a building may have serious consequences, such as the cracking of plastered walls, inability to close doors, or actual structural failure. There may be two somewhat different types of settlements with which the designer must contend. These are *contact settlement*, which takes place because of the compression of the soil immediately beneath a shallow foundation, and settlement due to the consolidation of underlying compressible soil layers. Since the latter type of settlement may be caused by the consolidation of compressible soils located deep beneath the structure it is sometimes termed *deep-seated settlement*.

*c.* Although it is sometimes difficult to assign quantitative values to them, several other general requirements must be met by foundations if they are to satisfactorily perform their intended function throughout the life of the structure. For example, frost action may be a factor in the design. Shallow structural foundations are commonly located so that the bottom of the foundation is below the line

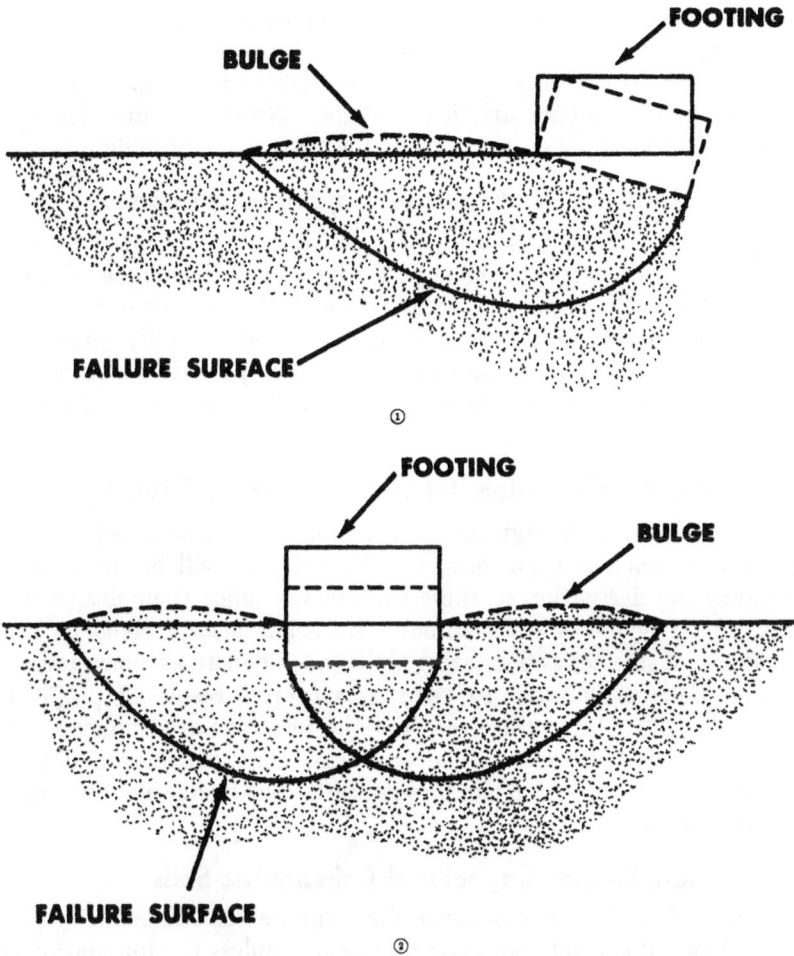

Figure 47. Typical failure surfaces beneath shallow foundations.

of frost penetration, thus eliminating this factor from consideration. Somewhat similarly, in some regions surface soils alternately swell and shrink during the wet and dry seasons; this situation is common in some areas of Texas, for example. Foundations should normally be placed below the level of such volume changes. When foundations are being designed for bridge piers and abutments, provisions must be made to protect them against being undermined by scour when the stream is in flood stage. Sometimes fluctuations in the water table are a source of foundation difficulties. A rise in the water table above a shallow foundation may cause it to be subjected to uplift pressures which may threaten its stability. A permanent drop in the water table may cause a foundation to settle because of increased intergranular pressure in compressible soil layers. In areas where it is likely that excavations may be made for the construction of adjacent structures at a later date, foundations must be placed deep enough to preclude failure from this cause. These and other factors are frequently very important in planning the foundations for a given structure, particularly if it is large. No further mention will be made of most of these items in the discussions which follow.

## 147. Considerations Relative to Ultimate Bearing Capacity

The ultimate bearing capacity of a soil may be estimated by the application of theoretical concepts, or by the performance of field loading tests. It is important at this stage in the discussion to notice that the theoretical concepts used are not intended to supply extremely precise results. Their application to soils yields approximate formulas which must be used in conjunction with a fairly large factor of safety in the design of foundations.

## 148. Theoretical Expressions for Ultimate Bearing Capacity

In the following paragraphs are presented formulas which are applicable to various types of soils. No attempt will be made here to present the derivation of these expressions, since their derivation is beyond the scope of this manual. Each has been derived by the comparison of stresses and shearing resistance upon an assumed failure surface. Each is believed to yield conservative results if properly applied. It must be emphasized that the formulas are applicable to homogeneous soil deposits. The effects of nonuniformity and particularly the presence of a buried stratum of weak soil, are discussed in paragraph 170.

## 149. Ultimate Bearing Capacity of Cohesionless Soils

a. For loads in the normal range the ultimate bearing capacity of a cohesionless soil is rarely a matter of concern, unless the foundation is very narrow, the soil is very loose, or the water table is very high. The following expression may be used for the ultimate bearing capacity ($q_0$) of a footing located at the surface of a cohesionless soil; $q_0 = \frac{\gamma b}{2} \tan^5$ ($45° + \phi/2$), wherein $\gamma$ is the unit weight of the soil, $b$ is the width of the footing (the smaller dimension, in the case of a rectangular footing), and $\phi$ is the angle of internal friction. As an example of the application of this formula, assume that it is desired to estimate the ultimate bearing capacity of a footing which is 10 feet square located at the surface of a homogeneous, medium sand deposit for which $\gamma = 100$ pounds per cubic foot, and $\phi = 30°$ (estimated). For this case, $\tan^5$ ($45° + \phi/2$) = about 15. Thus, $q_0 = \frac{100(10)(15)}{2} = \frac{15,000}{2} = 7500$ pounds per square foot. For values of $\phi$ above about 35° the equation gives very conservative results. Approximate values of $\tan^4$ ($45° + \phi/2$) and $\tan^5$ ($45° + \phi/2$) are given in figure 48.

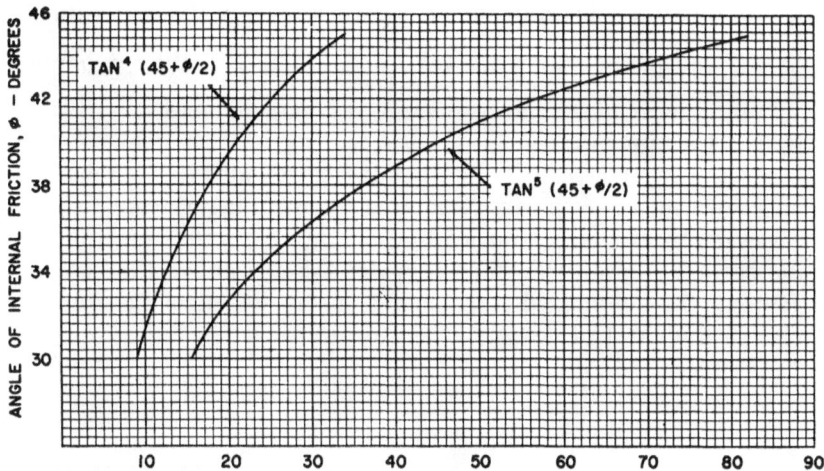

*Figure 48.* *Approximate values of trigonometric functions to be used in calculating the ultimate bearing capacity of cohesionless soils.*

*b.* As is evident from inspection of the formula above, the principal variables which affect the ultimate bearing capacity of a cohesionless soil are $b$, $\gamma$ and $\phi$. As the width of the foundation is increased, the ultimate bearing capacity is increased in direct proportion. This is the reason that there is very little possibility of a large raft foundation breaking into a soil of this type, unless the deposit is very loose. Very loose silt deposits, even though they may be classed as cohesionless, are not suitable for the support of shallow foundations. Both $\gamma$ and $\phi$ have a large influence on the ultimate bearing capacity. Since both may generally be increased if the soil is compacted, it is sometimes feasible to adopt elaborate construction measures to densify a loose cohesionless soil. To be effective in an extensive loose deposit of this type, the compaction would have to extend several feet below the surface. This process is generally expensive and would only be justified on major projects, where the advantages of the site outweighed the additional cost. Small increases in $\phi$ lead to comparatively large increases in $q_o$. In connection with the unit weight, it is important to note that a rise in the water table to the base of a shallow foundation will considerably decrease the ultimate bearing capacity, since the effective unit weight will be reduced to about half the unit weight above the water table. As a general rule, it is probably advisable to reduce the value of $q_o$ obtained by the equation of $a$ above to about $\frac{1}{2}q_o$, if the water table is within $b$ feet of the surface, or may be later.

*c.* The ultimate bearing capacity of a cohesionless soil is greatly increased by a surcharge, such as illustrated in figure 49 ①. If the approach used in deriving the approximate equation of $a$ above is

used, the ultimate bearing capacity is given by the expression:
$q_0=[q'+\frac{\gamma b}{2} \tan (45°+\phi/2)] \tan^4 (45°+\phi/2)$, where $q'$ is the surcharge.
Thus the influence of a small surcharge is greatly increased by being multiplied by $\tan^4 (45°+\phi/2)$. If the footing is located below the surface of the ground, so that the surrounding soil acts as a surcharge, the effect of depth may be approximated by assuming $q'=\gamma D_f$, where $D_f$ is the depth of the bottom of the footing below the surrounding ground surface (fig. 49 ②). If the water table is above the base of the footing, the submerged unit weight of the soil must be used in calculating the effect of depth.

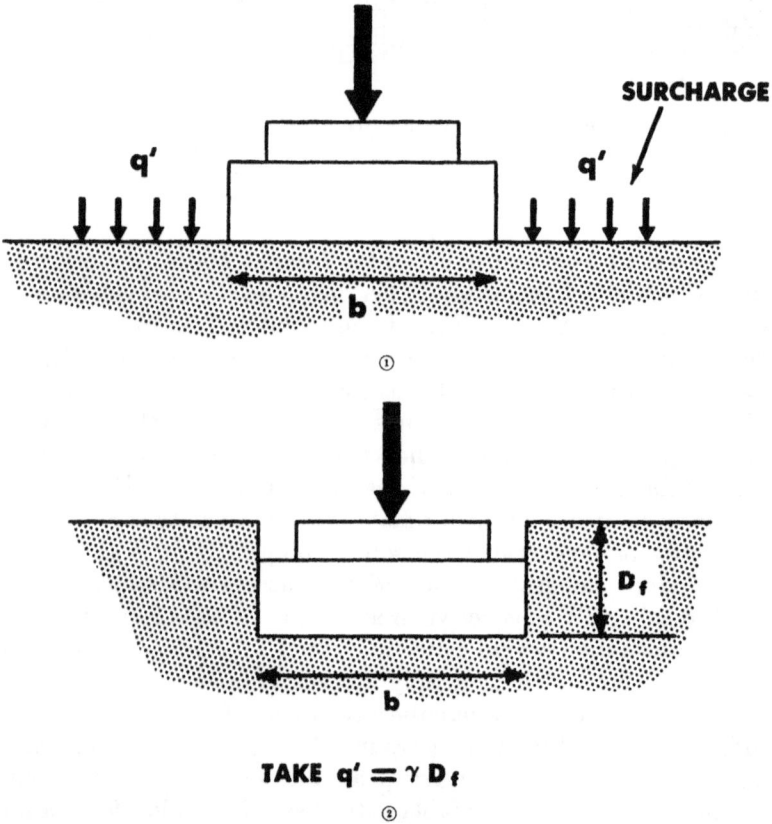

Figure 49. Effect of surcharge on bearing capacity of cohesionless soils.

## 150. Ultimate Bearing Capacity of Saturated Clays

a. For saturated clays which do not have structural defects, such as slickensides or cracks, a simplified analysis shows that the ultimate bearing capacity of a strip footing located at the surface of the ground is given by the expression, $q_0=4c$, where $c$ is the cohesion. Since the unconfined compressive strength is widely used as

120

a measure of the shearing resistance of a clay soil and is approximately equal to $2c$, $q_0=2q_u$, where $q_u$ is the unconfined compressive strength of the soil involved. This expression is based upon an assumption that the soil is homogeneous to a depth of at least $b$ below the bottom of the footing. In contrast to cohesionless soils, the ultimate bearing capacity of a saturated clay soil is essentially independent of the width of the foundation. The situation beneath a large footing is no better than beneath a small one and may be worse, if the strength of the soil decreases with depth. If the clay involved is insensitive, the formula gives results which are somewhat conservative. If the clay is highly sensitive, results will be somewhat on the unsafe side and require additional study.

$b$. The ultimate bearing capacity of a square footing founded on a homogeneous clay soil may be approximated by multiplying the value of $q_0$ for a strip footing by 1.3. Similarly, for a rectangular footing, the approximate expression is $2q_u$ $(1+0.3\ b/L)$, where $L$ is the longer dimension. It should be noted that these equations should be used only for the design of isolated footings.

$c$. The ultimate bearing capacity of a saturated clay soil is also increased by the presence of a surcharge, although the effect is much less than for a cohesionless soil. By comparison, the effect of surcharge is practically negligible. The effect of a surcharge may be approximated by simply adding the surcharge, $q'$, to the value of $q_0$ obtained from the equations given above. As before, the effect of depth may be taken into account by assuming $q'=\gamma D_f$.

## 151. Ultimate Bearing Capacity of Other Soils

In the case of soils which are intermediate between the limiting types of cohesionless materials and saturated clay, the ultimate bearing capacity is also generally between the values given above for the limiting cases. Rather elaborate equations have been derived for the case of partially saturated clays, for example. However, for practical purposes it would generally be sufficiently accurate and safe to assume the case of a saturated clay and use the value of $q_u$ from an unconfined compression test, assuming that one can be performed. On major projects such a procedure may not be economical and additional investigation may be required. The ultimate bearing capacity of such soils may be estimated from the results of a field load test, as described in paragraph 153.

## 152. Factor of Safety

The ultimate bearing capacity is the load which corresponds to failure of the soil. In selecting the allowable soil pressure to be used in design, values obtained from the theoretical expressions given in the preceding paragraphs must be divided by an appropriate factor of safety. Generally speaking, a value of the factor of safety of 3.0

should be used. Loads should be the maximum sustained loads that the structure will experience. As discussed later, considerations relative to settlement may govern the design.

## 153. Field Load Test

*a. Purpose.* A field load test may be employed as an aid in estimating the ultimate bearing capacity of a soil which is to receive loads from a structure. Results of the test may also be used to approximate the amount of contact settlement which may occur, as explained later. It may be desirable to use a test of this sort when a saturated clay has structural defects. On major projects a properly conducted load test may be regarded as a valuable check upon values estimated by theory.

*b. Procedure.* As a first step in conducting a field load test, a pit is dug to the level at which it is desired to estimate the allowable soil pressure. The plate which is to be used should be at least 1 foot square, and preferably 2 feet square or larger. In order to avoid the effects of surcharge, the pit should be at least 3, and preferably 5, times as wide as the plate which is to be used. The plate is placed in position and arrangements made for applying load, as indicated in figure 50. Some arrangement must also be made to measure the settlements which take place as the plate is loaded; a bench mark and an engineer's level may be employed for this purpose. The load is then applied in increments. Each load increment is maintained until the rate of settlement becomes very small. The ultimate settlement which corresponds to each increment of load is recorded. A load-settlement curve for the test may then be prepared, as indi-

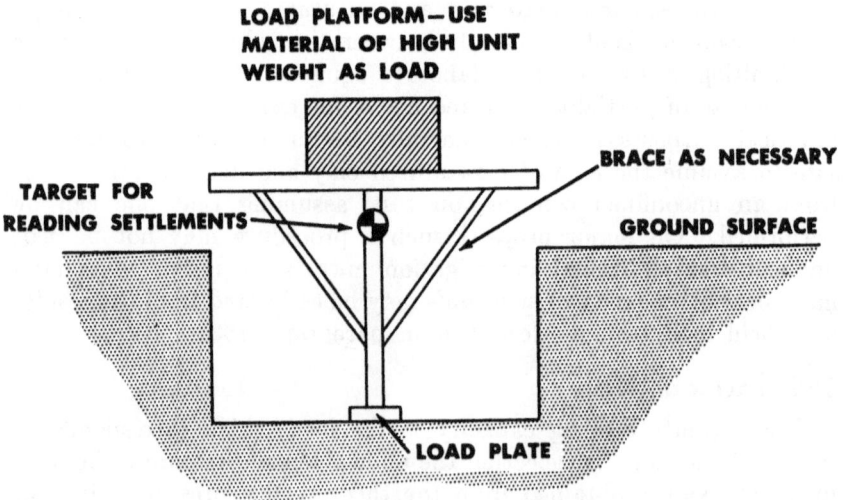

*Figure 50. Field loading test for estimating bearing capacity of soil to be used as a structural foundation.*

cated in figure 51. The test should be continued until failure occurs or the load reaches a value of three times that which will be used in the design of the foundation.

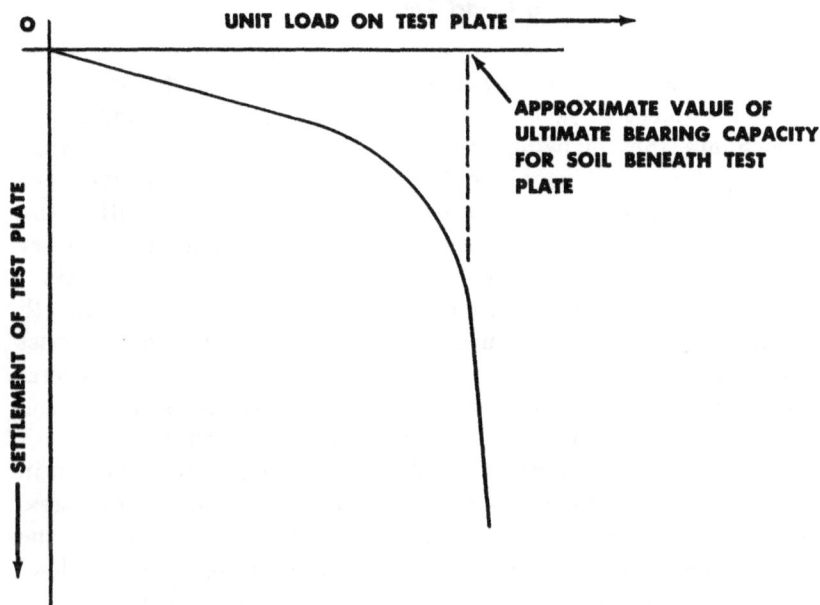

*Figure 51. Load-settlement curve obtained from a field load test.*

## 154. Interpretation of Load Test Results

As indicated in figure 51, a typical load-settlement curve will show a definite break or increase in steepness at some value of the load applied to the plate. If this occurs, the ultimate bearing capacity should be assumed to be the unit load corresponding to the break in the curve. In some cases there will be no definite break in the curve and the bearing capacity may then be selected to correspond to some definite value of the settlement. The value of $q_0$ selected from the test curve is strictly applicable only to a foundation of the same size, depth and location as the test plate. If the location is representative, then for a saturated clay soil $q_0$ determined from the plate is taken to be directly applicable to the foundation, since $q_0$ is essentially independent of the width of the loaded area; it is conservative to neglect the effects of surcharge, if present in the foundation. For cohesionless soil, however, the value of $q_0$ from the test is less than for the foundation, since $q_0$ is directly proportional to the width of the loaded area. The value of $q_0$ for the foundation may thus be increased by the ratio which the width of the proposed foundation bears to the width of the test plate, when a cohesionless soil is involved. Surcharge also greatly increases the bearing capacity of cohesionless soils, as previously noted. Values of $q_0$ determined from the load test are ultimate values, and a factor of safety must be applied, as before, if allowable values are desired.

### 155. Limitations of Field Load Test

The field load test is very popular among practicing engineers and justly so, if the test is properly conducted and interpreted. However, the test is subject to several limitations, among the most important of which are the following: First, the test is expensive and time consuming; Second, it may be difficult to select a representative place at which to conduct the test in order to be certain that critical soil conditions are evaluated. This consideration may make it necessary to perform several load tests at the site of a large structure. A third and very important consideration is that the *bulb of pressure* beneath the load test is generally much smaller than that which will exist beneath the actual foundation. The bulb of pressure is the general name given to a plot of contours of equal vertical stress in the soil beneath a loaded area. The pressures may be determined theoretically. The outer boundary of a bulb of pressure like those shown in figure 52 may be a line connecting points at which the vertical stress is equal to 5 percent of the unit load, or some similar figure. The general importance of this concept is illustrated in figure 52. A load test conducted at or near the surface of a soil deposit which is underlain by a weaker soil will not disclose the presence of the weaker layer, because of the small scale of the test. However, the weak layer may control the design of the actual foundation, as indicated in figure 52. In such a situation a load test is useless or, worse, even misleading. From the standpoint of ultimate bearing capacity, the soil must be

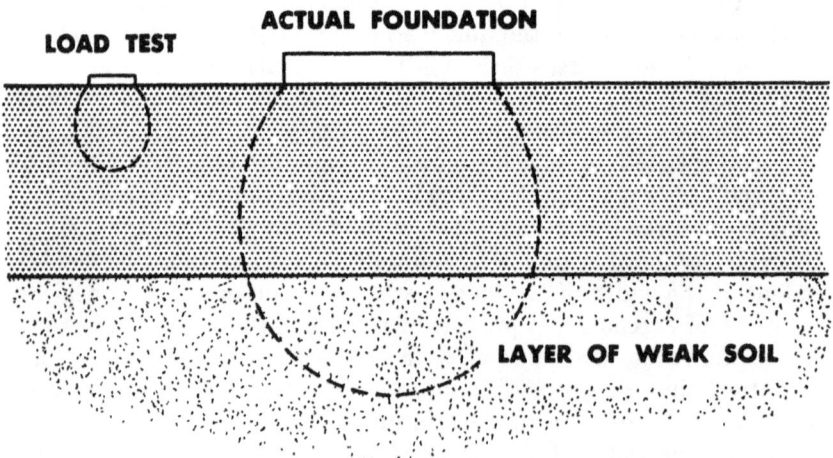

*Figure 52. Comparison between bulb of pressure for load test and that of actual foundation.*

reasonably homogeneous, or underlain by a stronger stratum, to a depth of about twice the width of the foundation for the load test to be valid. A final limitation is that the load test, as usually conducted, cannot disclose settlement which may result from the consolidation of compressible soils. This is true primarily because the time of the load test is very short, as compared with the time required for the consolidation of a thick layer of clay soil. The small scale of the load test is a contributing factor here also.

## 156. Settlement of Cohesionless Soils

Unless the deposit is quite loose or the water table high, the settlement of comparatively rigid footings on cohesionless soils is not likely to be a matter of practical concern. Settlements on this type of soil are apt to be quite variable, but quite small. They are also likely to take place quite rapidly as the load is applied. Principal variables which affect the settlement are the relative density and the position of the water table. If the relative density is low, the settlement will be comparatively great. Settlements will be much greater if the water table is at the base of the footing than if it is $b$ or more feet below the footing. The settlement also increases somewhat with an increase in width of the loaded area. No formulas will be given here for the settlement of shallow foundations on this type of soil. If the tabular values of table XI, appendix II, are used, detrimental settlements are not likely to occur. If the soil is loose, the water table high, or it is desired to use higher values of the allowable soil pressure than those given in the table, more detailed investigations which are beyond the scope of this manual should be undertaken. If the results of a load test are available, and the soil is homogeneous to a depth of at least twice the width of the loaded area, an approximate rule of some usefulness is that the settlement of a foundation which is 5 or more feet wide is approximately twice the settlement of a 2-foot square test plate under the same unit load. Similarly, the settlement of a foundation which is 5 feet wide is approximately three times that of a 1-foot square plate with the same load. If a raft foundation is being used on a cohesionless soil, settlement should not be a factor, unless the soil is very loose, and provided that the bottom of the raft is from 8 to 10 feet below the surrounding ground surface.

## 157. Contact Settlement of Saturated Clays

Shallow foundations which are located on compressible clays are subject to contact settlement. Formulas based on the theory of elasticity have been derived and are occasionally used in estimating this type of settlement. These expressions are not presented here, because the amount of contact settlement is generally quite small and relatively unimportant, as compared with the settlement which may be caused by consolidation of the clay. If the results of a load test are available, the contact settlement may be estimated by making use of the fact that on clay soils this type of settlement is directly proportional to the width of the loaded area.

## 158. Settlement Due to Consolidation

*a. General.* The consolidation of confined clay soils subjected to the weight of a structure may be the source of uniform settlement, or differential settlements, which may be very damaging. This behavior is to be expected when a heavy load is applied to, or above, a thick, soft, compressible clay layer. In the paragraphs immediately following, an approximate procedure is presented for estimating such settlements. It is important to note that the procedure presented is approximate in nature. Its principal usefulness is in emphasizing certain principles of soil action. The procedure presented should also be useful in field situations in indicating whether or not a shallow foundation will be satisfactory in a location where clay soils exist, or whether a pile foundation must be used. If a major structure is involved, a more accurate analysis of settlements may be necessary.

*b. Steps Involved in Estimating Consolidation Settlement.* It is assumed in this portion of the discussion that soil conditions at the site of the structure are reasonably well known to a depth of at least 1.5, preferably 2, times the width of the structure, and that the position of the water table has been established with reasonable accuracy. Loads and dimensions of the structure also must be known. Principal steps in an approximate settlement analysis then are—(1) an estimate of preloading pressures, (2) an estimate of the stresses which will be caused in the clay by the weight of the structure, and (3) a computation of the predicted settlement.

## 159. Computation of Preloading Pressure

It will be assumed for the purpose of this explanation that a normally loaded clay soil is involved. That is, the clay is completely consolidated under the weight of the overburden which is now on it. The computation of the preloading pressure is then simply a matter of estimating the vertical pressure on some plane, due to the weight of the overburden. The basic equation is $p = \gamma\, z$, where $p$ is the vertical pressure on a horizontal plane, $\gamma$ is the unit weight of the soil, and $z$ is the depth; below the water table the submerged unit weight must be used. In our approximate analysis, all pressures will be computed at the mid-depth of the clay layer. If several layers are involved, the pressure and settlement analyses would be made separately for each layer, and the effects added. It may be noted that, occasionally, deviations from the assumption of normal loading may be important. Such might be the case, for example, if a fill had been constructed recently at the site. The clay may not have consolidated completely under the weight of the fill. Similarly, a structure may settle because of the lowering of the water table in an area, and a consequent increase in overburden pressure on a clay stratum.

## 160. Stress Caused by the Weight of Structure

*a. Methods Based on the Theory of Elasticity.* Vertical stresses which are caused by the application of surface loads to soil masses are generally estimated by the use of mathematical equations derived by the use of the theory of elasticity. Although soils in nature seldom conform exactly to the assumptions made in the derivation of these equations, they give reasonably accurate results when used in settlement predictions and are widely used for that purpose. Formulas and charts for estimating vertical pressures caused by surface concentrated loads and distributed loads acting on various types of footings are available in standard soil mechanics textbooks.

*b. Approximate Method.*

(1) In figure 53 is shown an approximate method of computing the vertical stresses which result from uniform surface loads. This method is particularly useful in making a rough preliminary estimate of the settlement which a structure may undergo. Basis of the method is the assumption that the area over which the stresses caused by the surface load are distributed increases uniformly with depth. Assume that a total load $Q$ is carried at the surface by a rectangular footing with the dimensions $b$ times $L$. At depth $z$ below the surface the average vertical stress is $Q/(b+z)(L+z)$. This involves

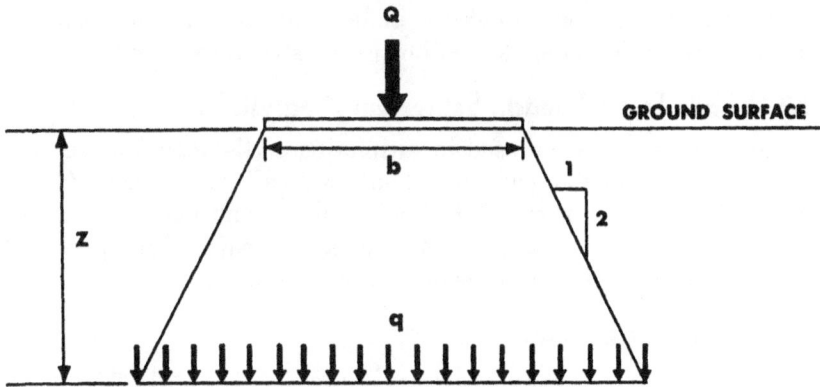

FOR SQUARE FOUNDATION, AVE $q = \dfrac{Q}{(b+z)^2}$

FOR RECTANGULAR FOUNDATION, AVE $q = \dfrac{Q}{(b+z)(L+z)}$

WHERE L = LENGTH

*Figure 53. Approximate method of computing vertical stresses beneath a shallow foundation.*

the assumption that the dimensions of the stressed area increase at a 2 to 1 slope, as indicated in the drawing.

(2) The maximum intensity of pressure at any depth may be estimated by multiplying the average stress at that depth by 1.5.

(3) If two or more footings are involved in the calculations, and the stressed areas overlap at the depth concerned, the calculation is usually made by simply assuming that the total, combined load is distributed over the entire stressed area.

## 161. Stress Release Due to Excavation

Construction of a shallow foundation usually involves the excavation of some of the surface soil. The removal of this soil causes a decrease in pressure, or *stress release*, at any depth below the foundation. The magnitude of this decrease in pressure may be estimated by the use of the same formulas that are used in estimating the pressure increase caused by a distributed load. If the approximate method of the preceding paragraph is used, the effect of excavation may be taken into account by deducting the weight of excavated soil from the total load which will be carried by the structure, including the weight of the foundation itself. If the excavation is shallow, it may simply be assumed that the weight of the excavated soil is offset by the weight of the concrete used to form the foundation. $Q$ would then be the net load carried by the structure. *One of the most effective ways of preventing settlement under any structure is to excavate a weight of soil which is equal to the weight of the structure.* If this is done very little settlement should take place, even though the underlying clay may swell slightly after the excavation is made and before the structure is built.

## 162. Loads To Be Used in Settlement Analysis

Loads which are to be used in estimating settlements due to consolidation should be those which will be sustained over a period of time. Loads of relatively short duration, such as unusual wind loads, normally need not be considered, as they act for such a brief period of time as to be ineffective in causing consolidation.

## 163. Settlement Estimate

*a.* One equation, which may be used to estimate the settlement of a clay layer in the field, is based upon a simple comparison between the change in thickness of a sample in a laboratory consolidation test and that of the field stratum. This equation is $\Delta H = H \Delta e / (1 + e_o)$, where $\Delta H$ is the ultimate decrease in thickness of a confined clay layer and, therefore, the settlement of a structure. Settlement calculations are based upon the generally unfavorable assumption of a completely flexible structure. $H$ is the thickness of the field layer. $\Delta e$ is the change in void ratio which accompanies a change in pressure from $p_0$, the initial pressure, to $p_1$, the final pressure. The increase in pressure from $p_0$ to $p_1$ is caused by the weight of the structure. This value must come from the *e-p* curve obtained from a laboratory consolidation test (fig. 37), and both $p_0$ and $p_1$ must be intergranular pressures. $e_0$ is the void ratio corresponding to the initial pressure. *This is the approach which must be used if the clay is not normally loaded.* It may also be used for normally loaded clays, if desired.

*b.* For normally loaded clays the following expression is applicable: $\Delta H = [HC_c/(1+e_0)] \log_{10} (p_1/p_0)$, where the symbols have the same meaning as in *a* above, and $C_c$ is the compression index (par. 119). For preliminary purposes, the approximate relationship between $C_c$ and the liquid limit presented in figure 38 may be sufficiently accurate to provide an estimate of the settlement. *This equation must be limited to clays which are normally loaded.* If the clay is precompressed, settlements obtained in this approximate fashion will be conservative. If the clay is highly sensitive, predictions made in this approximate fashion may be dangerously unsafe. Settlement predictions for such soils must be based upon the results of laboratory consolidation tests.

## 164. Time Rate of Settlement

As has been emphasized, the consolidation of clay soils is a slow process, requiring many years to complete in most field applications. Estimates of the time rate of settlement may be made using results of laboratory consolidation tests. Because of the theory involved and the fact that such computations are seldom necessary in practice, details of time-settlement calculations will not be presented here. The time required for the ultimate settlement to take place is primarily dependent upon the thickness of the layer, and the permeability. The thicker the layer and the lower the permeability, the longer will be the time required for consolidation, other things being the same. The time required is also a function of the drainage conditions of the clay layer. For example, other things being the same, approximately four times as long is required for the ultimate settlement of a clay layer which is bounded only on the top or bottom by a more pervious soil, e. g. sand, than if the clay is bounded on both top and bottom by a more pervious soil.

## 165. Numerical Example of Settlement Calculations

In order to illustrate the approximate approach which has been discussed, the settlement of the foundation shown in figure 54 will be estimated. This is a square concrete footing, 10 feet on a side. Contact settlement of the sand will be ignored, since the sand is of medium density. The clay will be assumed to be normally loaded. Stress calculations will be referred to the center of the clay layer. The overburden pressure, $p_0$, may be estimated as follows: $p_0 = 110(15) + (100 - 62.5)5 = 1650 + 188 = 1838$ pounds per square foot. The increase in stress in the clay, $\Delta p = 200(2000)/(10+20)(10+20)$ $=400,000/900=444$ pounds per square foot. Therefore, $p_1=1838+$ $444=2282$. From figure 38, $C_c=0.47$, since the liquid limit is 62 percent. From paragraph 163, $\Delta H=[HC_c/(1+e_0)]\log_{10}(p_1/p_0)$. All the quantities in this equation are now known with the exception of $e_0$, which may be estimated from the fundamental relationships of paragraph 24, assuming that the specific gravity of the clay is 2.75; $e_0=1.71$.

Then, 
$$\Delta H=10(0.47)/(1+1.71)\log_{10}(2282/1838)$$
$$\Delta H=(4.7/2.71)\log_{10}1.24=(4.7/2.71)\,0.093$$
$$\Delta H=0.162 \text{ feet}=1.94 \text{ inches.}$$

Thus, the settlement of this isolated spread footing may be expected to be on the order of 2 inches. Since the footing is isolated, such a settlement should not be serious, unless the structure which it supports can not withstand an ultimate settlement of this magnitude. Differential settlements are difficult to estimate by the approximate method. If total settlements are small, differential settlements will also be small. If total settlements are large, differential settlements may also be large, assuming a flexible structure.

## 166. Bearing Capacity of Buried Clay Layer

Another principle may be conveniently illustrated at this point. This is the fact that the ultimate bearing capacity of the clay layer may govern the design and should be checked. Assume that the clay has an unconfined compressive strength of 800 pounds per square foot, as indicated in figure 54. Using the approximate method, the maximum vertical stress at the top of the clay layer is $1.5(200)2000/(10+15)(10+15)=600,000/625=960$ pounds per square foot. Using the information in paragraph 150, the ultimate bearing capacity of a saturated clay is $4c+\gamma z=4(400)+100(15)=3100$ pounds per square foot. Thus, the factor of safety against a shear failure in the underlying clay layer is approximately 3100/960 or somewhat more than 3. The proposed foundation should be satisfactory from this standpoint for the assumed conditions.

## 167. Summary of Considerations Relative to Settlement

The approximate nature of the approach which has been suggested for evaluating settlements must again be emphasized. Information which has been presented should be adequate for a preliminary rough estimate of the magnitude of settlements which may occur when a shallow foundation is being contemplated on a reasonably uniform soil deposit. Such an estimate of settlement, used in combination with considerations relative to ultimate bearing capacity, should permit the engineer to decide if a shallow foundation will be satisfactory, or that a deep (pile) foundation must be used. The information which has been presented is undoubtedly not sufficient for the safe

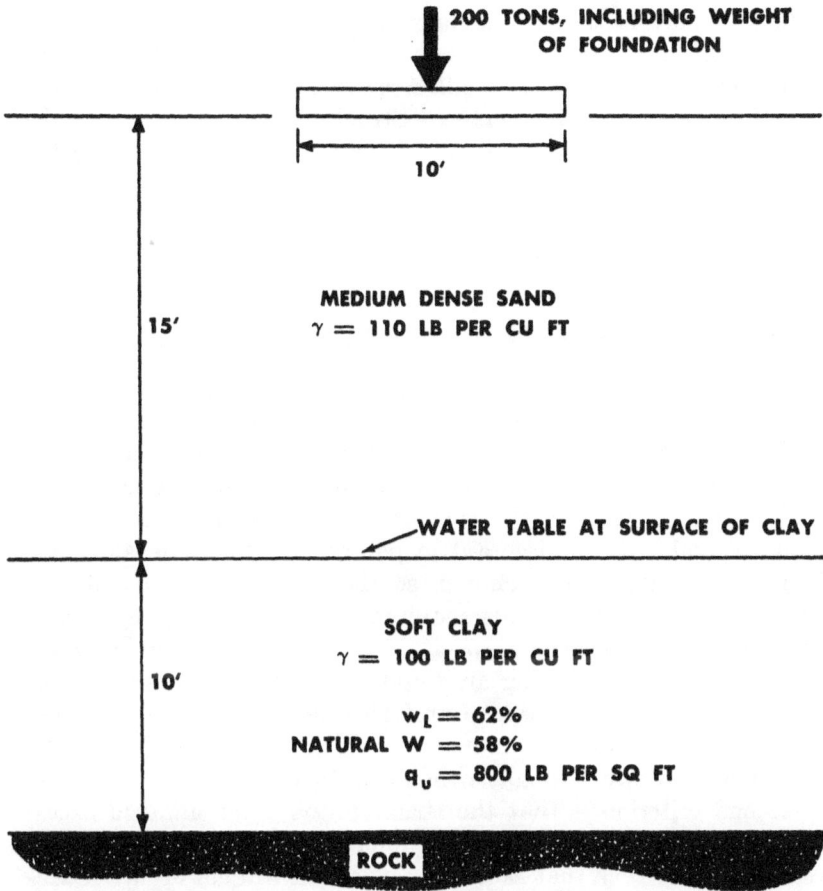

**200 TONS, INCLUDING WEIGHT OF FOUNDATION**

10'

MEDIUM DENSE SAND
$\gamma = 110$ LB PER CU FT

15'

WATER TABLE AT SURFACE OF CLAY

SOFT CLAY
$\gamma = 100$ LB PER CU FT

10'

$w_L = 62\%$
NATURAL $W = 58\%$
$q_u = 800$ LB PER SQ FT

ROCK

*Figure 54. Sketch accompanying numerical example of settlement calculations*

and economical design of the foundations of a heavy, important, or permanent major structure founded above compressible soils. In such cases, investigations which are beyond the scope of this manual may be required.

## 168. Considerations Relative to Other Soils

Obviously many soils do not fall into the categories of cohesionless soils or saturated clays, the limiting cases upon which the previous discussion has been based. Perhaps the best approach with these soils is to place them in one category or the other, depending on which type they most closely resemble. Soils which are essentially non-plastic may be expected to behave like cohesionless soils, while plastic soils may come closer to the case of saturated clays. If this does not seem feasible, the soil's behavior under both limiting classifications may be estimated, and compromise values decided upon. Non-plastic silts, such as rock-flour, may be analyzed as sands, while plastic silts may be treated as clays, although this presents a somewhat simplified picture of the behavior of silt soils. It must be emphasized that *loose or soft silt deposits are not normally suitable for the direct support of foundations.* Structures on such deposits must be founded on piles or piers, or on raft foundations which are located at such an elevation that the weight of the soil removed is approximately equal to the weight of the structure. *More highly organic soils, such as peat and muck, are similarly unsuited for the direct support of foundation loads.*

## 169. Rational Method of Estimating Allowable Soil Pressure

The engineer who is concerned with the selection and design of a shallow foundation is interested in the *allowable soil pressure*, the unit load which the soil beneath the foundation can support with safety and economy. As has been emphasized, two major criteria are applied in estimating the allowable soil pressure, together with certain general criteria discussed in paragraph 146. The allowable soil pressure must be low enough so that the foundation will not break into the ground. This requirement may be met by evaluating the ultimate bearing capacity by theoretical means, or by a load test, and then applying an ample factor of safety; a factor of safety of 3 is generally used. Careful consideration must be given to the selection of the maximum load which will be carried by the foundation, and which might contribute to a failure of this type. The second criterion is that the structure must not undergo excessive settlement, either uniform or differential. The allowable soil pressure must be such that excessive settlement will not occur. Careful attention again must be given to the selection of loads which will contribute to settlement, and to the amount of settlement which is tolerable to the structure concerned. When columns on spread footings are used at normal spacings, tolerable differential settlements may range from a low of perhaps one-half of an inch for a reinforced concrete structure up to perhaps 3 inches for a light steel structure which is comparatively flexible. With these two criteria in mind, the engineer must then select the allowable soil pressure which satisfies both requirements and which, in addition, provides an economical design. Circumstances will occur in which a shallow foundation will not be satisfactory. Consideration must then be given to the use of the deep foundation described in paragraphs 175 through 196.

## 170. Effect of Nonuniform Soil Profile

*a.* Most of the material which has been presented in this discussion of shallow foundations has dealt with uniform soil deposits — deposits that are reasonably homogeneous to a considerable depth. Many soil deposits are nonuniform in character, and their nonuniformity may be extremely important in the design of shallow foundations. Some deposits are extremely erratic in nature, containing lenses of soft or weak soil and marked by extreme variations in soil type and condition within a small area. Approximate methods of analysis which have been presented are of little value in such situations except to indicate extremes of behavior which may occur in different portions of the deposit. Some idea of the behavior of a shallow foundation may be obtained by imagining that the area of greatest stress beneath the foundation and the zone of weakest soil coincide.

*b.* Soil deposits are frequently nonuniform in the sense that there are alternate layers of materials of different strengths, one on top of the other. For example, a comparatively soft clay layer may be underlain by stronger material, such as sand or rock. This is generally a favorable situation as far as the engineer is concerned in that the weakest soil is visible and readily detected. The engineer is made sharply aware of any problem which the weak soil may present. A much more serious situation has already been noted in figure 54, in which a layer of comparatively strong, stiff soil is underlain by a much weaker one. Such a situation is not uncommon and may have serious consequences if the weaker layer is not detected and its possible effect on the structure recognized. Stresses produced by loads at the surface may cause detrimental settlement due to consolidation of the buried layer, or may actually cause a shear failure if the bearing capacity of the buried layer is exceeded. In some situations layers of weak and strong soils occur alternately with depth. Computations such as have been presented previously are useful in such cases in judging if a shallow foundation will be satisfactory, or in judging the depth to which piles should be driven.

## 171. Tabular Values of Allowable Soil Pressure

*a.* Many tabulations of allowable pressures on various classes of materials are available in engineering handbooks and building codes. Information of this type is contained in table XI, appendix II. The values given in the table are maximum values which are suggested for use in the absence of a load test or other evidence which would permit the use of a higher value. The descriptive terms in the table are self-explanatory, except for the term *soft rock*, which is taken to include shale, decomposed schist, and similar materials which are somewhat disintegrated and softened, with some cracks allowed.

*b.* A number of soils, including muck, peat, organic silt, loose inorganic silt, and soft clay are not included in the tabulation, since they are regarded as unsatisfactory materials for the direct support of foundations.

*c.* Values given in table XI should be satisfactory for the design of shallow foundations which are located on homogeneous deposits, or in cases in which the foundation rests on a comparatively weak soil which lies above a stronger one. *They are not applicable if the soil upon which the foundation rests is underlain by a weaker soil. For the design of a major structure the use of the tabular values must be tempered by judgment, experience, and frequently laboratory testing.*

## 172. Foundation Characteristics of Groups of Unified Soil Classification System

Column 11 of table VII, appendix II, gives general information relating to the value of foundations of soils classified under the Unified Soil Classification System. The information given in the table is quite general, and should be used with a full understanding of the principles which have been discussed previously in this section.

## 173. Observation of Existing Structures

Much information may be gained from the intelligent examination of foundations which are providing satisfactory support for existing structures in the area where a new structure is to be built. Because of the variable nature of soil deposits and the importance of variations within a small area, caution must be exercised in applying this knowledge to the design of a particular structure. If the opportunity is available, a discussion of local soil problems with a skilled person can be very helpful.

## 174. Effects of Vibration

The bulk of the discussion contained in this section has dealt with static loads. Vibratory loads have not been considered. As noted previously, vibrations have relatively little effect upon cohesive soils but may have a great effect upon loose cohesionless soils. The relative density of a loose cohesionless soil deposit may be greatly increased by vibration and vibratory methods are frequently used to compact soils of this type. Principal methods used for structural foundations include pile driving and the vibroflotation process, which employs jets of water in combination with steel vibrators. Sometimes cohesionless soil deposits are compacted inadvertently during construction operations, as by pile driving or blasting. Unexpected settlements which may result may be of serious consequence. Little knowledge is available concerning the design of foundations which must support heavy vibratory loads, as from vibrating machinery. The design of a major foundation of this type should be regarded as a special problem beyond the scope of this manual.

## Section III. DEEP FOUNDATIONS

### 175. Introduction

When tne surface soils which exist at the site of a proposed structure are too weak and compressible to provide adequate support, deep foundations are frequently used to transfer the load to underlying suitable soils. Two types of deep foundations are in common use. These are *pile foundations* and *pier foundations*. Most of the discussion which follows is concerned with piles and pile foundations, since these are very commonly used, both in military and civil construction. Piers, which are normally used only for the support of very heavy loads, are much less common.

### 176. Piles and Pile Foundations

By common usage, a pile is a load-bearing member which may be made of timber, concrete, steel, which is generally forced into the ground. Piles are used in a variety of forms and for a variety of purposes, as indicated below. A *pile foundation* is a group of piles which is used to support a pier or column, a row of piles under a wall, or a number of piles which are distributed over a large area to support a mat foundation. A foundation of this type normally is used for the support of vertical loads, although it may also be used to support inclined or lateral forces.

### 177. Bearing Piles

Piles which are driven vertically and used for the direct support of vertical loads are commonly called *bearing piles*. Bearing piles may be used to transfer the load through a soft soil to an underlying firm stratum; these are called *end-bearing piles*. They may also be used to distribute the load through relatively soft soils which are not capable of supporting concentrated surface loads; these are *friction piles*. A bearing pile may sometimes receive its support from a combination of end bearing and friction. Bearing piles also may be used in situations in which it is likely that a shallow foundation would be undermined by scour, as in the case of some bridge piers. Bearing piles are illustrated in figure 55.

*a.* A typical illustration of an end-bearing pile is one which is driven through a very soft soil, such as a loose silt or the mud of a river bottom, or water to bearing on a firm stratum beneath. The firm stratum may, for example, be rock or a firm statum of sand or gravel. In such cases the pile derives practically all its support from the underlying firm stratum.

*b.* A friction pile develops its load-carrying capacity entirely, or principally, from *skin friction* along the sides of the pile. The load is transferred to the adjoining soil by friction between the pile and the surrounding soil. The load is thus transferred downward and laterally to the soil. The soil surrounding the pile or group of piles, as well as that beneath the points of the piles, is stressed by the load.

*c.* Some piles carry load by a combination of friction and end bearing. A pile of this sort may pass through a fairly soft soil which provides some frictional resistance, and then into a firm layer which develops load-carrying capacity through a combination of friction over a relatively short length of embedment and end bearing.

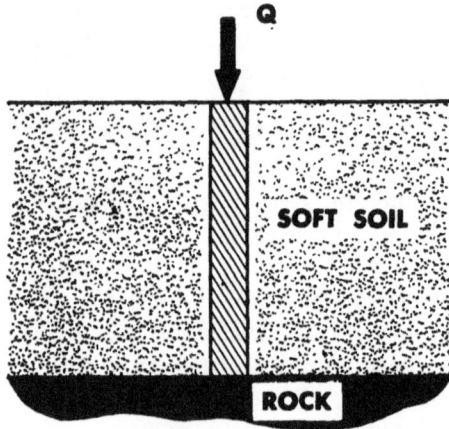

Figure 55. Bearing piles.

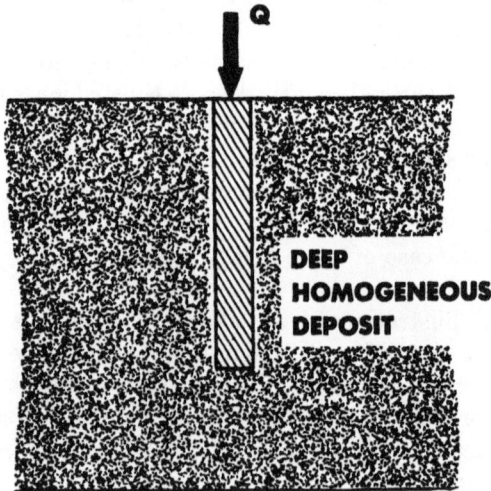

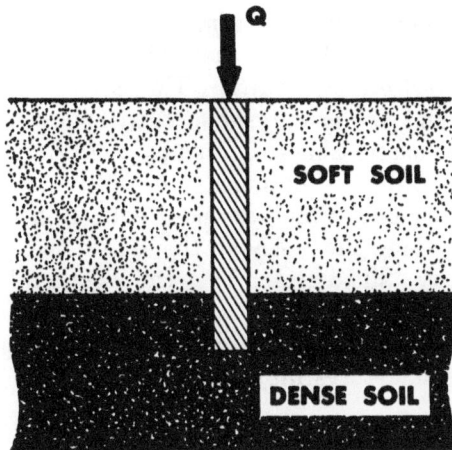

**COMBINATION END-BEARING & FRICTION**

*Figure 55*—Continued.

## 178. Other Uses of Piles

Piles are used for many purposes other than supporting vertical loads. Piles which are driven at an angle with the vertical are commonly called *batter piles*. They may be used to support inclined loads, or to provide lateral stability to pile foundations which support structures subjected to lateral loads. Piles are sometimes used to support lateral loads directly, as in the *pile fenders* which may be provided along waterfront structures to take the wear and shock of docking ships. Piles are sometimes used to resist upward, tensile forces; these are frequently called *anchor piles*. Anchor piles may be used, for example, as anchors for bulkheads, retaining walls, or guy wires. Vertical piles are sometimes driven for the express purpose of compacting loose cohesionless deposits. Closely spaced piles, or *sheet piles* may be driven to form a wall or a bulkhead which restrains a soil mass.

## 179. Timber Piles

A timber pile is nothing but a straight tree trunk which has been trimmed of branches. It is driven with the small end (tip) down. The diameter of the tip should not be less than 5 to 8 inches. Timber piles are widely used in areas where suitable timber is readily available and are frequently used under expedient conditions. Their comparative cheapness, ready availability and ease of handling are advantages accompanying their use. Timber piles, normally are used in lengths up to about 70 feet, although the use of piles 200 feet in length has been reported. The load-carrying capacity of any pile is dependent upon the soil into which it is driven, but timber piles normally are used for the support of comparatively light loads, generally from 10 to 20 tons per pile. Timber piles are somewhat flexible and resilient in nature; these properties make them useful in certain applications, as in fender piles on waterfront structures. They are comparatively strong for their weight and are usually able to withstand normal driving. However, they are susceptible to injury by overdriving. The tips of the piles may be broken, "broomed", or split under hard driving conditions. Timber piles which are used entirely beneath the water table have a very long life. Above the water line, untreated piles are subject to decay and to attack by termites and other insects. In salt water untreated timber piles are subject to attack by marine borers. If timber piles are given proper preservative treatment, their life in waterfront structures may be extended to 30 years or more.

## 180. Concrete Piles

Concrete piles are made in two forms; precast and cast-in-place. One of their principal advantages is the ready availability of materials for the concrete. Under comparable conditions, they can support greater loads than timber piles of comparable size. Another advantage is their durability and immunity from attack by biological organisms. The latter is a particular advantage in waterfront structures.

*a. Cast-in-Place Concrete Piles.* Many different commercial types of cast-in-place concrete piles are available in civil practice. In some commercial types a thin metal shell is driven into the ground to the desired length by the use of a mandrel which fits inside the shell. After the shell has been driven, the mandrel is withdrawn and the shell filled with concrete. Both tapered and cylindrical sections are used. In other types a hole is formed in the ground by drilling, or by driving a steel casing which is later withdrawn: the hole is then filled with concrete. A further variation is seen in the formation of an enlarged lower section, or bulb, by various means. Cast-in-place concrete piles have become very popular in many sections of the United States in recent years, particularly on large projects where many piles are required. They are commonly used in lengths of from 30 to 80 feet, and much longer ones have been used in particular situations. Again depending upon the soils involved, piles of this sort have load-carrying capacities up to about 40 tons per pile in normal applications. Probably the principal disadvantage of the usual pile of this type, which uses the steel shell, is that special pile-driving equipment is required.

*b. Precast Concrete Piles.* Precast concrete piles generally are made with parallel sides, pointed ends, and either square or octagonal cross sections. They are heavily reinforced to withstand driving and handling stresses. Piles of this sort are constructed in lengths up to 50 or 60 feet in the usual case and are capable of supporting loads up to 80 tons in some situations. They are very heavy for their length and are difficult to handle for this reason. They are very durable if made of good quality concrete. Their principal disadvantage is that the required length of pile must be determined accurately before the pile is driven, since it is difficult to either cut off the pile to shorten it, or to add extensions to it.

## 181. Steel Piles

Rolled steel structural shapes and steel pipe sections are both widely used as piles, particularly where the pile must be very long and to support heavy loads.

*a.* The structural shapes which are most widely used are the H-beam and WF (wide-flange) sections. These members have great strength in both compression and tension and relatively small cross-sectional areas, so that resistance to driving is comparatively slight. They are perhaps best adapted to use as end-bearing piles on rock. They are easily transported, and have the advantage of comparative ease of cutting to shorten them in the field. They are also easily spliced by simply welding on another section. Lengths of 100 feet are not uncommon with this type of pile, and they are capable of supporting very heavy loads, when driven to firm bearing on rock or a dense soil stratum. Their cost is comparatively greater than that of other pile types, as is the cost of steel pipe piles.

*b.* Steel pipe sections make excellent piles. They are most frequently used as bearing piles, either end-bearing or friction. They are most frequently driven with closed ends. After being driven they are filled with concrete. Open-end pipe piles are also used. The soil generally is cleaned out and they, too, are filled with concrete. These piles have the advantage that they can be inspected before the concrete is poured. Pipe sections are less flexible than other structural shapes, which may be important under hard driving conditions, as they are not as likely to be deflected from a vertical position. Pipe piles have been used in lengths up to 200 feet. They are commonly designed to support loads of from 50 to as high as 200 tons per pile, where soil conditions permit.

## 182. Composite Piles

*Composite piles*, which are made up of one material in the lower section and another in the upper section, are another common type. The most commonly used composite pile combines timber, which is used below the water line, and concrete, which is used above. This type of pile combines the economy of timber with the durability of concrete. Composite piles also are sometimes made up of a steel pipe or H-section in the lower portion and concrete in the upper.

## 183. Pile Driving Equipment

*a. Pile-Driving Rig.* Piles are commonly driven into the ground by the use of a pile-driving rig or a crane fitted with pile-driving attachments.  Crane units may be truck-mounted or on crawler treads. This type of rig works on the ground or on the completed part of the structure.  The pile and the hammer are supported and guided by the pile leads.  Under emergency conditions a frame may be constructed to support the pile leads.  In some situations manpower may be employed to raise the hammer.

*b. Pile Hammers.* The simplest type of pile hammer is the *drop hammer*, which usually consists of a block of cast steel weighing from 500 pounds to a ton or more.  The hammer is raised a few feet in the leads, generally by a winch, and allowed to fall on top of the pile.  The pile is forced into the ground under the energy imparted to it by the falling hammer.  This method has the advantage of simplicity, but is quite slow when compared with modern *steam hammers*.  In a steam hammer, steam pressure is employed to raise the striking part or "ram".  If the ram is then allowed to fall by gravity, it is called a *single-acting steam hammer*.  If the downward stroke of the ram is aided by steam pressure, it is called a *double-acting steam hammer*. Commonly used single-acting steam hammers have a ram which weighs from 3000 to 5000 pounds and a stroke of about 3 feet; these hammers deliver about 60 blows per minute.  Double-acting steam hammers deliver blows of comparable energy, but are generally more rapid in action.  They deliver a light, sharp blow and are most useful in driving steel sheet piling and other comparatively light piles.  In general, the lighter hammers are used in driving timber piles and the heavier ones for steel and concrete piles.  The top of the pile frequently is protected by a cast steel driving head which contains a wooden cushion block which receives the direct blows of the hammer.

## 184. Behavior of Pile During Driving

*a.* During the first few feet of driving, or if the pile is being driven through very soft soils, the pile will penetrate several inches, or perhaps several feet, under each hammer blow.  As driving continues or when stiffer soils are encountered, the penetration per blow will decrease. Piles are frequently driven to a resistance which is measured by the number of blows required to effect the last inch of penetration.  When hard driving is encountered, care must be taken to prevent overdriving of the pile and consequent damage.  This is particularly important when a heavy hammer is being used to drive light timber piles.  In such cases driving should be stopped when more tban 4 or 5 blows are required to accomplish the final inch of penetration.  For heavier piles, 8 or 10 blows per inch should be regarded as the limit.

*b.* It is a commonly observed phenomenon that friction piles driven into some clays will encounter comparatively little resistance to driving.  However, if driving is stopped for some time, say 3 or 4 days, the resistance to driving will be found to be much greater.  This is due to the remolding of the clay during driving and the time effects associated with the low permeability of these soils. While the soil is remolded during driving, compressive stresses are also set up due to the displacement of soil by the pile and the pressure beneath the tip of the pile. These forces cause consolidation over a period of time, which tends to somewhat offset the effects of remolding. As emphasized again later, this is the principal reason why pile-driving formulas based upon the behavior of the pile during driving can not be used to predict static loads which will be carried by piles driven in clay. Similarly, static load tests should be performed after the elapse of some time after a pile is driven in cohesive soils. A similar action may be observed in driving piles in silt and silt-clay soils. Sometimes the recovery in strength is very slow.

## 185. Driving of Piles in Groups

Piles which are driven in loose sand frequently will decrease the volume of the sand and cause some settlement of the surface of the deposit. In driving groups of piles in these soils, the center pile should be driven first, followed by the peripheral piles working outward from the center. In this way the outer piles may be somewhat shorter than the inner ones, because of the increased density of the sand. If piles are driven into dense sand, penetration will usually be quite difficult and some heaving of the surface may result from displacement of the soil by the piles. If piles have already been driven nearby, they may be raised by the driving of an additional pile. If they are friction piles the heaved ones need not be redriven. However, if they are end-bearing piles the heaved piles should be redriven to satisfactory bearing. A deposit of clay can not be made more dense by driving piles into it.

## 186. Other Methods of Placing Piles in Position

In situations in which piles must be driven through relatively dense layers of sand and gravel in order that the pile may be carried through underlying soft strata to bearing beneath, jetting is frequently used. In this process a jet of water is discharged along the sides of the pile, or at the point, in order to loosen the sand or gravel by causing a temporary quick condition. The pile then may penetrate of its own weight or with very light driving. After jetting, the pile must be driven to bearing. If the pile has to be driven through a thin, dense layer of clay or soft rock, a hole is sometimes broken through the layer by means of a hard metal point. The pile may then be driven more easily. This operation is called "spudding". Another practice which is sometimes used, particularly when piles are to be driven through a plastic clay to reach firm bearing beneath, is to remove a portion of the soil prior to placing the pile. This may be done by boring, or by driving a casing and then cleaning out the soil on the inside. Piles are occasionally jacked into position, particularly in cities where the disturbances accompanying pile driving may have harmful effects on adjacent structures.

## 187. Determination of Allowable Load on a Single Pile

There are three methods which may be used to estimate the safe load which may be carried by a single bearing pile. The first of these is the *dynamic method*, in which the behavior of the pile during driving is used to predict the static load which can be carried. This procedure is applicable only to cohesionless soils. A second procedure is the so-called *static method*, in which soil properties are used to compute the allowable load. This approach is best used in saturated clay soils. The third method is the performance of a *pile load test*.

## 188. Dynamic Method

*a.* The dynamic method is based upon the use of *pile-driving formulas*. There are many pile-driving formulas in existence, some of which are very elaborate. The best known is the simple *Engineering News Formula*, which is $Q_a = 2W_h H/(s+C)$. The symbols in the formula have the following meanings: $Q_a$ is the allowable load on the pile in pounds; $W_h$ is the weight of the ram in pounds; $H$ is the height of fall of the ram in feet; $s$ is the penetration of the pile in inches under the final blow from the hammer; and $C$ is a constant, the value of which is 1.0 for a drop hammer and 0.1 for a steam hammer. The formula includes a theoretical factor of safety of 6, although the actual factor of safety has been found to vary from less than one to more than 10 in actual tests. On the average, the factor of safety is about 4. *The formula is not applicable to cohesive soils.* It is best applied to timber piles driven with a drop hammer into sandy soils. The value of $s$ for such conditions should not be less than about one inch. The value $s$ is often based upon the average penetration per blow during the last six blows of the hammer. Caution should be exercised in the interpretation of the formula in driving through alternate layers of soft and stiff soils, as a portion of the driving resistance may come from friction which will disappear with time.

*b.* As an example of the use of the equation, consider that a timber pile is driven into a sandy soil with a drop hammer which weighs 2500 pounds and has a fall of ten feet. Penetration during the last six blows was 5.4 inches. The allowable load on the pile is equal to $2(2500)10/(0.9+1.0) = 50,000/1.9 = 25,300$ pounds, or about 12.5 tons.

## 189. Static Method

The ultimate bearing capacity of a single, cylindrical, friction pile driven in clay soil may be estimated by multiplying the embedded surface of the pile by the shearing strength of the clay. The shearing strength of the clay may be taken to be one-half the unconfined compressive strength. The ultimate value is then divided by an appropriate factor of safety, usually 3, to determine the allowable load on the pile. This approach does not indicate the settlement which may be expected of a friction pile in clay. This factor is discussed when pile groups are considered. This approach has also been used for piles driven into other soil types. Allowable values of the skin friction for other soils and various types of piles are available in some references. This is not recommended here because of the many uncertainties involved and the difficulty of field application.

## 190. Pile Load Tests

*a.* The only reliable method of determining the safe load which can be placed on a pile is by means of a load test. The *pile load test* may be conducted in a fashion which closely resembles that of the plate bearing test shown in figure 50. A platform may be built on top of the pile and some material of high unit weight used to provide dead load. A more common arrangement is that which is illustrated in figure 56. Here a hydraulic jack reacting against a beam, which in turn reacts against anchor piles, is used to apply load. The load is commonly applied in increments, and some time is allowed to elapse between each increment of load. The final load is generally required to be at least twice the load which is contemplated in the design of the foundation.

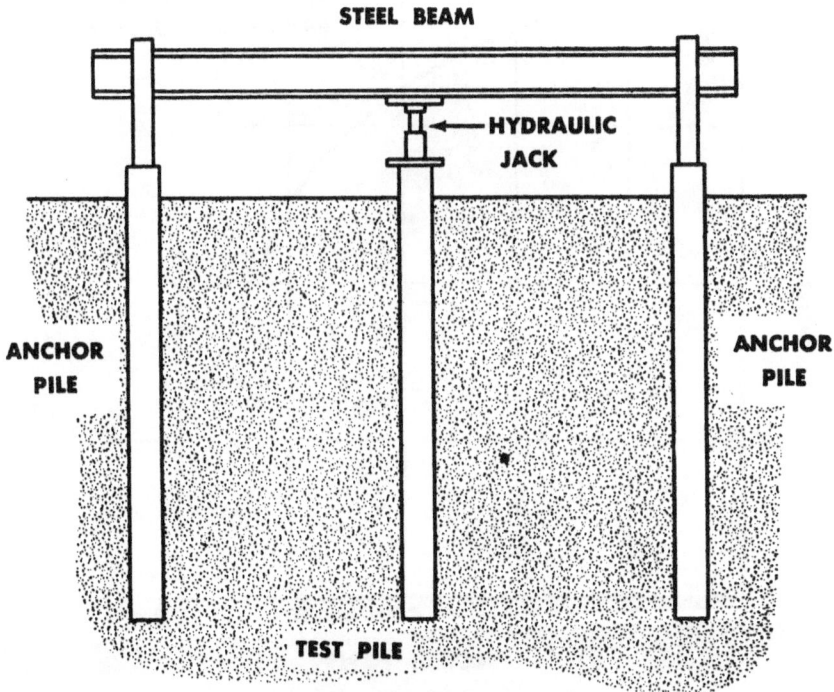

*Figure 56. Pile load test.*

*b.* The relationship between load and settlement may be plotted as indicated in figure 57. One of the curves ① is a typical load-settlement curve for a pile which is driven through very soft soils to an underlying dense stratum. In this case there is no sharp break in the curve; thus the allowable load must be selected to correspond to the maximum allowable settlement. A typical requirement is that the maximum allowable pile load be equal to one-half that which causes a net settlement of not more than one-hundredth of an inch per ton of total test load, or one-half that which causes a gross settlement of one inch, whichever is less. The curve of figure 57 ② is typical of those obtained in load testing friction piles in clay. The allowable load may be taken to be from one-third to one-half of the failure load on the pile. It must be noted that the load test is not normally conducted over a long enough period of time to disclose settlement due to the consolidation of compressible soils.

*c.* It must be emphasized that, unless the piles are driven to bearing on rock, time is required for the development of ultimate bearing capacity. The time may vary from a few hours to 2 or 3 days for permeable, cohesionless soils, to a month or more in silt and clay soils. To be valid, the load test must be conducted after an appropriate waiting period.

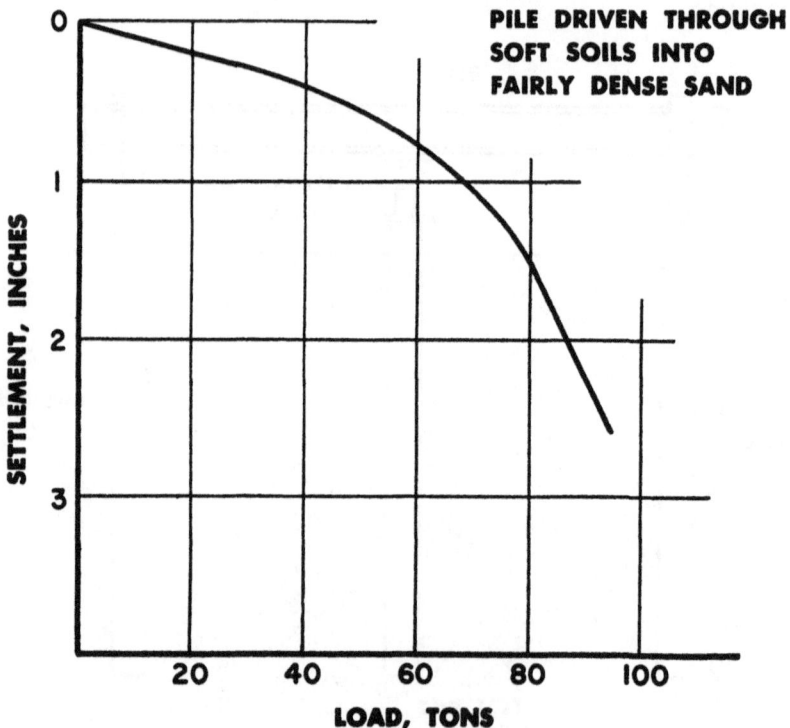

① Pile driven through soft soils into fairly dense sand

*Figure 57. Settlement curves for pile load tests.*

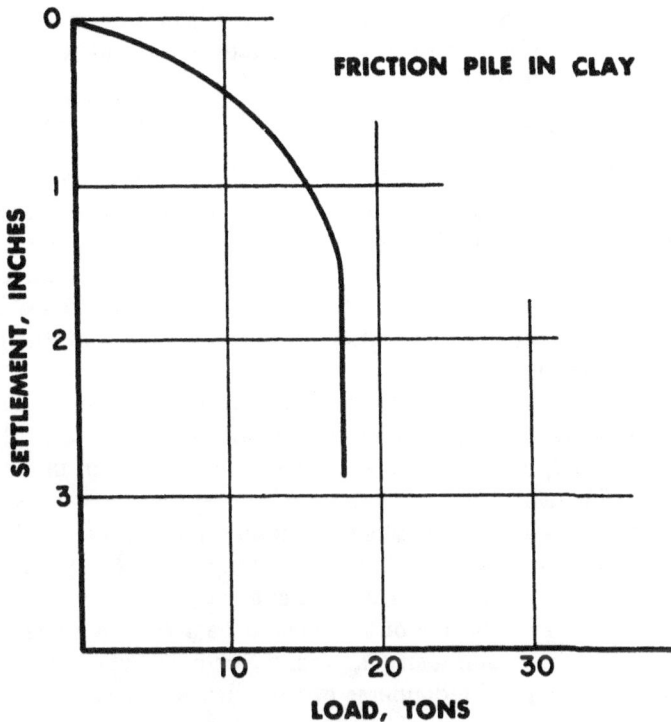

FRICTION PILE IN CLAY

SETTLEMENT, INCHES

LOAD, TONS

③ Friction pile in clay

*Figure 57*—Continued.

## 191. Miscellaneous Considerations Relative to Pile Loads

*a.* The allowable load on a group of piles may or may not be equal to the allowable load on a single pile multiplied by the number of piles in the group, as discussed in more detail in paragraph 192.

*b.* It is rarely necessary to design a pile as an unbraced column, even in soft soils. Exceptions occur when an appreciable length of the pile extends through air, water, or an extremely soft soil. Otherwise, the pile behaves as though it were fully supported and braced, because of the restraining effect of soil which surrounds it.

*c.* Allowable load which can be supported by piles which are subjected to tensile, lateral, or inclined loads can generally be determined accurately only by appropriate loading tests. Pull-out tests may be used to evaluate the tensile load which may be carried by an anchor pile; in clay soils the allowable load may be approximated by the static method. Friction piles in clay soils which are subjected to considerable lateral or inclined loads are likely to undergo lateral movement, or creep, over a period of time. This difficulty will generally not be experienced if batter piles are end-bearing in a firm stratum. Vertical piles have a limited ability to withstand lateral loads; the resistance may be estimated from the passive pressure discussed in chapter 12.

As a minimum value and except for very soft silt or clay soils, vertical bearing piles should be able to withstand a lateral force of 1000 to 1500 pounds without difficulty.

## 192. Action of a Group of Piles

*a*. Since pile foundations normally involve a group of piles acting together to provide support for a structure, it is important that the fundamentals of group action be thoroughly understood. It should never be assumed that the capacity of a pile group is equal to the capacity of an individual pile multiplied by the number of piles in the group, unless it is certain that soil conditions are such that the group will behave in the assumed fashion.

*b*. In judging the behavior of a pile foundation, it must be realized that the load-carrying capacity of the foundation is dependent upon the soils involved, since all loads are transmitted to the underlying soil layers. The same general criteria of soil action must be applied to pile foundations as were applied to shallow foundations. That is, the foundation must be safe against breaking into the ground and must also be free from excessive settlement.

*c*. The number and spacing of the piles to be used in a given case is obviously dependent upon many factors. Normal spacings vary from about 2.5 to 3.5 times the diameter of the piles which are to be used. A symmetrical pattern is usual. The minimum spacing is dependent upon practical matters of construction, since piles can not be driven too close together without interfering with one another. The maximum spacing is determined largely by the economical design of the foundation or pile cap. The *pile cap* is usually of reinforced concrete designed to be rigid enough so that all the piles aid in carrying the load of the structure.

## 193. Design of Pile Foundations

*a*. Adequate soil exploration to a sufficient depth is essential for the design of a pile foundation for anything except a crude, hasty structure.

*b*. The type of pile to be used in a given pile foundation is also a function of many variables. Included among them are the available materials, loads to be carried, probable length of the piles, soil and ground-water conditions, type of pile-driving equipment available, and cost.

*c*. The length of pile to be used may frequently be estimated from the soil profile, particularly when end-bearing piles are to be used. If friction piles are to be driven in clay, the required length may be estimated by the static method. In the majority of cases, the required pile length can be determined only by the driving and loading of test piles in the field.

## 194. Pile Foundations in Sand

*a.* Piles which are driven through soft soils into a layer of dense sand of great depth behave in approximately the same manner as if the entire load were resting on the sand at the level of the pile tips. The load-carrying capacity of a pile group may be taken to be the product of the number of piles times the allowable load on each pile, as determined by a load test or approximately be a pile-driving formula. Settlement is usually not a critical factor, as the settlement occurs quickly and would seldom exceed 1 inch.

*b.* If the sand stratum into which the piles of *a* above are driven is underlain by a compressible layer of soil, settlement may occur because of consolidation. The amount of settlement may be approximated by assuming the load to be distributed over the area of the foundation at the level of the pile tips, and using the approximate method of figure 53. A situation of this type is a classic example of one in which piles may do as much harm as good and should be carefully analyzed. If the sand stratum is thin, the possibility exists that long piles driven through it may be more satisfactory.

*c.* When piles are driven through alternately stiff and soft layers to bearing in sand, care should be taken in the interpretation of load test results, since some of the resistance may be due to frictional resistance along the portion of the pile which is above the sand. After the structure is complete, this frictional resistance may disappear, and the load will be carried by the point resistance of the pile in the sand. If necessary, the point resistance may be evaluated by performing two load tests, one on a pile driven to 3 or 4 feet above the top of the sand and the other on a pile driven into the sand.

*d.* In some cases piles are driven through a recent fill to bearing on firm sand. As the fill consolidates the load on the piles will be increased, due to what is called "negative skin friction." The piles must be able to support this additional load plus the weight of the structure. A similar situation may occur if a fill is placed around the upper portion of the piles after they have been driven.

*e.* Piles may be driven into a loose deposit of sand to compact it by vibration and lateral displacement. After the piles are driven, the group of piles contained in the area will form the sand into a block of dense sand, which serves as a foundation resting upon the loose sand. Other methods of compaction may be more efficient.

## 195. Pile Foundations in Clay

*a.* Friction piles driven into a deep clay deposit should be checked from the standpoint of ultimate bearing capacity and for settlement.

    (1) For the purpose of estimating the safe bearing load, a group of piles of this sort may be visualized as forming a giant block resting on the soil at the elevation of the pile tips.

This is illustrated in figure 58. As indicated, the bearing capacity of the group is comprised of two things; the shearing resistance which acts on the surface of the periphery of the pile group, and the ultimate bearing capacity beneath the assumed block, the area of which is assumed to be the same as the area of the pile group. The shearing resistance on the sides may be estimated by multiplying the area over which it acts by one-half the unconfined compressive strength. The bearing capacity of the clay beneath the group may be estimated by the methods of paragraph 150. An appropriate factor of safety must be used to obtain the allowable load on the group. The spacing of the piles should be such that the full capacity of each pile is realized, if possible. It will be noted that, within the limits of economy, considerable benefit will be gained from the use of longer piles.

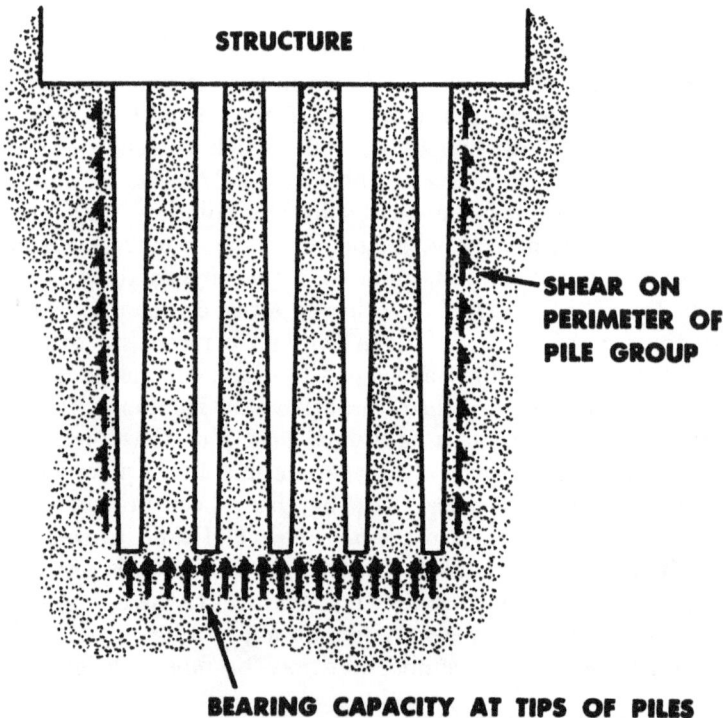

Figure 58. Bearing capacity of a group of friction piles in clay.

(2) The settlement of a group of friction piles in clay may be estimated by assuming that the total load is located on a plane one-third of the length of the piles above their tips. The approximate method of paragraph 165 may then be

used to estimate the settlement due to consolidation of the soil below this plane. It may be noted that longer piles generally offer an advantage here also, in that the compressibility of clay soils frequently decreases with depth.

*b.* Piles may be driven into clay in two other general circumstances. One is the case in which relatively soft compressible soils lie over a stiff, hard layer of clay. Piles may then be driven to bearing in the stiff clay layer. Another situation is one in which piles are driven into a clay deposit which increases in stiffness with depth. In both these situations the load-carrying ability of a group of piles should be practically the same as the product of the number of piles times the load-carrying capacity of a single pile. Settlement may be a problem in both cases, and is difficult to estimate by approximate methods.

## 196. Pier Foundations

A *pier foundation* is a large, deep foundation which serves the same function as a pile foundation. The principal differences between a pier and a pile are the size (a pier is usually 18 inches or more in diameter) and the method of construction used. Piers are constructed by the open-shaft method or by the use of caissons, which may be open or pneumatic. Generally speaking, pier foundations may be designed in the same way as footings. Both ultimate bearing capacity and settlement must be considered; settlement is frequently the governing factor in design.

# CHAPTER 9

# COMPACTION

## 197. Introduction

By *compaction* is meant the process of artificially densifying a soil. Densification is accomplished by pressing the soil particles together into a closer state of contact, with air or water being expelled from the soil mass in the process. Compaction, as here used, implies dynamic compaction or densification by the application of moving loads to the soil mass. This is in contrast to the consolidation process described in chapter 6, in which a soil is made more dense as the result of the application of a static load. With relation to compaction, the density of a soil is normally expressed in terms of dry density or dry unit weight, common units of measurement being pounds per cubic foot. Occasionally, use is made of the wet density or wet unit weight.

## 198. Advantages Gained from Compaction

*a. Introduction.* Certain advantages which accompany the compaction of soils have made the process of compaction a standard procedure in the construction of earth structures, particularly embankments, earth-fill dams, subgrades and bases for highway and airport pavements. There is no single construction process which can be applied to natural soils which produces so marked a change in their physical properties at so low a cost as compaction, when it is properly controlled to produce the desired results. Principal soil properties which are affected by compaction include settlement, shearing resistance, movement of water, and volume change. Compaction does not improve the desirable properties of all soils in the same degree, and in certain cases the engineer must give careful consideration to the effect of compaction upon the various properties listed. For example, with certain soils the desire to hold volume change to a minimum may be more important than just an increase in shearing resistance.

*b. Minimizing of Settlement.* One of the principal advantages which results from the compaction of soils used in embankments is that settlement which might be caused by the consolidation of the soil within the body of the embankment is reduced to a minimum. This is true because compaction and consolidation both bring about a closer arrangement of soil particles. Artificial densification by compaction will prevent later consolidation and settlement of an embankment, but this does not necessarily mean that the embankment will be free from settlement, since its weight may cause consolidation of compressible soil layers which may form the embankment foundation.

*c. Increase in Shearing Resistance.* Increasing density by compaction increases shearing resistance. This effect is highly desirable in that it may make possible the use of a thinner pavement structure over a compacted subgrade or the use of steeper side slopes for an embankment than would otherwise be possible. Other things being the same, the shearing resistance of a given soil increases with an increase in density. In addition to density, shearing resistance depends upon water content. For the same density the highest strengths are frequently obtained by the use of greater compactive efforts and with water contents somewhat below optimum moisture content (par. 199). Large-scale experiments conducted by the Corps of Engineers indicated that the unconfined compressive strength of a clayey sand could be doubled by compaction, within the range of practical field compaction procedures.

*d. Effects on Water Movement.* When soil particles are forced together by compaction, both the amount of voids contained in the soil mass and the size of the individual void spaces are reduced. This change in voids has an obvious effect upon the movement of water through the soil. One effect is to reduce the permeability, thus reducing the seepage of water. Somewhat similarly, if the compaction is accomplished with pioper moisture control, the movement of capillary water is minimized, thus reducing the tendency for the soil to take up water and suffer later reductions in shearing resistance.

*e. Volume Change.* Volume change (shrinkage and swelling) is an important soil property which is particularly critical when soils are used as subgrades for roads and airport pavements. Fundamental considerations relative to the shrinkage and swelling of soils is contained in paragraphs 38 and 39. Generally speaking, volume change is not a great matter of concern in relation to compaction, except for clay soils. Compaction has a marked influence on the volume change of clay soils. For these soils, the greater the density, the greater the potential volume change, unless the soil is restrained. An expansive clay soil should be compacted at a moisture content and to a density at which swelling will be a minimum. Similarly, the soil should be compacted so that shrinkage will be a minimum. Although the conditions corresponding to minimum swell and minimum shrinkage may not be exactly the same, soils in which volume change is a factor generally may be compacted so that these effects are minimized. The effect of swell on bearing capacity is important and is recognized by the standard method used by the Corps of Engineers in preparing samples for the California Bearing Ratio test.

## 199. Moisture-Density Relationships

*a. Definitions.* Nearly all soils exhibit a similar relationship between moisture content and dry density when subjected to a given compactive effort. This relationship is indicated in figure 59. For each soil there is developed a maximum dry density at an optimum moisture content for the compactive effort used. The *optimum moisture content*, at which *maximum density* is obtained, is the moisture content at which the soil has become sufficiently workable under the given amount of compactive effort to cause the soil particles to become so closely packed as to expel most of the air. For most soils (except cohesionless sands) when the moisture content is less than optimum the soil is more difficult to work, and thus to compact. Beyond optimum, most soils become more workable, but are not as dense under a given compactive effort because the water interferes with the close packing of the soil particles. Beyond optimum and for the stated conditions, the air content of most soils remains essentially the same, even though the moisture content is increased. The moisture-density relationship shown in figure 59 is indicative of the workability of the

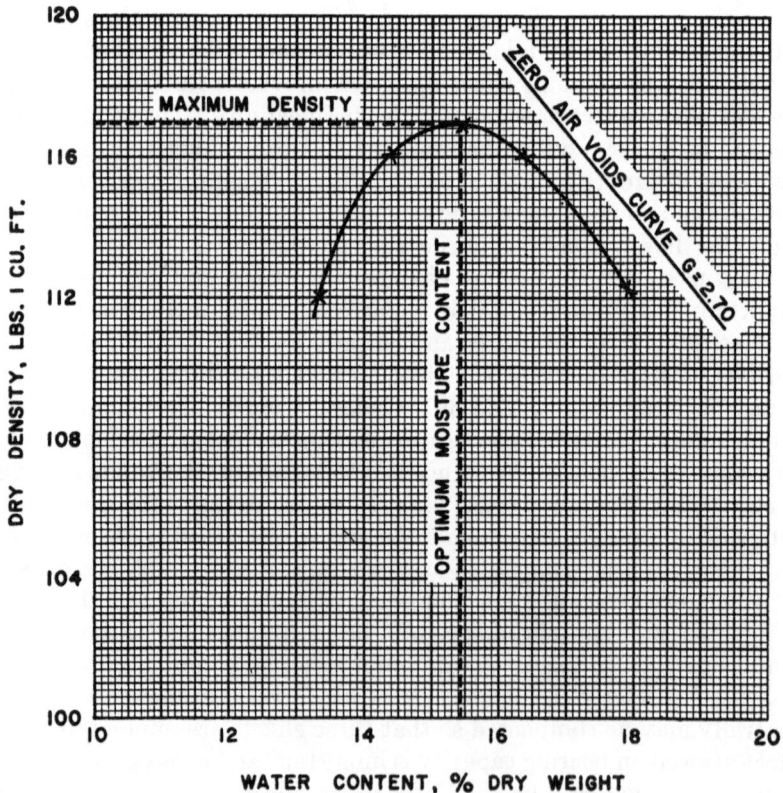

*Figure 59. Typical moisture-density relationship.*

soil over a range of water contents for the compactive effort used. The relationship is valid for laboratory compaction, and for field rolling. The maximum density is frequently visualized as corresponding to *100 percent compaction* for the given soil under the given compactive effort.

*b. Laboratory Compactive Efforts.* The curve of figure 59 is valid only for one compactive effort; it was established in the laboratory. Two standardized laboratory compactive efforts are widely used in this country. They are the *Standard AASHO* (American Association of State Highway Officials) or *Standard Proctor compaction procedure,* which is widely used among highway engineers, and the *Modified AASHO compaction procedure,* which is most often used in airfield and earth dam work, and which has been adopted by the Corps of Engineers. The compactive effort involved in the Standard AASHO procedure is 12,400 foot-pounds per cubic foot. It involves 25 blows of a 5½-pound hammer, with a striking face 2 square inches in area, falling 12 inches upon each of three layers of soil in a mold having a volume of 1/30 cubic foot. The Corps of Engineers modification of the AASHO method represents a much greater amount of energy put into the soil (56,200 foot-pounds per cubic foot), and involves 25 blows of a 10-pound hammer, with 2 square inches of striking face, falling 18 inches upon each of five layers of soil in a mold having a volume of 1/30 cubic foot. The moisture-density relationship shown in figure 59 was established in the laboratory by compacting several samples of the same soil with different moisture contents in exactly the same way. The wet unit weights and moisture contents were determined, and the dry unit weight corresponding to each moisture content calculated. Detailed procedures for performing the Modified AASHO compaction test are given in TB 5–253–1.

*c. Zero Air Voids Curve.* On the plot of figure 59 is shown the *zero air voids curve* for the soil involved. This curve is obtained by plotting the dry densities corresponding to complete saturation at different moisture contents. The dry density corresponding to a given water content is a function only of the specific gravity of the soil particles. It may be calculated by the expression, $\gamma d = 62.4 G / (1 + \frac{wG}{100})$, where $w$ is the moisture content in percent. The zero air voids curve represents theoretical values which are practically unattainable, because it is not possible to remove all the air contained in the voids of the soil by compaction alone. Typically, at moisture contents beyond optimum the actual compaction curve closely parallels the theoretically perfect compaction curve. Any values of the dry density curve which plot to the right of the zero air voids curve are in error. The error may be in the test measurement, the calculation, or the specific gravity.

## 200. Effect of Variation of Compactive Effort

*For each compactive effort which is used in compacting a given soil there is a corresponding optimum moisture content and maximum density.* If the compactive effort is increased, the maximum density is increased and the optimum moisture content is decreased. This fact is illustrated in figure 60, in which are shown moisture-density relationships for two different soils, each of which was compacted at two different compactive efforts in the laboratory. If the same soil is compacted under several different compactive efforts, it is possible to develop a relationship between density and compactive effort for that soil. This information is of particular interest to the engineer who is preparing specifications for compaction and to the inspector, who must interpret the results of density tests made in the field

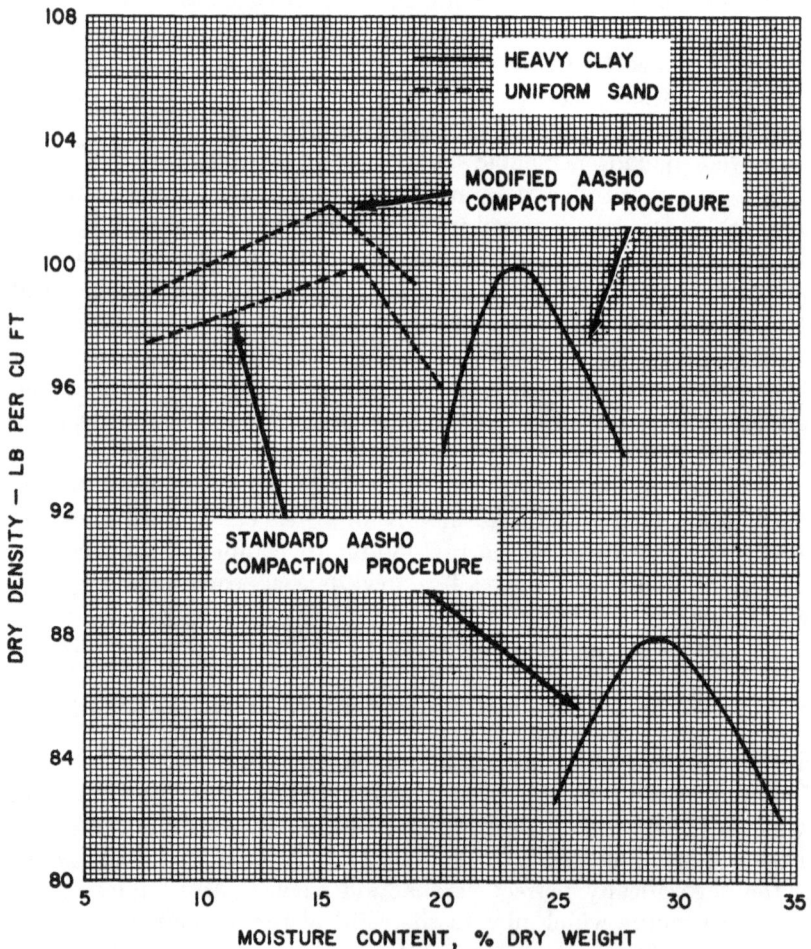

Figure 60.  *Effect of variation in compactive effort on moisture-density relationships of two soils.*

during rolling. The relationship between compactive effort and density is not linear, and a considerably greater increase in compactive effort may be required to increase the density of a clay soil from 90 to 95 percent of AASHO maximum density than is required to effect the same change in density of a sand. The effect of variation in the compactive effort is as significant in the field rolling process as it is in the laboratory compaction procedure. In the field the compactive effort is a product of the drawbar pull, which is a function of the weight of the roller, and the number of passes for the width and depth of the area of soil which is being rolled. Increasing the weight of the roller or the number of passes increases the compactive effort. Other factors which may be of consequence include lift thickness, contact pressure, and in the case of sheepsfoot rollers, the size of the tamping feet. An understanding of the basic relationship between compactive effort and density (and shearing resistance) leads to the conclusion that there must be careful correlation between laboratory and field compaction, if best results are to be achieved.

## 201. Compaction Characteristics of Various Soils

*a.* Obviously the nature of the soil itself has a great effect upon the density which is obtained under a given compactive effort. Soils which are extremely light in weight, such as diatomaceous earths and some volcanic soils, may have maximum densities under a given compactive effort as low as 60 pounds per cubic foot. Under the same compactive effort, the maximum density of a clay may be in the range of 90 to 100 pounds per cubic foot, while that of a well-graded, coarse granular soil may be as high as 135 pounds per cubic foot. Moisture-density relationships for seven different soils are shown in figure 61. These curves were established in the laboratory using the Standard AASHO compaction procedure.

*b.* Information relative to the compacted dry unit weights of the soil groups of the Unified Soil Classification System is given in tables VI and VII, appendix II. Unit dry weights given in column (13) of table VI are based upon compaction at optimum moisture content for the Modified AASHO compactive effort: those in column (10) of table VII upon the Standard AASHO compactive effort.

*c.* The curves of figure 61 indicate that the different soils react somewhat differently to compaction at moisture contents which are somewhat less than optimum. Moisture content is less critical for the heavy clay than for the slightly plastic sandy and silty soils. Heavy clays may be compacted through a relatively wide range of moisture contents below optimum with comparatively small changes in dry density. On the other hand, the granular soils which have better grading and higher densities under the same compactive effort react sharply to small changes in moisture content, with slight changes in moisture producing sizable changes in dry density. The relatively

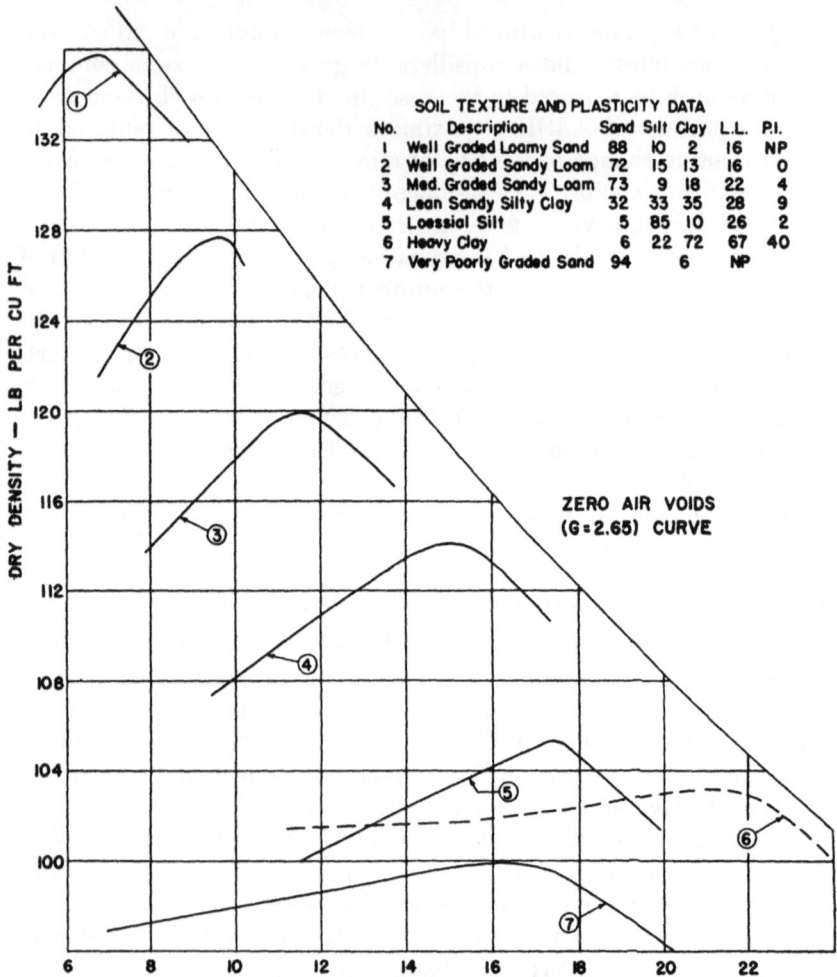

*Figure 61. Moisture-density relationships for seven soils compacted according to the Standard AASHO compaction procedure.*

clean, poorly graded, nonplastic sands also are relatively insensitive to changes in moisture.

*d.* There is no generally accepted and universally applicable relationship between the optimum moisture content under a given compactive effort and the limit tests which have been described previously (ch. 2). Optimum moisture content varies from about 12 to 25 percent for fine-graded soils, and from 7 to 12 percent for well-graded aggregate containing soil binder. For some clay soils the optimum moisture content and the plastic limit will be approximately the same.

*e.* The presence of coarse particles, gravel or stone, in a soil has an effect upon its compaction characteristics. Since the laboratory compaction test is normally performed upon the portion of the soil which passes a No. 4 sieve, this effect is not included. TB 5–253–1 gives an approximate procedure for correcting the optimum moisture content of soils which contain more than about 25 percent by weight of gravel.

## 202. Other Factors Which Influence Density

Several factors, in addition to those which have been listed, have an influence on soil density, although to a smaller degree. For example, temperature is a factor in the compaction of soils which have a high clay content: both density and optimum moisture content may be changed by a great change in temperature. Some clay soils are sensitive to manipulation. That is, the more they are worked the lower the density for a given compactive effort. Manipulation has little effect on the degree of compaction of silty or sandy soils. Curing, or drying of a soil following compaction, may increase the density of subgrade and base materials, particularly if cohesive soils are involved.

## 203. Maintenance of Soil Density after Construction

Soil densities which are obtained by compaction during construction may be changed during the life of the structure. Such considerations of greatest concern to the engineer who is engaged in the construction of semi-permanent installations, although they should be kept in mind during the construction of any facility in order to insure satisfactory performance. The two principal factors which tend to change soil density are climate and traffic. As far as embankments are concerned, normal embankments retain their degree of compaction, unless subjected to unusual conditions and except in their outer portions, which are subjected to seasonal wetting and drying, and frost action. Subgrades and bases are subject to more severe climatic changes and traffic than are embankments. Climatic changes may bring about seasonal or permanent changes in soil moisture and accompanying changes in density which may cause distortion of the pavement surface. High-volume change soils are particularly susceptible and should be compacted to meet conditions of minimum swelling and shrinkage. Granular soils retain a large measure of their compaction under exposure to climatic conditions, while other soils may be affected to some degree, particularly in areas of severe seasonal changes such as semi-arid regions where long, hot, dry periods may occur, or in humid regions where deep freezing occurs. Frost action may change the density of a compacted soil, particularly if it is fine grained. With particular regard to subgrades and bases for airfields, heavy traffic may bring about an increase in density over that secured during construction. This increase in density may cause the rutting of a flexible pavement or the subsidence of a rigid pavement. The protection which a subgrade soil receives after construction is complete has an important effect upon the permanence of compaction. The use of good shoulders, the maintenance of tight joints in a concrete pavement, and adequate drainage all contribute toward maintaining the degree of compaction achieved during construction. In general statement, compliance with the compaction requirements given in paragraphs 205 and 206 should insure satisfactory behavior of most earth structures which incorporate compacted soils.

## 204. General Construction Procedures

Military structures in which soil compaction is an essential part of the construction process include embankments, subgrades and bases for roads and airfields, and backfills around retaining walls and other earth-retaining structures. Considerations relative to earth-fill dams are presented in chapter 11, while the backfills behind earth-retaining structures are discussed in chapter 12. Soil compaction procedures and requirements for the other structures mentioned are discussed in the remainder of this chapter. The general construction process which is followed in the building of a rolled-earth embankment requires that the fill be built up on relatively thin layers or "lifts," each of which is rolled until a satisfactory degree of compaction is obtained. The subgrade in a fill section would generally be the top lift in the compacted fill, while the subgrade in a cut section generally also would be compacted. Soil bases normally are compacted to a high degree of density. Compaction requirements frequently stipulate the attainment of a certain minimum density, which for military construction is generally a specified minimum percentage of Modified AASHO maximum density for the soil concerned. The moisture content of the soil is maintained at or near optimum, within the practical limits of field construction operations. Principal types of equipment used in field compaction are sheepsfoot, three-wheel (smooth), and pneumatic-tire rollers.

## 205. Compaction Requirements, Embankments

*a.* The general requirement for the compaction of soils used to form embankments to support roads or airfield pavements is that the soil should be compacted to at least 90 percent Modified AASHO maximum density. This requirement prevails at depths below those indicated in figure 62 and discussed in detail in paragraph 206.

*b.* Experience gained in the design and construction of highway fills is embodied in table XII, appendix II. Two conditions of exposure are noted in the table. Condition 1 relates to fills which are not subject to inundation after construction, while condition 2 relates to fills which may be inundated at certain periods. It must be noted that the compaction requirements given in table XII are based upon the Standard AASHO compaction procedure, while the requirement of *a* above is based upon the Modified AASHO effort.

## 206. Compaction Requirements, Subgrades and Bases Subjected to Normal Compaction

In order to prevent detrimental settlement under traffic a definite degree of compaction of subgrade and base courses is needed, depending upon the wheel load and the depth below the surface. In figure 62 is shown the degree of compaction which should be obtained at a given depth below the surface in situations in which the Modified AASHO compaction test procedure is applicable. One set of curves is shown for cohesionless sands, and another set for other materials. On each set of curves the solid, curved line indicates the depth to which a degree of compaction equal to 100 percent Modified AASHO maximum density should extend for a given wheel or assembly load and tire pressure. Similarly, in each case the dashed, curved line indicates the depth to which 95 percent Modified AASHO maximum density should extend. Below the depth indicated by the dashed line, fills should be compacted to 90 percent Modified AASHO maximum density. The compaction requirements of figure 62 should be used as a guide in the construction of certain full-operational airfields and all special airfields, especially when rigid pavements are to be constructed. For other airfield construction and most road construction in a theater of operations greater settlement can be accepted, although the amount of maintenance will generally be increased. In these cases, the following minimum compaction requirements should be met. For airfields, the base course and subgrade should be compacted to 95 percent Modified AASHO maximum density, except that the top 6 inches of all cohesionless-sand subgrades should be compacted to 100 percent Modified AASHO maximum density. For roads the subgrade and base course should be compacted to 90 percent Modified AASHO maximum density. Equipment and construction procedures which may be used to attain the desired degree of compaction are discussed in the paragraphs which follow. Some subgrade soils cannot be compacted to the design densities given above. The treatment of such soils is as indicated in paragraph 207.

## 207. Subgrade Compaction of Special Soils

*a. Expansive Clays.* As has been previously indicated, soils which have a high clay content (particularly CH, MH, OH) may expand in detrimental amounts, if compacted to a high density at a low moisture content, and then exposed to water. Such soils are not desirable as subgrades and are difficult to compact. If they must be used, they cannot be compacted to the design densities given above, but to the maximum density obtainable with the minimum moisture content which will result in a minimum amount of swelling. This method requires detailed testing and careful control of compaction. In some cases it may be desirable to construct a base of sufficient thickness to insure against the harmful effects of expansion.

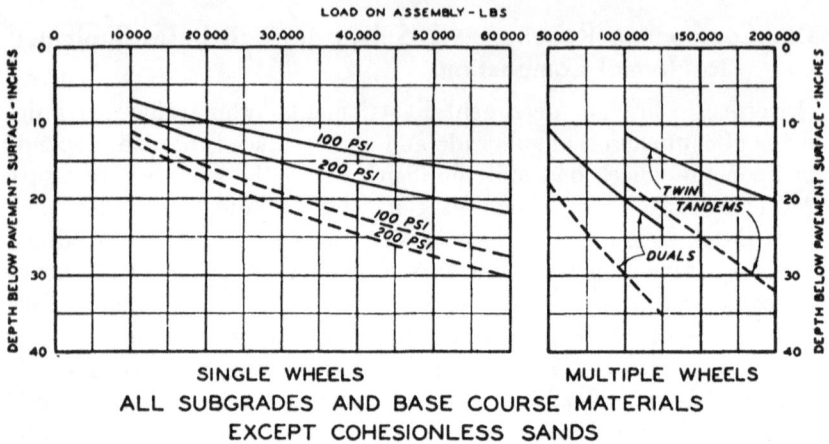

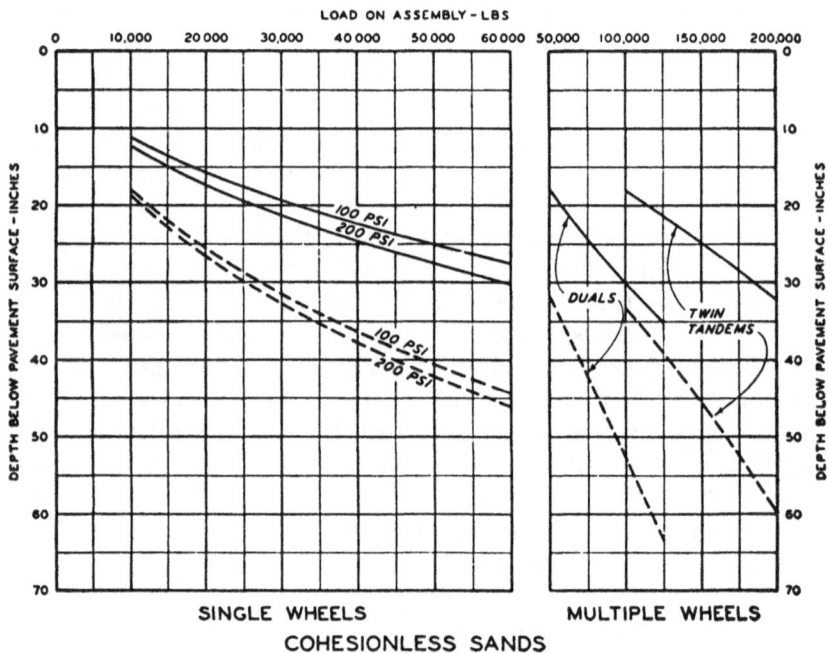

Figure 62. Subgrade and base course compaction requirements.

*b. Silts That Become Quick When Remolded.* Experience has shown that some silts and very fine sands (predominantly ML and SC soils), when compacted in the presence of a high water table, will pump water to the surface and become "quick", with a consequent loss of shearing strength. These soils can not be properly compacted unless they are dried. If they can be compacted at the proper moisture content, their shearing resistance is reasonably high, and every effort should be made to lower the water table so that compaction can be done properly. If trouble occurs with these soils in localized areas, they can be removed and replaced with more suitable ones. If removal, or drainage and later drying, can not be accomplished, these soils should not be disturbed by attempting to compact them. Instead, they should be left in their natural condition and an additional thickness of base used to prevent the subgrade from being overstressed.

c. *Clays Which Lose Strength When Remolded.* As has been noted previously, certain clay soils lose strength when remolded. This is particularly true of some CH and OH soils. Some of these soils have high strengths in their undisturbed condition, but scarifying, reworking, and compacting them in cut areas may reduce their shearing strengths, even though they are compacted to design densities. When these soils are encountered, their sensitivity may be detected by performing unconfined compression tests upon the undisturbed soil, and upon the remolded soil compacted to the design density at the design moisture content. If the undisturbed value is the higher, no compaction should be attempted and construction operations should be conducted so as to produce the least possible disturbance of the soil. Pavement design should be based upon the bearing value of the undisturbed soil.

## 208. Compaction Equipment

Equipment which is normally available to the military engineer for the compaction of soils includes rollers of the sheepsfoot, pneumatic-tire, and smooth, 3-wheel types. Good use may also be made of other construction equipment in certain instances, particularly crawler-type tractor units and loaded hauling units, including rubber-tired scrapers.

## 209. Sheepsfoot Rollers

a. Variables in a sheepsfoot roller which influence compaction include the weight of the roller, the area and shape of the feet, and the spacing of the feet. In military construction the first of these variables is frequently the only one which is under the control of the engineer in the field, as far as the roller itself is concerned. Standard sheepsfoot rollers may be varied in weight by filling the drum with water or sand. Other variables which influence the rolling process are soil type, moisture content, initial density of the soil, and the thickness of lift.

b. The contact pressure of the roller feet on the soil should be as large as possible without greatly exceeding the bearing capacity of the soil. If the bearing capacity of the soil is exceeded the roller will sink deeper until greater contact area reduces the pressure to that which the soil can carry. The bearing capacity increases with density, which is the reason why a sheepsfoot roller "walks up" as rolling proceeds, provided only that the contact pressure is not too great. Figure 63 indicates what is meant by the "walking up" or "walking out" of a sheepsfoot roller. In general, contact pressures which result in most efficient rolling should be relatively low for friable, silty, and clayey sandy soils which depend largely on friction for the development of bearing capacity, somewhat higher for clayey silts, clayey sands and lean clay soils which have low plasticity, and still higher for medium to heavy clays. Clean cohesionless soils normally can not be compacted successfully by sheepsfoot rollers.

c. Considerations relative to the thickness of layer which can be compacted most efficiently with a sheepsfoot roller are given later. However, the maximum thickness of lift which can be effectively compacted with conventional rollers of this type is 9 inches.

d. The number of passes of the roller has a great effect upon the density which is obtained. The relationship between density and the number of passes is typically a straight-line relationship when plotted on semilogarithmic paper. However, rolling beyond a certain number of passes with a sheepsfoot roller may not be economical. The number of passes which is required to compact the soil effectively on a given job should be determined by trial, as indicated in e below. Under normal conditions compaction to 95 percent Modified AASHO maximum density can be achieved in 10 to 12 passes of a conventional sheepsfoot roller.

e. When compaction operations are begun on a sizeable project it is desirable to conduct tests on trial lifts in order to determine the best rolling procedure. If there is no choice of equipment, test rolling is limited to determining the best lift thickness, the number of passes required for compaction of each of the major soil types encountered, and the need for increasing or decreasing foot pressures. Test rollings of this sort should include a minimum of variables, and the soil should be at optimum moisture. Density tests will indicate the most effective combination of lift thickness and number of passes. If the roller walks up too fast and densities are not great enough, the foot pressure may need to be increased or the lift thickness reduced, or both. On the other hand, if the roller does not walk up, or sinks deeper into the soil with an increasing number of passes, the shear strength of the soil is being exceeded, and the foot pressures must be decreased by removing ballast from the roller. In either case, the moisture content of the soil may have to be adjusted.

**1**

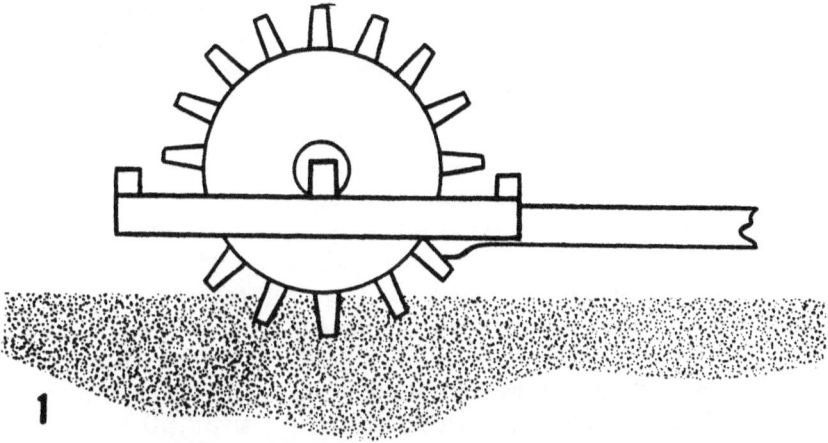

**ROLLER FEET EMBEDDED TO
WITHIN TWO INCHES OF THE DRUM**

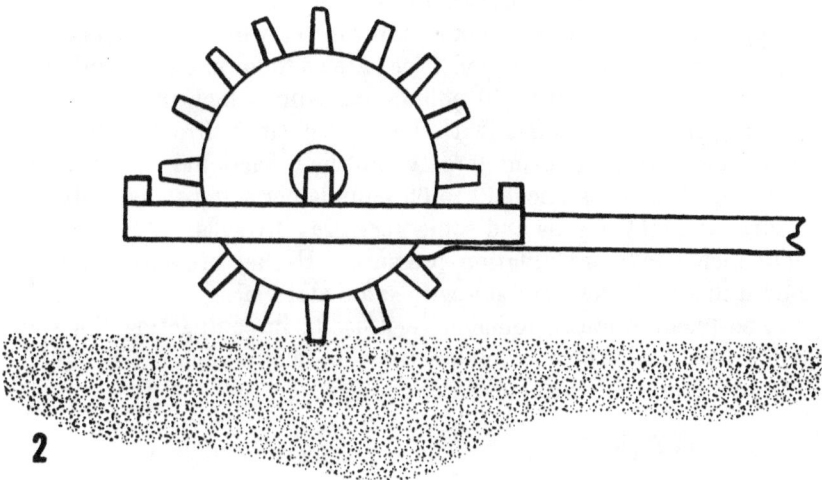

**2**

**ROLLER AFTER IT HAS "WALKED OUT"**

*Figure 63.   Compaction by a sheepsfoot roller.*

## 210. Pneumatic-Tire Rollers

*a.* Pneumatic-tire rollers, as a type, can be used in compacting any type of soil, provided the values of contact pressure and wheel load are proper for the soil which is being compacted. They are least effective with fine-grained soils of high compressibility. Rollers of this type which may be available for military construction include the 13-wheel rubber-tired roller and the 50-ton pneumatic-tire roller.

*b.* Compaction by a pneumatic-tire roller is dependent upon area of contact pressure, thickness of lift, and number of coverages. The contact pressure is equal to the inflation pressure plus some pressure due to sidewall thickness. The area of contact, contact pressure, and total wheel load are interrelated. If the contact pressure is constant for given tires, increasing the total load will not increase the density obtained by rolling. However, increasing the total load will increase the size of the loaded area and, therefore, the effective depth of compaction. With cohesionless soils, the larger the size of the tire, the greater is the contact area and the greater the confining effect. The contact pressure is a major factor in obtaining densities, and the wheel load and the number of passes are important in determining the most economical lift thickness for a given roller.

*c.* With this type of roller, the lift thickness can be the maximum which can be compacted satisfactorily, as determined in field tests. The 50-ton pneumatic-tire roller can often compact lifts which are 18 inches or more in thickness, while the lift thickness for compaction with the 13-wheel rubber-tired roller will not exceed 12 inches.

*d.* Approximate ranges of contact pressures which are effective in compacting different general soil types are as follows: clean sands and some gravelly sands, 20 to 40 pounds per square inch inflation pressure, the greater pressures with large size tires; friable silty and clayey sands which depend largely on their frictional qualities for developing shear resistance, 40 to 65 pounds per square inch inflation pressure; and clayey soils and some very gravelly soils, 65 pounds per square inch and more inflation pressure. Higher pressures may be required in airfield construction with some materials.

*e.* The pneumatic-tire roller is very useful in compacting the thin layer of loose soil which is left on the top of an embankment which has been compacted with sheepsfoot rollers.

## 211. Smooth Rollers

*a.* Smooth-wheel rollers may be used in the compaction of soils, although that is not the purpose for which they are best suited. Both tandem and 3-wheel types are used in the final rolling of subgrades and the rolling of base courses. If a smooth-wheel roller is used in earth work the 3-wheel type is generally preferred, because of the greater pressures exerted by the driving (rear) wheels.

*b.* When 3-wheel rollers are used, the soil generally should be spread in relatively thin layers, up to about 6 inches loose thickness in the usual case. With the heavier rollers with total weights of 10 to 12 tons, thicker lifts may sometimes be rolled satisfactorily, particularly in friable fine-grained soils. They are not effective on clean cohesionless sands, but may be used in compacting gravelly and clayey soils. In compacting clayey soils, care must be taken to roll a thin enough layer of soil so that compaction will be to the full depth, otherwise compaction may be limited to a surface crust.

## 212. Other Equipment

Crawler tractors are practical compacting units, particularly for cohesionless gravels and sands. The material should be spread in thin layers, on the order of 3 or 4 inches, and is compacted by the combination of static pressure and vibration. When clean sands and sandy gravels are being compacted, good results may be achieved by saturating the material with water by ponding, or by jetting, if water under pressure is available. Items of rubber-tired construction equipment, particularly scraper units, may be used for compaction under expedient conditions. Proper routing of construction equipment over the working area may serve to decrease the amount of rolling which is necessary.

## 213. Recommended Equipment To Be Used With Soil Groups of Unified Soil Classification System

Column (12) of table VI, appendix II, gives information relative to compaction equipment which is recommended for use with each of the groups of the Unified Soil Classification System. Similar information is given also in column (9) of table VII, appendix II.

## 214. Construction of Embankments

In paragraph 215, certain facts relating to the construction of embankments for roads and airfields are discussed. Much of this discussion also applies to the construction of earth dams. Other items which are of importance for dams are discussed in chapter 11. Many of the things which are discussed are directly applicable to the construction of subgrades and bases, which receive additional attention in paragraphs 222 and 223.

## 215. Selection of Materials

*a.* Soils which are to be used in the construction of fills generally come from cut sections of the road or airfield concerned, provided that this material is suitable. If the material excavated from cut sections is not suitable for use in fills, or there is not enough of it, then it is obtained from borrow sources. Except for highly organic soils, nearly any soil can be used in fills. However, some soils are more difficult to compact than others, and some require flatter side slopes for stability. Certain soils, as indicated in chapter 11, require elaborate protective devices to maintain the fill in its original condition. These considerations and others may make it desirable to thoroughly investigate cost, compaction characteristics, and shear strengths of soils which are to be used in major fills, when time is available. Under expedient conditions the military engineer must simply make the best possible use of the soils at hand.

*b.* In general terms, the coarse-grained soils of the Unified Soil Classification System are desirable for fill construction, ranging from excellent to fair. The fine-grained soils are less desirable, being more difficult to compact and requiring more careful control of the construction process. More specific information concerning the suitability of these soils is given in column (7) of table VII, appendix II.

## 216. Removal of Unsuitable Material

Materials which are unsuitable for use in embankments, such as top soil, excessive organic material, and vegetation are generally removed by stripping operations to a depth of about one foot. In swampy areas, much deeper pockets of highly organic soils, such as peat and muck, may be encountered in both cut and fill sections, or the surface soils over an entire area may be unsuitable. In fill areas it may be necessary to remove isolated pockets of unsuitable soils, to build over them in expedient situations, or to remove them to a considerable depth and replace them with more suitable materials. In excavation areas it is sometimes possible to blend unsuitable materials with other soils for use in an embankment, or the unsuitable material may be wasted.

## 217. Dumping and Spreading

Since the majority of fills are built up of thin lifts to the desired height, it is important that the soil for each lift be spread in a uniform layer of the desired thickness. In typical operations the soil is brought in, dumped, and spread by scraper units; careful attention must be given to the adjustment of the scrapers to accomplish this objective. Materials may also be brought in by trucks or wagons and dumped at properly spaced locations, so that a uniform layer may be easily spread by blade graders or bulldozers. It is possible for bulldozers to form very short and shallow fills working alone. End-dumping of soil material to form a fill without compaction is rarely permitted in modern embankment construction. An exception occurs when a fill is being built over very weak soils, as in a swamp. The bottom layers may then be end-dumped until sufficient material has been placed to allow hauling and compacting equipment to operate satisfactorily. As has been indicated, the best thickness of layer to be used with a given soil and given equipment cannot be determined exactly in advance, but is best determined by trial during the early stages of rolling on a project.

## 218. Adding Water to Soil and Mixing

*a.* It is frequently necessary to add water to soils which are being incorporated in embankments (and those in subgrades and bases) in order to obtain the desired degree of compaction and to achieve uniformity. The soil can be watered in the pit or on the grade. After the water is added it must be thoroughly and uniformly mixed with the soil. Even if additional water is not needed, mixing may still be desirable in order to insure uniformity. In the processing of granular materials, best results are generally obtained by sprinkling and mixing on the grade; any good mixing equipment should be satisfactory. The more friable sandy and silty soils are easily mixed with water. They may be handled by sprinkling and mixing, either on the grade or in the pit. Mixing can be done with cultivators and rotary speed mixers to depths of 8 inches or more without difficulty. These soils may be mixed with a blade grader, although blade mixing is generally too slow to be desirable if other mixing tools are available. Water also may be added to these soils by diking or ponding the pit and flooding until the desired depth of penetration has taken place, if time is available. Several days will generally be required by this method to accomplish uniform moisture distribution. Medium clayey soils can be worked in the pit or on the grade as conditions dictate. Best results are obtained by sprinkling and mixing with cultivators and rotary speed mixers. They can be worked in lifts up to 8 inches or more without great difficulty. Heavy clay soils present many difficulties, as they are difficult to pulverize and it is hard to incorporate water uniformly. The most effective approach is generally sprinkling, followed by mixing on the grade. Heavy disc harrows may be needed to break up dry clods and to aid in incorporating water, followed by heavy-duty cultivators and rotary speed mixers. Time is required to effect uniform distribution of water, and it is difficult to work lifts in excess of about 6 inches loose thickness.

*b.* It may be noted that the length of the section which is being rolled may have a large effect upon densities in hot weather, when water evaporates quickly. When this condition occurs, quick handling of the soil may mean the difference between obtaining adequate density with a few passes or the effort of adding and mixing in water.

## 219. Handling Wet Soils

When the moisture content of a soil which is to be compacted is markedly in excess of that which is necessary to obtain the desired density, some of the water must be removed. In some cases it may be possible to use excessively wet soils in ways in which the excessive moisture content is not detrimental, as in the outer sections of embankments where they will not endanger the stability of the section beneath the roadway and where they will dry sufficiently to attain the necessary stability as construction proceeds. Most often they must be dried, and this can be a slow and costly process. Drying is usually accomplished by manipulating and exposing the soil to aeration and to the rays of the sun. Manipulation is most often done by the use of cultivators, plows, and rotary speed mixers. Rotary speed mixers, with the tail-hood sections raised, permit good aeration and are very effective in the drying of excessively wet soils. A rotary speed mixer engaged in this type of operation is shown in figure 64. An excellent method which may be useful when both wet and dry soils are available is simply to mix them together. In some cases, also, it has been necessary to use alternate layer construction, in which a layer of wet soil is covered by a thin layer of dry stable soil, and the two are compacted together. Layers must be thin enough to insure adequate compaction, and the method is not recommended except in emergency situations.

*Figure 64. Rotary speed mixer engaged in aeration of an excessively wet soil.*

## 220. Compaction of Embankments

*a.* Considerable information relating to the compaction of soils in embankments has been presented in previous paragraphs. Further information is presented below.

*b.* If the fill consists of cohesive or plastic soils, the embankment generally must be built up of uniformly compacted layers not greater than 6 inches in thickness, with the moisture content carefully controlled. Rolling should be done with sheepsfoot rollers. Bonding of a layer to the one placed on top of it is aided by the thin layer of loose material which is left on the surface of the rolled layer by the roller feet. Rubber-tired or smooth-wheel rollers may be used to provide a smooth, dense, final surface. Rubber-tired construction equipment may provide supplemental compaction, if properly routed over the area.

*c.* If the fill material is clean sand or sandy gravel, moisture control is not too important. These soils may be compacted effectively either very dry or very wet. Since soils are not usually found in a very dry condition, these soils are best compacted when saturated. They may be saturated by ponding or jetting. Rubber-tired equipment may be used in compacting these soils. Very effective compaction may be obtained from the combination of saturation and the vibratory effects of crawler tractors, particularly when tractors are operated at fairly high speeds, so that vibration is increased.

*d.* Sands and gravels which have silt and clay binders require effective control of moisture for adequate compaction. The need for close control is especially great with certain soils of the GM and SM groups. Pneumatic-tire rollers are best for compacting these soils, although sheepsfoot rollers may also be used effectively.

*e.* Rock is sometimes used in fills, particularly in the lower portion. In some cases, the entire fill may be composed of rock layers with only a cushion layer of soil for the subgrade. The thickness of such rock layers should not be more than the diameter of the largest rock fragments. Compaction of this type of fill is difficult, but may generally be done by vibration from the passage of track-type equipment over the fill area.

## 221. Finishing

Finishing operations in embankment construction include all the things which are necessary to complete the earth-work operations. Included among these are such items as the trimming of side and ditch slopes, where necessary, and the fine-grading operations which are needed to bring the embankment section to its final grade and cross section. Actually, most of these are not separate operations which are performed after all other operations are complete, but are carried along as the work progresses. The tool which is used most often in finishing operations is the motor grader, while scraper and dozer units also may be used. The provision of adequate drainage facilities is an important part of the work.

## 222. Compaction of Subgrades

*a.* Compaction requirements for subgrades and bases have been given in paragraph 206 and figure 62.

*b.* In fill sections, the subgrade is the top layer of the embankment, which is compacted to the required density and brought to desired grade and section. For subgrades, plastic soils should be compacted at moisture contents which are close to optimum. It is recognized that moisture content cannot always be carefully controlled during military construction, but certain practical limits must be recognized. Generally, plastic soils cannot be compacted satisfactorily at moisture contents which are more than about 25 percent (of optimum) above or below optimum. Much better results will be obtained if the moisture content can be controlled within 10 percent of optimum. For the cohesionless soils moisture control is not as important, but some sands tend to bulk at low moisture contents; compaction should not be attempted until this situation has been corrected.

c. In cut sections, particularly when flexible pavements are being built to carry heavy wheel loads, subgrade soils which gain strength with compaction should be compacted to the requirements of figure 62. This may make it necessary to remove the soil, replace it, and compact it in layers in order to obtain the required densities at greater depths. In most construction in theaters of operation the subgrade soil in cut sections should be scarified to a depth of about 6 inches and rolled. This procedure is generally desirable in the interests of uniformity. If the natural subgrade has a density which is equal to or greater than the required percentage of Modified AASHO maximum density, no rolling is necessary, except to provide a smooth surface.

## 223. Compaction of Bases

Selected soils which are used in base construction must be compacted to the requirements of figure 62. The thickness of layers must be within the limits which will insure proper compaction; it is generally from 3 to 6 inches, depending on the material and method of construction. Smooth-wheel rollers are recommended for use in compacting hard, angular materials with a limited amount of fines or stone screenings, while pneumatic-tire rollers are recommended for softer materials which may break down (degrade) under the action of a steel roller.

## 224. Field Control of Compaction

As has been emphasized in previous paragraphs, specifications for adequate compaction of soils used in military construction generally require the attainment of a certain minimum density in field rolling. This requirement is most often stated in terms of a specified minimum percentage of Modified AASHO maximum density; with many soils the close control of moisture content is necessary to achieve the stated density with available equipment. In many instances careful control of the entire compaction process is necessary, if the required density is to be achieved with ease and economy. Control generally takes the form of field checks of moisture and density to determine if the specified density is being achieved, to control the rolling process, and to permit adjustments in the field rolling to be made as required. In the discussion which follows, it is assumed that the laboratory compaction curve is available for the soil which is being rolled, so that the maximum density and optimum moisture content are known, that the soil compacted in the laboratory and being rolled in the field are actually the same, and that the required density can be achieved in the field with the equipment available.

## 225. Determination of Moisture Content

*a. General.* It may be necessary to make checks of the moisture content of the soil during field rolling for two principal purposes. First, since the specified density is in terms of dry unit weight and the density measured directly in the field (par. 227) is generally the wet unit weight, it is necessary that the moisture content be known in order that the dry unit weight can be calculated. Second, the moisture content of some soils must be maintained close to optimum if satisfactory densities are to be obtained. Adjustment of the field moisture content can only be done if the moisture content is known. The determination of density and moisture content is often done in one overall test procedure; they are described here separately for convenience.

*b. Visual and Manual Examination.* Experienced engineers who have become familiar with the soils which are encountered on a particular project can frequently judge moisture content closely by field examination. Friable or slightly plastic soils usually contain enough moisture at optimum to permit forming a strong cast by compressing the soil in the hand. As noted, some clay soils have optimum moisture contents which are close to their plastic limits; thus a plastic limit or "thread" test conducted in the field may be highly informative.

*c. Field Drying.*

    (1) The moisture content of a soil (par. 25) is best and most accurately determined by drying in an oven at controlled temperature. Methods of determining the moisture content of a soil in this fashion are described in TB 5–253–1, using the equipment of Soil Testing Set No. 1.

    (2) The moisture content of the soil may be determined by drying the soil in the air in the sun. Frequent manipulation will speed up the drying process. From a practical standpoint this method is generally too slow to be of much value in the control of field rolling.

    (3) Several quick methods may be used to determine approximate moisture contents under expedient conditions. For example, the sample may be placed in a frying pan and dried over a hotplate or a field stove. The temperature is difficult to control in this procedure and organic materials may be burned, thus causing a slight to moderate error in the results. On large-scale projects where many samples are involved, the quick method may be used to speed up determinations by

comparing the results obtained by this method with comparable results obtained by oven-drying.

(4) Another quick method which may be useful is to mix the damp soil with enough denatured grain alcohol to form a slurry in a perforated metal cup, igniting the alcohol, and permitting it to burn off. The alcohol method, if carefully done, will produce results equivalent to those obtained by careful laboratory drying. For best results, it is suggested that the process of saturating the soil with alcohol and burning it off completely be repeated three times.

## 226. Numerical Example, Determination of Water To Be Added to Attain Optimum Moisture

If the moisture content of the soil is less than optimum, the amount of water which must be added for efficient compaction generally is computed in gallons per 100 feet of length (station). The computation is based upon the dry weight of soil contained in a compacted layer. For example, assume that the soil is to be placed in layers 6 inches in compacted thickness at a dry unit weight of 120 pounds per cubic foot. The moisture content of the soil is determined to be 5 percent, while the optimum moisture content is 12 percent. Assume that the strip to be compacted is 40 feet wide. Compute the amount of water which must be added per 100-foot station to bring the soil to optimum moisture, allowing for 10 percent loss by evaporation. The weight of water required per cubic foot of soil (no evaporation) is equal to 120 $(0.12-0.05)=120(0.07)=8.4$ pounds. To allow for the loss by evaporation this must be divided by $100-10$ percent$=0.90$. The weight of water per cubic foot of soil then is $8.4/0.90=9.33$ pounds. Since water weighs 8.34 pounds per gallon, the number of gallons per cubic foot of soil$=9.33/8.34=1.119$. The volume of compacted soil per station$=40$ (100) $6/12=2000$ cubic feet. The amount of water required per station to insure compaction at optimum moisture content then is 2000 $(1.119)=2238$ gallons.

## 227. Density Determinations

a. Two principal methods are used in checking the density which is being obtained by field rolling. These methods are the undisturbed sample method and the disturbed sample method. Both methods are dependent upon the determination of the wet unit weight of the compacted soil, the measurement of the moisture content, and the calculation of the dry unit weight for comparison with the specified density.

b. The undisturbed method depends upon the measurement of the wet weight of soil corresponding to a known volume. The unit wet weight may then be calculated. Suitable undisturbed samples may be secured by forcing a cylindrical sampler into the soil, by trimming an undisturbed sample to a regular geometric shape the volume of which can be calculated, or volume-weight measurements of a chunk sample of irregular shape. Expedient methods for determining the wet unit weight by these techniques are described in TB 5–253–1.

*c.* The disturbed sample method depends upon the weighing of the wet soil which is taken from a roughly cylindrical hole which is dug in the compacted layer. The volume of the hole may be determined by one of several methods, including the use of heavy oil of known specific gravity, water which is poured into a thin sheet of rubber which lines the inside of the hole, calibrated sand, and the rubber-balloon density apparatus. The first three of these methods are described in TB 5–253–1.

*d.* If the soil which is being compacted contains gravel the density measured in the field must be corrected for the presence of these coarse particles. This is because the coarse particles, larger than a No. 4 sieve, are removed from the soil before the maximum density and optimum moisture content are determined in the laboratory. A method for computing this correction is given in TB 5–253–1.

## 228. In-Place CBR Test

In some situations it may be desirable to judge the adequacy of compaction by performing the in-place CBR test upon the compacted soil of a subgrade or base. The CBR thus obtained can then be compared with the design CBR which is desired, provided that the design was based upon CBR tests on unsoaked samples. If the design was based on soaked samples, the results of field-in-place CBR tests must be correlated with the results of laboratory tests performed on undisturbed mold samples of the in-place soil, subjected to soaking.

## 229. Frequency of Density Checks

If the density determined by the methods described above is equal to or greater than that required, then rolling generally may be judged to be satisfactory and the placing of another lift may proceed. If the density is lower than that required, additional rolling may be necessary or the moisture content may have to be adjusted. If these measures fail, then the weight of the roller may have to be increased, the thickness of lift reduced, or some other measure taken to obtain adequate compaction. The possibility that the soil which is being rolled in the field is not the same one which was tested in the laboratory should never be overlooked. Under normal field conditions the number of density and moisture checks required should not be very great after the initial period of adjustment, assuming that the work is proceeding smoothly and uniform soils are being compacted. The engineer may quickly learn to judge the moisture content of the soil by appearance and feel. If adequate densities are being obtained and the proper moisture content is being maintained, the job of inspection may then be principally one of deciding on the number of passes of the roller which will achieve the desired result and seeing that this number of passes is actually made. Under such conditions only two or three density checks per day may be required on a fill of moderate length. Where conditions are more variable, density and moisture checks may be needed as often as once an hour for a fill of moderate length. The exact number of checks needed can only be determined by the engineer in charge of the job.

# CHAPTER 10

# SOIL STABILIZATION FOR ROADS
# AND AIRFIELDS

---

## Section I. INTRODUCTION

### 230. Soil Stabilization Defined

*Soil stabilization* may be defined as a method of processing soil in order to render it suitable for the use for which it is intended under the prevailing traffic and climatic conditions. In other words, the objective of soil stabilization, as related to roads and airfields, is to produce a firm soil mass which is capable of withstanding the stresses imposed upon it by traffic loads in all kinds of weather without excessive deformation. The term is further generally applied to the processing of a layer of soil of predetermined thickness which is to serve as a base or wearing surface. The construction of bases or wearing surfaces from stable materials, such as crushed rock, is not generally regarded as soil stabilization.

### 231. Purposes of Soil Stabilization

The purpose for which soil stabilization is most frequently used is to improve the existing subgrade soil so that its bearing capacity is increased sufficiently that the thickness of base course may be reduced, or to eliminate the need for a base course. A stabilized soil mixture is often used to provide a base course for the support of a relatively thin bituminous wearing surface. The stabilized soil layer may serve as a wearing surface in some applications, although it generally should be provided with a wearing surface to resist the abrasive effects of traffic and to reduce the infiltration of surface water. Soil stabilized mixtures which are discussed in this chapter should be used as wearing surfaces only if they are to withstand very light traffic or under expedient conditions.

### 232. Importance of Drainage and Compaction

The importance of adequate drainage and compaction in contributing to soil stabilization cannot be overemphasized. Many subgrade soils which, when first examined, appear to be unsatisfactory, may be made to satisfactorily fulfill their intended function by the provision of adequate drainage facilities and proper compaction.

Soil stabilization procedures are frequently expensive and time-consuming. They should not be undertaken unless it is certain that drainage and compaction alone can not accomplish the intended objective. Similarly, adequate drainage and proper compaction are essential to the proper construction and functioning of soil stabilized mixtures, when it is determined that soil stabilization is necessary.

## 233. Types of Soil Stabilization

Several types of soil stabilization are of consequence in military construction. By far the most important of these is *mechanical stabilization*, which generally involves the blending of the existing subgrade soil with other available soil materials in order to produce a stabilized soil mixture which has the desired properties. Commercial materials are also used in soil stabilization, particularly under conditions such that soil materials needed for blending with the existing subgrade soil are not easily or economically available in the area. The most important commercial materials which may be available to the military engineer are bituminous materials (asphalt or tar) and Portland cement. Soil stabilization using bituminous materials is termed *bituminous soil stabilization*. When cement is used as a stabilizing agent, the mixture which is produced is called *soil-cement*. A number of chemicals have been used for stabilization in the field or laboratory. Most of these are not important to the military engineer from a practical standpoint, since the chemicals are not normally available in field situations. Perhaps the most important of these chemical materials, in the sense that it is frequently available in the field, is lime, which is useful in the treatment of certain cohesive soils.

## Section II. MECHANICAL STABILIZATION

## 234. Introduction

*a. Definition.* A mechanically-stabilized soil is one which has resistance to deformation under load because of the mechanical interlocking and friction of the aggregate particles and, to some extent, because of the cementing action of the fines after compaction.

*b. Fundamental Consideration.*

(1) The three essentials for obtaining a properly stabilized soil mixture are proper gradation, a satisfactory binder soil, and proper control of the moisture content. To obtain uniform bearing capacity, uniform mixing and blending of all materials is essential. The mixture must be compacted at or near optimum moisture content in order to obtain satisfactory densities.

(2) The primary function of the portion of a mechanically-

stabilized soil mixture which is retained on a No. 200 sieve is to contribute internal friction. Practically all materials of a granular nature which do not soften when wet or pulverize under traffic can be used, although the best aggregates are those which are made up of hard, durable, angular particles. The gradation of this portion of the mixture is important, as the most suitable aggregates, generally speaking, are those which are well graded from coarse to fine. Well-graded mixtures are preferred because they generally have greater stability when compacted than do poorly graded mixtures, can be compacted more easily, and have greater increases in stability with corresponding increases in density. Materials which are satisfactory for this use include crushed stone, crushed and uncrushed gravel, sand, and crushed slag. Many other locally available materials have been successfully used, including disintegrated granite, talus rock, mine tailings, caliche, coral, lime-rock, tuff, shell, clinkers, cinders, and iron ore. When local materials are used, requirements relative to proper gradation can not always be met. If conditions are encountered in which the gradations obtained by blending local materials are such as to be either finer or coarser than the specified gradation, it is desirable to satisfy the size requirements of the finer fractions and neglect the gradation of the coarser sizes.

(3) The portion of the soil which passes a No. 200 sieve functions as a filler for the rest of the mixture and supplies cohesion, which will aid in the retention of stability during dry weather. The swelling of this material, if clay, serves somewhat to retard the penetration of moisture during wet weather. Clay is most commonly used as a binder. The nature and amount of this finer material must be carefully controlled, since an excessive amount of it will result in an excessive change in volume with change in moisture content and other undesirable properties. The properties of the soil binder are usually controlled by controlling the plasticity characteristics, as evidenced by the liquid limit and plasticity index. These tests are performed upon the portion of the material which passes a No. 40 sieve. The amount of fines is controlled by limiting the amount of material which may pass a No. 200 sieve. When the stabilized soil is to be subjected to frost action, this factor must be kept in mind when designing the soil mixture.

## 235. Typical Uses of Mechanically-Stabilized Soil Mixtures

Mechanical soil stabilization may be used in preparing soils to function as subgrades, bases, or surfaces. Several commonly encountered situations may be visualized to indicate the usefulness of this method. One of these situations occurs when the surface soil is a loose sand which is incapable of providing support for wheeled vehicles, particularly in dry weather. If suitable binder soil is available in the area, it may be brought in and mixed in the proper proportions with the existing sand to provide an all-weather surface for light traffic or under expedient conditions. This would be a *sand-clay road*. The same thing may be done in some cases to provide a "working platform" during construction operations. A somewhat similar situation may occur in areas where natural gravels which are suitable for the production of a well-graded sand-aggregate material are not readily available. Crushed stone, slag, or other materials may then be stabilized by the addition of suitable clay binder to produce a satisfactory base or surface. A common method of mechanically stabilizing an existing clay soil is by adding gravel, sand, or other granular materials. The objectives here are to increase the drainability of the soil, increase stability, reduce volume changes, and generally control the undesirable effects which are associated with clays.

## 236. Objective and Importance of Mechanical Stabilization

*a.* The overall objective of mechanical stabilization is simply to blend available soils in such a fashion that, when properly compacted, they will give the desired stability. In one sense the goal is to achieve a soil combination which is as near the top of table V, appendix II, as possible. In certain areas, for example, the natural soil at a selected location may be of low supporting power because of an excess of clay, silt, or fine sand. Within a reasonable distance there may exist suitable granular materials which may be blended with the existing soil to effect a marked improvement in stability of the soil at a very much lower cost in manpower and materials than is involved in applying imported surfacing, and may even produce better results in the end.

*b.* The importance of the possibility of mechanical stabilization of soils in military construction cannot be overemphasized. What is needed is an awareness of the possibilities of this type of construction, an understanding of the principles of soil action previously presented, and a flexible viewpoint, open to the possibilities of making full use of locally available materials.

## 237. Limitations of Mechanical Stabilization

Without minimizing the importance of mechanical stabilization, limitations of this method should also be realized. The principles of mechanical stabilization have frequently been misused, particularly in areas where frost action is a factor in the design. For example, clay has been added to granular subgrade materials in order to "stabilize" them, when in reality all that was needed was adequate compaction to provide a strong, easily-drained base which would not be susceptible to detrimental frost action. An understanding of the compaction which can be achieved by modern rolling equipment should prevent a mistake of this sort. Somewhat similarly, poor trafficability of a soil during construction because of a lack of fines should not necessarily provide an excuse for mixing in clay binder. It may be possible to solve the problem by applying a thin surface treatment or some other expedient method.

## 238. Requirements for Mechanically-Stabilized Soil Bases

*a.* Grading requirements relative to mechanically-stabilized soil mixtures which are to serve as base courses are given in table X, appendix II. Experience in civil highway construction has indicated that best results will be obtained with this type of mixture if the fraction passing the No. 200 sieve is not greater than two-thirds of the fraction passing the No. 40 sieve. The size of the largest particles should not exceed two-thirds of the thickness of the layer in which they are incorporated. The mixture should be well graded from coarse to fine.

*b.* A basic requirement of soil mixtures which are to be used as base courses is that the plasticity index should not exceed 6. Under certain circumstances, this requirement may be relaxed if a satisfactory bearing ratio is developed. Experience also indicates that under ideal circumstances the liquid limit should not exceed 28. It may be possible to relax these requirements in theater-of-operations construction. The requirements may be lowered to a liquid limit of 36 and a plasticity index of 10 for full-operational airfields, and to a liquid limit of 45 and a plasticity index of 15 for emergency and minimum-operational airfields, when good drainage is provided.

*c.* If the base is to function satisfactorily, other requirements than those relating to the soil mixture alone must be met. The base must normally have a high bearing value. Density requirements and those relating to frost action are also of particular importance.

## 239. Requirements for Mechanically-Stabilized Soil Surfaces

*a.* Grading requirements for mechanically-stabilized soils which are to be used directly as surfaces, usually under emergency conditions, are generally the same as those indicated in table X, appendix II. Preference should be given to mixtures which have a maximum size of aggregate equal to 1 inch, or perhaps 1½ inches, as experience has indicated that particles larger than this tend to work themselves to the surface over a period of time under the action of traffic. Somewhat more fine soil is desirable in a mixture which is to serve as a surface, as compared with one for a base, so that it will be more resistant to the abrasive effects of traffic, more resistant to the penetration of precipitation which falls directly upon the surface, and can, to some extent, replace by capillarity moisture which is lost by evaporation.

*b.* Emergency airfields which have surfaces of this type require a mixture which has a plasticity index between 5 and 10. Civil experience indicates that road surfaces of this sort should have a liquid limit not in excess of 35, and that the plasticity index should be between 4 and 9. The surface should be made as tight as possible and good surface drainage should be provided. It is to be noted that, for best results, the plasticity index of a stabilized soil which is to function first as a wearing surface and then as a base, with a bituminous surface being provided at a later date, should be held within very narrow limits.

*c.* Considerations relative to compaction, bearing value, and frost action are as important for surfaces of this type as for bases.

## 240. Rule-of-Thumb Proportioning

Satisfactory mixtures of this type are difficult to design and build satisfactorily without laboratory control. A rough estimate of the proper proportions of available soils in the field is possible, and depends upon manual and visual inspection. Suppose that a loose sand is the existing subgrade soil, and it is desired to add silty clay from a nearby borrow source to achieve a stabilized mixture. Each soil should be moistened to the point where it is moist, but not wet; in a wet soil the moisture can be seen as a shiny film on the surface. What is desired is a mixture which will feel gritty and in which the sand grains can be seen. Also, when the soils are combined in the proper proportion a cast formed by squeezing the moist soil mixture in the hand will not be either too strong or too weak; it should just be able to withstand normal handling without breaking. Several trial mixtures should be made until this consistency is obtained, being careful to note the proportions of each of the two soils. If gravel is available, this may be added, although there is no real rule-of-thumb to tell how much should be added. It is always desirable to have too little gravel rather than too much.

## 241. Use of Local Materials

The essence of mechanical soil stabilization is the use of locally available materials. Desirable requirements for bases and surfaces of this type have been given above. It is quite possible, especially under emergency conditions, that mixtures of local materials will give satisfactory service, even though they do not meet the stated requirements. Many stabilized mixtures have been made using shell, coral, soft limestone, cinders, marl, and other materials previously listed. Reliance must be placed upon experience, an understanding of soil action, the qualities that are desired in the finished product, and other factors of local importance in proportioning such mixtures in the field.

## 242. Blending

It is assumed in this discussion that an existing subgrade soil is to be stabilized by the addition of a suitable borrow soil to produce a base course mixture which will meet the specified requirements. The mechanical analysis and limits of the existing soil will generally be available from the results of the subgrade soil survey (ch. 13); similar information is necessary concerning the borrow soil. The problem is then to determine the proportions of these two materials which should be used to produce a satisfactory mixture. In some cases more than two soils must be blended to produce a suitable mixture. This situation is to be avoided when possible because of the difficulties frequently encountered in getting a uniform blend of more than two local materials. Trial combinations are usually made on the basis of the mechanical analysis of the soils concerned. In other words, calculations are made to determine the gradation of the combined materials and the proportion of each component adjusted so that the gradation of the combination will fall within specified limits. The plasticity index of the selected combination is then determined and compared with the specification. If this value is satisfactory then the blend may be assumed to be satisfactory, provided that the desired bearing value is attained. If the plasticity characteristics of the first combination are not within the specified limits then additional trials must be made. The proportions finally selected then may be used in the field construction process.

## 243. Numerical Example of Proportioning

This process of proportioning will now be illustrated by a numerical example. Two materials are available, material B in the roadbed and material A from a nearby borrow source. The mechanical analysis of each of these materials is given in *a* below, together with the liquid limit and plasticity index of each. The desired grading of the combination is also shown, together with the desired plasticity characteristics.

*a. Results of Laboratory Tests.*

### Mechanical Analysis

| Sieve Designation | Percent passing, by weight | | |
|---|---|---|---|
| | Material A | Material B | Specified |
| 1-inch | 100 | 100 | 100 |
| ¾-inch | 92 | 72 | 70–100 |
| ⅜-inch | 83 | 45 | 50–80 |
| No. 4 | 75 | 27 | 35–65 |
| No. 10 | 67 | 15 | 20–50 |
| No. 40 | 52 | 5 | 15–30 |
| No. 200 | 33 | 1 | 5–15 |

### Plasticity Characteristics

| | Percent by weight | | |
|---|---|---|---|
| Liquid Limit | 32 | 12 | Not more than 28. |
| Plasticity Index | 9 | 0 | Not more than 6. |

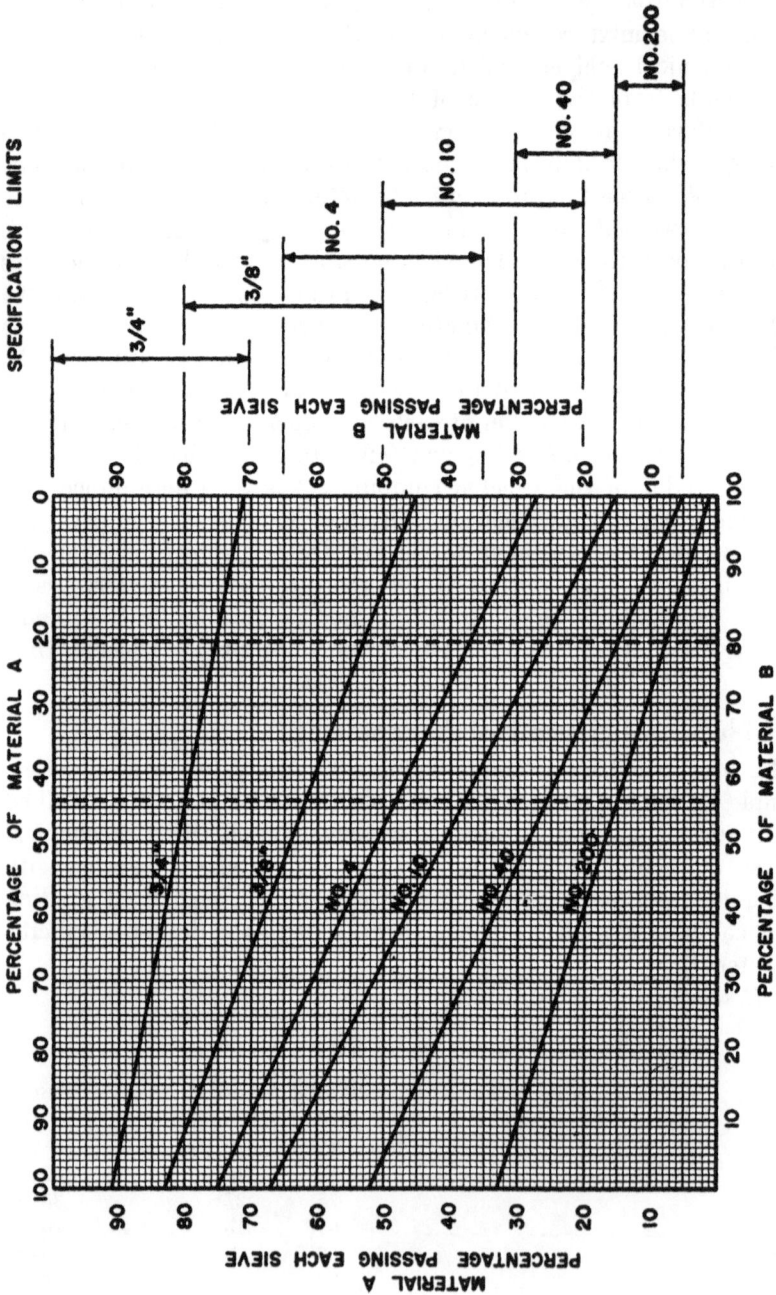

Figure 65. Graphical method of proportioning two soils to meet gradation requirements.

*b. Proportioning To Meet Specified Gradation.* Proportioning of trial combinations may be done either arithmetically or graphically. A graphical method is illustrated in figure 65. The two dashed vertical lines on the chart represent the outer limits of combinations of the two materials which will meet the specified gradation requirements. The boundary on the left represents the combination of 44 percent material A and 56 percent material B; the position of this line is fixed by the upper limit of the requirement relating to the material passing the No. 200 sieve (15 percent). The boundary on the right represents the combination of 21 percent material A and 79 percent material B; this line is established by the lower limit of the requirement relative to the fraction passing the No. 40 sieve (15 percent). Any mixture falling within these limits will satisfy the grading requirements. For purposes of illustration, assume that a combination of 30 percent material A and 70 percent material B is selected for a trial mixture. A similar diagram can be prepared for any two soils. The proportioning can also be done arithmetically. Graphical methods for proportioning more than two soils are more complex and are not included here.

*c. Proportioning To Meet Plasticity Requirements.* An approximate method of determining the plasticity index and liquid limit of the combined soils will serve to indicate if the proposed trial mixture is satisfactory, pending the performance of laboratory tests. This may be done either arithmetically or graphically. A graphical method for getting these approximate values is shown in figure 66. The values shown in figure 66 require additional explanation, as follows: Consider 500 pounds of the mixture tentatively selected (30% A, 70% B). Of this 500 pounds, 150 pounds will be material A and 350 pounds material B. Within the 150 pounds of material A, there will be $150(0.52)=78$ pounds of material passing the No. 40 sieve. Within the 350 pounds of material B, there will be $350(0.05)=17.5$ pounds of material passing the No. 40 sieve. The total amount of material passing the No. 40 sieve in the 500 pounds of blend$=78+17.5=95.5$ pounds. The percentage of this material which has a P. I. of 9 (material A) is $(78/95.5)100=82$. As shown in figure 66, the approximate P. I. of the mixture of 30% material A and 70% material B is 7.4 percent. By similar reasoning, the approximate liquid limit of the blend is 28.4 percent. These values are somewhat higher than permissible under the specifications. An increase in the amount of material B will somewhat reduce the plasticity index and liquid limit of the combination.

## 244. Field Proportioning

In the field, the materials used in a mechanically stabilized soil mixture probably will be proportioned by loose volume. Assume that a mixture incorporates 75 percent of the existing subgrade soil,

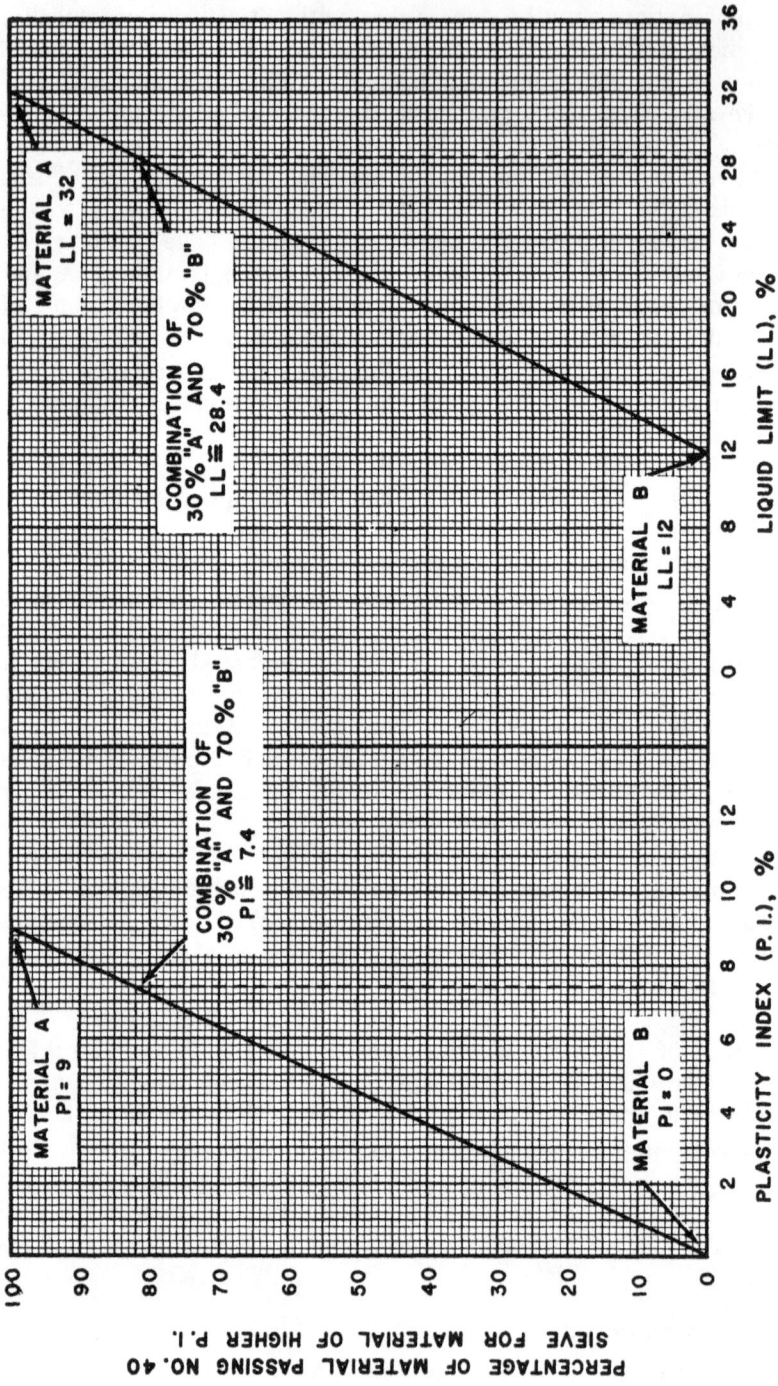

Figure 66. Graphical method of estimating plasticity characteristics of a combination of two soils.

184

while 25 percent will be brought in from a nearby borrow source. It is desired to construct a layer which has a compacted thickness of 6 inches. It is estimated that a loose thickness of 8 inches will be required to form the 6-inch compacted layer; a more exact relationship can be established in the field as construction proceeds. Of the 8-inch loose thickness, 75 percent, or $0.75(6)=6$ inches, will be the existing soil. The remainder of the mix will be the borrow soil, which will be spread in a 2-inch loose layer on top of the existing soil. The two materials then will be mixed thoroughly to a depth of 8 inches, and compacted by rolling. The proportions may be more accurately controlled by weight, if weight measurements can be made under existing conditions.

## 245. Construction Procedure

*a.* The steps followed in the construction of a mechanically-stabilized soil base or wearing surface will depend upon the equipment available and the soils. It is assumed in this discussion that preliminary steps of cutting drainage ditches, shaping of the road to the required crown and grade, removal of organic materials and undesirable soils, and similar operations have been performed. Since subgrade soils are typically nonuniform, both in nature and extent, sufficient sampling must be carried out so that variations within the top few inches will be detected and accounted for in the design of the various sections.

*b.* If the existing soil is clay, it is generally the smaller proportion of the mixture. Thus, it is generally desirable to blade the required amount of clay to the side, bring in the granular material, and spread it uniformly on the subgrade in the required amount. The clay is then bladed back on top of the granular layer and the two materials uniformly mixed. If the existing soil is sand, it may be scarified and pulverized to the desired depth. The binder soil is then brought in, spread uniformly, and the two materials thoroughly mixed. If more then two materials are to be blended, similar operations are conducted. It is generally desirable to place coarser material, like gravel, near the middle of the layers to be processed, in order to avoid segregation.

*c.* Thorough pulverization and blending of the soil mixture is an absolute necessity for the success of this method. Tools used for this purpose include blade graders, harrows, discs, and rotary tillers. When mixing is complete, the layer has a uniform color and appearance throughout the processed depth. If field conditions permit, frequent samples should be taken of the combined materials in order to detect any sections which deviate from the desired mixture. Some variation is to be expected in this type of construction and should be corrected by adding the missing soil elements before construction is continued.

*d.* When the necessary equipment is available mixing operations may be performed with a traveling plant or, occasionally, at a central mixing plant.

*e.* When the mixing operation is finished the soil is compacted, using the type of equipment which is best suited to the soil involved. The soil should be compacted at or near optimum moisture and in accordance with the compaction requirements of paragraphs 222 and 223. If the soil is too dry, water is added by sprinkling and thoroughly mixed with the soil, using the tools indicated in *c* above. If too wet, the soil must be dried until the moisture content reaches the desired amount. Sheepsfoot rollers are frequently used for initial compaction, with final rolling being done with pneumatic-tire rollers.

*f.* After compaction the surface is shaped by light blading and rolled with three-wheel rollers, accompanied by sprinkling as necessary, until the surface is true to line and grade and has a tight, uniform surface texture.

## Section III.   BITUMINOUS SOIL STABILIZATION

### 246. Definition

Bituminous soil stabilization refers to a process by which a controlled amount of bituminous material is thoroughly mixed with an existing soil or aggregate material to form a stable base or wearing surface. The bituminous material used may be either asphalt or tar, although asphalts are more frequently used. Asphalts which are liquid at normal temperatures and require no, or very little, heating at the time of application generally are preferred for this use. The bituminous material may be used for one, or both, of two general purposes. First, to increase the resistance to deformation by supplying cohesion through the cementing action of the bituminous material. Second, in cases where exposure to water is a problem, a degree of waterproofing is achieved by the incorporation of the bituminous material, thus aiding in retention of a higher shearing strength than would otherwise be normal.

### 247. Uses and Conditions for Use

*a. Uses.* Bituminous soil stabilized mixtures have been extensively used in road and airfield construction by both military and civilian agencies. General uses include the making of an otherwise unsatisfactory base course material into a satisfactory one, the creation of a "working floor" on cohesionless sand subgrades, and the production of low-cost wearing surfaces to carry light traffic. This type of construction has been used to some extent at permanent military airfields and auxiliary landing fields in the United States, particularly in areas where suitable granular materials for mechanical stabilization were not economically available. It has also been used under emergency conditions for both road and airfield construction, and for the stabilization of beach sands at landing points.

*b. Conditions for Use.* Bituminous soil stabilization should not be considered for use except in situations in which suitable base course materials are not economically available, or where it is not feasible to mechanically stabilize the existing soil to produce the desired bearing value. The production of this type of mixture requires that the soil be thoroughly pulverized and mixed with the bituminous material. This requirement generally limits the economical use of this method to the treatment of essentially cohesionless soils, including sands, sandy and gravelly soils, and the more friable silts. Cohesive soils have been successfully stabilized with bitumen. However, because of the difficulties usually encountered in the pulverization and mixing processes, this practice is not recommended. For successful results, the moisture content of the soil must be reduced to fairly low values before the bituminous material is incorporated, and the soil-bitumen mixture must be thoroughly cured before compaction. These requirements, in general, limit the use of this type of construction to periods of good weather.

## 248. Considerations Relative to Materials

*a. Soil.* As indicated above, soils which are most suitable for stabilization with bituminous materials are those which can be easily pulverized and mixed with available equipment. Soils with high silt and clay content are not regarded as being suitable for this type of stabilization. Friable soils that are inherently somewhat stable and those that can readily be improved by the addition of other soils are best suited for bituminous stabilization. Experience on civilian airfields in the United States indicates that soils for bituminous stabilization should not contain more than 45 percent of material passing a No. 200 sieve, the liquid limit should be less than 30, and the plasticity index less than 10. Soils containing a large amount of mica are not suitable for this type of construction.

*b. Bituminous Materials.* It is generally desirable to use bituminous materials which require little or no heating in this work. Of the asphaltic products, rapid-curing cutback asphalts, grades RC–1 to RC–4, and medium-curing cutback asphalts, grades MC–1 to MC–4, are most frequently used. Grades RC–1 and RC–2 have been extensively used in the stabilization of cohesionless sands. Slow-setting asphalt emulsions are also used frequently and are sometimes used with moist soils. Slow-curing liquid asphalts or "road oils" have been used to some extent, particularly with soils which contain an excess of fines; their very long curing time makes them less useful than the cutback asphalts. Tars have been used, particularly grades RT–3 to RT–7, in areas where these materials are readily available; their use is not recommended in surface courses. The grade of bituminous material used will depend upon those available, the soil, and the mixing equipment. Lighter grades are generally preferred with soils which contain little or no fines and for in-place mixing with commonly available tools.

*c. Mixture.* Satisfactory bituminous soil stabilized mixtures generally contain from 4 to 9 percent bitumen by weight, exclusive of the volatile materials. The amount of bitumen generally increases with an increasing amount of fines in the soil. Although laboratory procedures have been developed for determining the proper amount of a given bituminous material to be used with a given soil, these methods have not been standardized and are not generally available to the military engineer in the field. There are no easy tests to be used in field determinations of the proper amount of bitumen, and this amount must generally be based upon judgment and experience.

## 249. Construction Procedure

*a.* Preliminary steps in this type of construction are generally the same as those described for mechanical soil stabilization.

*b.* Prior to application of the bituminous material, the soil must be pulverized and mixed to a uniform condition to the desired depth of treatment. When cutback asphalt or tar is used, the soil must be thoroughly dry before it is applied. A general requirement is that the moisture content of the soil should be less than 2 percent, although some sands can be handled successfully at moisture contents slightly in excess of this. Otherwise the water will adhere to the soil particles and the bitumen will not coat the material properly. When asphalt emulsion is used, the moisture content of the soil may be somewhat greater, depending on the soil involved. After the soil has been aerated and the moisture content reduced sufficiently, the bituminous material is applied by means of a pressure distributor, if mix-in-place methods are being used. If a traveling plant is used the bituminous material is incorporated in the plant.

*c.* The bituminous material must be thoroughly and uniformly mixed with the soil, using equipment like plows, harrows, cultivators, blade graders and rotary tillers. This operation may be done in a traveling plant, as illustrated in figure 67. It may also, on occasion, be done in a central mixing plant. Plant mixtures are generally more uniform than those produced by road mixing, and the operation is less affected by adverse weather conditions.

*d.* After mixing has been completed, aeration must be continued until virtually all of the volatile materials have been removed. This is absolutely essential in this type of construction in order for the mixture to "cure" and harden properly. The mix may then be compacted, using the type of equipment best suited to the operation. When cohesive soils are stabilized in this fashion, water is sometimes

*Figure 67. Traveling plant engaged in bituminous soil stabilization.*

added to facilitate mixing and compaction, but the compacted mixture must still meet the requirements of *e* below. Finishing operations are similar to those described under mechanical stabilization.

*e.* If the mixture is to be used as a base course, an additional curing time may be necessary before the surface is sealed or a wearing surface constructed. The base should not be sealed or surfaced until the moisture content has dropped to the range of 2 to 4 percent.

## Section IV.  SOIL-CEMENT

### 250. Definition

Soil-cement is an intimate mixture of pulverized soil and Portland cement. It is moistened, compacted, and cured during construction, and hardens into a semirigid material which possesses considerable compressive strength. The amount of cement which is used is generally not large, being from 7 to 14 percent by volume of the compacted mixture.

### 251. Uses and Conditions for Use

*a. Uses.* Soil-cement has been used extensively in this country in the construction of roads and airfields. It has also been used under emergency conditions for both purposes. It is principally used in the construction of base courses, since it can not withstand successfully the abrasive effects of traffic and direct exposure to climatic changes over an extended period of time. Soil-cement may serve as a wearing surface under emergency conditions for a brief period of time. Perhaps the most frequent use for soil-cement mixtures is as base courses which are covered by relatively thin bituminous surface treatments. It can also serve as a base for higher-type pavements which will be subjected to heavy traffic.

*b. Conditions for Use.* As with bituminous soil stabilization, soil-cement should not be considered for military construction unless it is certain that materials are not economically available for mechanical stabilization. Practically all soils can be stabilized with cement, but not all soils can be stabilized easily and economically by this method. Thorough pulverization of the soil and thorough mixing with cement and water are prerequisites to success. Hence, for military construction, soil-cement stabilization is limited to granular and friable soils which can be pulverized readily. Soils which have a high silt or clay content require more cement and present difficulties during construction, particularly during pulverizing and blending. Soils which contain appreciable amounts of organic matter are not suitable for soil-cement stabilization. Construction of soil-cement is limited to use as a base course directly under a bituminous wearing surface for pavements subject to gross loads of not more than 30,000 pounds.

## 252. Considerations Relative to Materials

*a. Cement.* The cement used is standard Portland cement.

*b. Water.* Any normal source of water may be used, although it should be free from excessive amounts of organic material, acids, and alkalies.

*c. Soil.* The maximum size of aggregate used in soil-cement is 3 inches and the soil should not contain more than about 50 percent of material which is retained on a No. 4 sieve. Experience in highway construction indicates that, for successful and economical soil-cement stabilization, the soil should not contain more than 50 percent material passing a No. 200 sieve, should have a liquid limit less than 40, and a plasticity index less than 18. An important factor in determining the suitability of a given soil for soil-cement construction is the ease with which the soil can be pulverized.

*d. Soil-Cement Mixture.* The amount of cement which is necessary for use with a given soil for satisfactory and economical results can not be determined by a simple procedure. Accurate determination of the proper cement content is based upon a rather elaborate series of laboratory tests, including the determination of volume changes and losses caused by alternate freezing and thawing, and alternate wetting and drying. Such testing is beyond the normal range of facilities available to the military engineer in the field and will not be presented here. As previously indicated, most soils which are suitable for soil-cement can be adequately hardened by the use of from 7 to 14 percent cement by volume of the compacted mixture. Granular soils which contain little binder soil and are non-plastic or feebly plastic in nature can generally be stabilized by cement contents in the lower end of the indicated range, as can many clean sands. Silts and the more plastic soils will require higher cement contents. In field situations in which economy is not the principal factor involved, the engineer should be certain that enough cement is used to insure hardening of the mixture. Laboratory determinations of the optimum moisture content and maximum density of the soil-cement mixture are necessary in order that compaction may be adequately controlled. The optimum moisture content of soil-cement will generally be from 2 to 3 percent greater than that of the untreated soil. Enough water must be present to insure hydration of the cement.

## 253. Construction Procedure

Soil-cement bases have been most frequently constructed with a compacted thickness of 6 inches. Assuming that the subgrade has been brought to the desired condition, principal steps in the construction process include pulverization of the existing soil, spreading of cement, dry mixing of the soil and cement, addition of water and mixing, compaction and finishing, and curing.

*a. Pulverizing Existing Soil.* The existing soil is scarified to the depth of processing, usually by the use of a rooter or a scarifier attachment on a blade grader. The soil is then pulverized by the use of disc harrows, plows, cultivators, or rotary tillers. It is generally desirable that the soil, with the exception of gravel or stone, be pulverized until at least 80 percent will pass a No. 4 sieve. The moisture content of soils which contain considerable silt or clay may have to be controlled in order to facilitate pulverization. For many of these soils, the correct moisture content will be close to optimum.

*b. Spreading of cement.* The proper amount of cement may be spread upon the surface, either by hand or by mechanical means. If cement is in sacks, these are generally spotted along the surface in rows of predetermined spacing. The number of sacks which is required may be calculated as follows. Assume that 10 percent of cement is to be used. The soil-cement layer is to be 6 inches thick, and the width of the processed strip is 24 feet. The volume of cement required per 100-foot station is then $24(100)(6/12)\ 0.10 = 120$ cubic feet or 120 sacks of cement, since the volume of one sack of cement may be taken to be one cubic foot. The sacks are then opened and the cement spread in uniform transverse rows. Spreading may be completed by use of a spike-toothed harrow or similar tools.

*c. "Dry" Mixing of Soil and Cement.* The mixing of soil and cement may be accomplished by the use of the tools listed in *a* above. Careful control is necessary during this operation, in fact throughout the entire procedure, in order that the mixing will be carried to the proper depth and that a uniform mixture is secured. Available equipment will generally be operated in a "mixing train," with units following one another in close sequence. Additional units are added to the train to apply water and accomplish wet mixing.

*d. Addition of Water and "Wet" Mixing.* After the dry materials have been thoroughly mixed, water is added by a distributor to bring the soil to the proper moisture content for compaction. Mixing is continued until the soil, cement, and water are thoroughly blended. Speed is essential in this phase of the process, as the remaining steps must be completed before the cement hydrates and the mixture hardens. Several passes of the mixing train will generally be required to obtain a uniform mixture.

*e. Compaction and Finishing.* Initial compaction of a soil-cement base is usually done by sheepsfoot rollers, although pneumatic-tire rollers may be necessary in very sandy soils. After rolling is completed, the surface is brought to final shape. Final shaping may be done with the aid of a spike-toothed harrow or a nail drag to remove the compaction planes left by the rollers, followed by a broom drag. A rubber-tired roller may then be used again, the surface shaped by a blade grader, and compaction planes again removed by a broom drag. Final compaction is then achieved by rollers with steel wheels. Rolling is continued until a tightly packed surface is secured. It is essential that the moisture content be very close to optimum during all the rolling operations which have been described. Frequent checks of moisture and density are usually necessary.

*f. Curing.* The water which is contained in a soil-cement mixture is necessary to the hardening of the cement, and steps must be taken to prevent the loss of moisture from the completed base by evaporation. A protective covering of from 2 to 4 inches of soil may be placed over the base, or hay or straw may be used. A light application of bituminous material may be used for the same purpose; if the base must be opened to traffic immediately after construction this is the best solution. It is desirable that curing be continued for 7 days before being opened to traffic. However, in emergency situations the base can be opened to traffic immediately.

## Section V. CHEMICAL STABILIZATION

### 254. General

A number of chemicals, other than bituminous materials and Portland cement, have been used successfully to stabilize soils in the laboratory and in the field. Certain of these materials hold great promise for future use in soil stabilization in both civilian and military applications. However, because most of these materials are not normally available to the military engineer in the field and are still in the research and development stage as far as practical application is concerned, their use will not be discussed here. The only additional material which may be of practical importance to the military engineer is lime, which may be used to improve the plasticity characteristics of some fine-grained soils.

## 255. Lime-Soil Stabilization

*a.* Lime may be used to reduce the plasticity and improve the mixing, compaction and strength characteristics of some plastic soils. The amount of lime which has been used is quite small, generally being from 2 to 10 percent by weight.

*b.* Experience has shown that for the more plastic soils (plasticity index more than 15), the plasticity index is reduced by the addition of lime. The soil is thus easier to pulverize and easier to mix.

*c.* Under a given compactive effort, the density of a lime-treated soil is less than that of the raw soil. However, the soil may be easier to compact and densities obtained in the field more uniform.

*d.* However, the strength of compacted lime-soil mixtures is greater than that of similar raw soils. The increase in soil strength is brought about principally by changes in water films surrounding the clay particles. There is some evidence that the strength increases with curing and age.

*e.* In general, lime-soil mixtures are susceptible to frost action and deteriorate rapidly under alternate freezing and thawing. A typical mixture of this type has poor weathering characteristics.

*f.* The construction of a lime-soil subgrade or base should present no particular difficulties, although the lime must be thoroughly mixed with the soil. It is generally spread on the soil layer and mixed in the same fashion as soil-cement. The moisture content must be brought to optimum and the mixture compacted in the same way as the natural soil. Time is not a critical factor from the standpoint of the reaction between the lime and the soil, since the lime will not harden within the normal construction day.

## Section VI. PAVEMENT DESIGN

## 256. Definitions and Fundamental Considerations

Although terms relating to pavements which are used for roads and airfields have been used in previous discussions, it is desirable to re-define them here. In this paragraph, also, are presented general facts as to the nature and function of each element of the pavement structure.

*a. Pavement or Pavement Structure.* In this manual the *pavement* is a structure which consists of one of more layers of processed materials. The term *pavement structure* is also used, and pavement is sometimes taken to have the same meaning as wearing surface. Elements of typical flexible and rigid pavement structures are shown in figure 68. The typical flexible pavement consists of a wearing surface, base course, and subgrade; a subbase course may also be used, although this layer is not indicated in figure 68. A rigid pavement is usually made of Portland cement concrete, with only two layers; the "pavement" itself, which fulfills the functions of both a wearing surface and a base, and the subgrade. Sometimes a base or subbase course may be used between a rigid pavement and the subgrade.

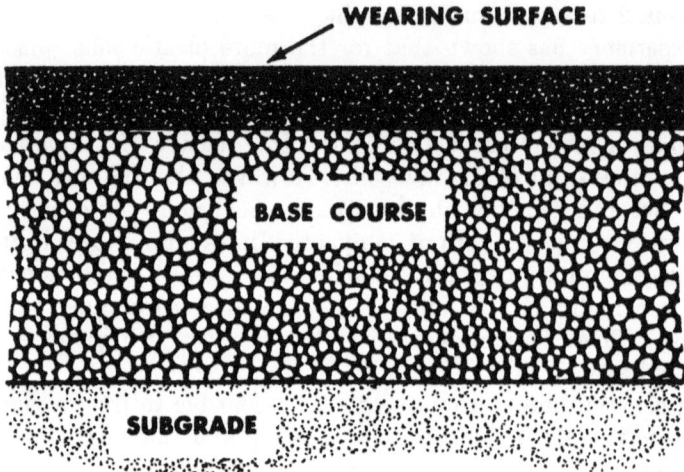

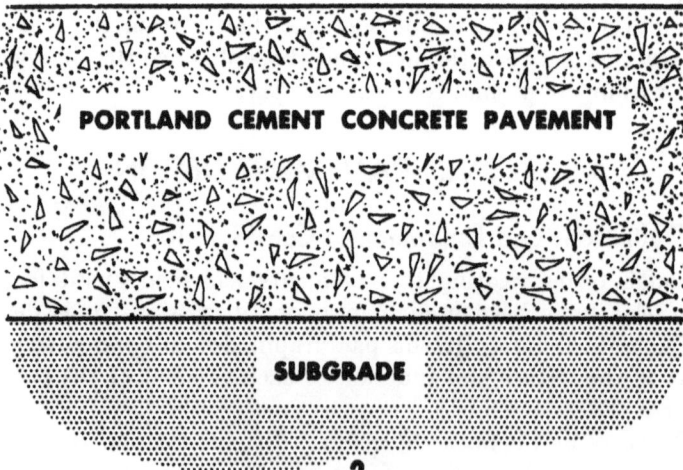

Figure 68. Typical pavement structures.

*b. Wearing Surface.* The upper, or surface, layer of a pavement is the *wearing surface.* In the case of flexible pavements, the wearing surface is usually a bituminous treatment or mixture of some sort. It may range in thickness from less than an inch, as in the case of a bituminous surface treatment, to several inches of high-type paving mixture, like asphaltic concrete. The wearing surface has four principal functions.

(1) It protects the base against the abrasive and disintegrating effects of traffic.

(2) It distributes the load to the underlying layers of the pavement structure.

(3) It prevents the penetration of excessive amounts of surface water into the base.

(4) It provides a smooth riding surface.

In the case of a rigid pavement, the combined wearing surface and base is at least 6 inches of Portland cement concrete.

*c. Base Course.* The *base course* is a layer of processed material which is directly beneath the wearing surface. Its principal function is to distribute the load to the underlying layers of the pavement structure. The base must be of such a character that it is not susceptible to damage by capillary water or frost action. The base is usually a layer of high density and stability. Many different materials may be successfully used as base courses, as discussed in paragraph 258.

*d. Subbase Course.* On occasion an additional layer of granular material is needed between the base and the subgrade. This layer is called a *subbase course.* It may be regarded simply as an additional thickness of base course, in which case a separate designation is not needed. Its principal function is to further distribute the load so that the subgrade will not undergo excessive deformation or displacement. A subbase is principally used over a very weak subgrade soil or in areas of deep frost penetration. It generally has a higher bearing value than the subgrade, but lower than the base course.

*e. Subgrade.* The *subgrade* is the foundation layer which must eventually support all the loads which come onto the pavement. In some cases this will simply be the natural soil. More frequently it will be the compacted soil in a cut or an embankment section. The combined thickness of the other elements of the pavement structure must be great enough so that stresses created in the subgrade will not cause excessive deformation or displacement.

## 257. Subgrade

The stability of a subgrade depends upon a number of different factors, including such things as the presence of an excessive amount of moisture or organic material, susceptibility to frost action, shrinkage and expansion, bearing value, and the traffic which the pavement is being designed to carry. Density is an important factor; compaction requirements relating to subgrades are given in paragraph 222. With a good subgrade soil, favorable climatic conditions, and light traffic, a minimum thickness of pavement structure will be required. With a poor subgrade soil, unfavorable climatic conditions, and heavy traffic, a very thick pavement structure will be needed. In the interests of economy, ease of construction, and satisfactory performance, poor subgrade soils are to be avoided in the location of a road or airfield, whenever it is feasible to do so. Column (7) of table VI, appendix II, gives information relative to the value as a foundation, when not subjected to frost action, of the groups of the Unified Soil Classification System. Values shown are for subgrades and base courses, except for base courses directly beneath a bituminous pavement.

## 258. Base Courses

a. Crushed rock, gravel, and slag are widely used for base course materials, when available. Uncrushed gravels and sands, gravelly and sandy soils, and stabilized soil mixtures are also frequently used. Many other materials, such as limerock, coral, and shells, are satisfactory for base courses when used singly, or in combination with other materials. Rubble may be used under expedient conditions. When higher-type bituminous pavements, like asphaltic concrete, are used the base frequently is an asphaltic mixture; in this type of construction a *binder course* may be used between the base and wearing surface. The selection of the type of base to be used under given conditions is a function of the materials and equipment available, and weather conditions. Information relative to base courses is contained in TM 5–255. Additional information concerning bases (and wearing surfaces) will be contained in separate technical manuals now under preparation. When printed, these manuals will be distributed automatically by The Adjutant General to organizations concerned.

b. In column (8) of table VI, appendix II, is given information concerning the value of the soils of the Unified Soil Classification System when used as bases directly beneath bituminous pavements (wearing surfaces). It will be noted that the term "excellent" has been reserved for bases constructed of high quality processed crushed stone. Compaction requirements relative to base courses have been given previously in paragraph 223.

## 259. Wearing Surfaces

a. Wearing surfaces of natural or processed soil and granular materials are widely used in military construction under expedient conditions. Included among these are natural soil surfaces and the base course materials mentioned in paragraph 258 including soil stabilized mixtures and locally available granular materials. In general, these surfaces are not satisfactory for all-weather, heavy-duty traffic without heavy maintenance.

*b.* A wide variety of bituminous surfaces are used for both roads and airfields. Principal categories include surface treatments, penetration macadam, road-mix surfaces, plant-mix surfaces, and natural rock asphalts. Within these categories there are many types which will give satisfactory service. Selection should be based upon traffic conditions, availability of equipment, personnel and materials, and construction time.

*c.* Rigid pavements, the most important of which is plain concrete, are built to serve as military roads and airfields. However, they are normally used only for important installations of a more permanent character, or which are subjected to very heavy traffic. From a soils standpoint, it may be noted here that fine-grained soils which contain more than 45 percent clay, with a liquid limit greater than 50 and a plasticity index of more than 25, are not suitable for the direct support of concrete pavements; these soils should be covered with a granular blanket course or subbase before the concrete is placed. Other types of rigid pavements include brick and cement-penetration concrete.

*d.* Rigid and flexible pavements are sometimes used in combination with one another, as when a bituminous surface is built over a concrete base course. Other combinations are possible.

*e.* Urgent conditions in theaters of operation frequently require the temporary use of portable road and airfield surfaces, such as metal landing mats, wood mats, various types of treadways, and prefabricated bituminous surfacing.

## 260. Thickness Design of Pavements

*a.* A very important part of the design of flexible and rigid pavements is the determination of the thickness of pavement structure which is required under given conditions. It is desired that the thickness be adequate to insure satisfactory behavior of the pavement during its service life, and that the design be economical. If more than one layer is used above the subgrade, the design is also concerned with determining the thickness of each layer.

*b.* Complete criteria for the thickness design of both flexible and rigid pavements for roads and airfields are contained in separate technical manuals now under preparation. When printed, these manuals will be distributed automatically by The Adjutant General to organizations concerned.

*c.* The design of flexible pavements for roads is based upon the California Bearing Ratio. Design curves have been established which give the relationship between the design CBR of the subgrade, design wheel load, and combined thickness of pavement and base. The design of rigid pavements for roads involves three principal variables: design wheel load, flexural strength of the concrete, and the subgrade modulus, $k_s$ (par. 142). Empirical curves give the relationship between these variables and the required design thickness. In some situations, the design of a pavement may have to be based upon considerations relative to frost action.

*d.* The design of flexible pavements for airfields where frost action is not a factor is also based upon the CBR. Design curves have been established which give the relationships between design CBR, assembly load, operational conditions of the airfield, and the combined thickness of pavement and base for single, dual, and tandem wheels at different inflation pressures. Similarly, design curves for different wheel assemblies and tire pressures have been developed for rigid pavements where frost action is not a factor; these show the relationships between flexural strength of the concrete, subgrade modulus, and wheel load. Criteria have also been established relative to the design of rigid-over-rigid pavement overlays, flexible-over-rigid pavement overlays, and for frost action in the subgrade soil.

# CHAPTER 11

# EARTH SLOPES, EMBANKMENTS, DIKES AND DAMS

## Section I. INTRODUCTION

### 261. Scope

This chapter is devoted to certain aspects of the design of earth slopes, embankments, dikes, and dams. Certain considerations relative to the seepage of water through earth masses are presented in chapter 4; additional information is contained in this chapter. The important subject of control of seepage, particularly as related to water-retaining earth structures, is covered. Considerable attention is given to the stability of slopes, and recommendations are made as to slopes which should be used under average conditions in cuts and embankments. A separate section is devoted to the location, design, and construction of dikes and small earth dams, such as may be used in developing a water supply for military purposes.

### 262. Limitations

Material contained in this chapter is of primary usefulness to the military engineer who is engaged in the design of relatively small structures under conditions in which time and available materials are limited. The design of large earth dams, in particular, is a highly specialized subject, and this chapter does not contain sufficient information for the adequate design of a major structure of this type. From a theoretical standpoint, the design of large earth structures is highly complex; only enough theory has been included here to emphasize the importance of the practical considerations presented.

## Section II. SEEPAGE

### 263. General

*Seepage* is the movement of free water through a soil mass, generally under the force of gravity. All water-retaining structures are subject to seepage through, under, or around them. Similarly, cut slopes and, occasionally, embankments which are not primarily designed to retain water may be subject to seepage. If uncontrolled, seepage may affect the stability of a structure because of the existence of excess seepage pressures, erosion, or associated effects. An excessive amount of seepage through or beneath an earth dam which is constructed for water supply purposes may destroy the usefulness of the structure.

## 264. Permeability of Soils

Quantitative determinations of the amount of seepage flow depend upon a determination of the coefficients of permeability, $k$, of the soils concerned. Permeability has been discussed previously in paragraph 82. Typical values of the coefficient of permeability of the soils included in the groups of the Unified Soil Classification System are given in column (8) of table VII, appendix II. These values should be adequate for field use, although where circumstances permit and for major projects, permeabilities should be determined by test. In analyzing seepage flow, variations in permeability due to stratification are often of great importance.

## 265. Flow Net

The *flow net* is a graphical representation of the direction of flow and the hydraulic head existing at any point in a cross section of soil through which water is flowing. The information which is given in the flow net is indispensable to the safe and economical design of major water-retaining structures. Basic assumptions which are made in the construction of a flow net include a homogeneous soil, laminar flow of the water, that the water completely fills the soil voids, and that there is no change in the size of the voids. All these conditions would seldom, if ever, be met with in practice; nevertheless the flow net is still extremely useful. It is not expected that the military engineer in the field would draw many flow nets. However, some understanding of the nature, properties, and uses of a flow net is desirable to the understanding of the effects of seepage flow and measures which may be undertaken to control seepage.

## 266. Properties of the Flow Net

The flow net is made up of two families of curves, one set of which is known as *flow lines* and the other as *equipotential lines*. The flow lines define the direction of flow, while the equipotential lines are contours of equal head (pressure and elevation heads, since velocity head is neglected). The important properties of the conventional flow net may be summarized as follows:

*a.* Flow lines and equipotential lines intersect at right angles to form similar rectangles. In the conventional flow net the infinite number of possible flow lines and equipotentials is reduced to a small number so selected as to form "squares". The sides of the squares are curved, but true square areas can be approached if the figures are subdivided far enough.

*b.* The quantity of water flowing between any two adjacent flow lines is the same throughout the flow net and is equal to a constant fraction of the total seepage.

*c.* The head loss between any two adjacent equipotential lines is the same throughout the flow net and is equal to a constant fraction of the total loss in head.

*d.* The velocity of flow and the hydraulic gradient are inversely proportional to the spacing of the flow net lines. Therefore, seepage forces are at a maximum at points where flow lines or equipotential lines tend to converge.

*e.* Every flow net has 4 boundaries, 2 of which are flow lines and 2 equipotentials. In cases where impervious boundaries exist the boundaries are easily established, as may be seen by examination of figure 69. In this case, one of the flow lines is that which follows down the left side of the sheet-pile wall and up the right side; the other is along the top of the impervious rock layer. One of the equipotentials is the natural ground surface to the left of the sheet-pile wall ($h=15$); the other is the natural ground surface to the right of the wall ($h=0$). In other cases the boundary conditions are not completely defined. This is the case when flow through dams and dikes is considered, since the location of the upper boundary of flow is not directly known but must be determined mathematically, by model studies, or graphically.

## 267. Methods of Obtaining Flow Nets

Three principal methods are available for obtaining flow nets for given conditions. These include mathematical solutions, the use of models, and the graphical method. Mathematical solutions are available only for very simple cases, cases which are seldom encountered in practice. However, mathematical solutions are useful in indicating the general proportions of more complex flow nets and for other purposes. Model tests conducted in the laboratory are very useful when time and facilities are available for performing them. The most useful method is the graphical one, since the flow net may be sketched for any combination of circumstances. In the sketching method, the flow net is simply drawn to conform to the boundary conditions and to fit the imposed mathematical requirements. Flow nets are rather easily drawn for homogeneous soils in cases in which boundaries are clearly defined. Other cases are more difficult and require greater experience. More complicated cases include those in which the top flow boundary is at atmospheric pressure, as in the case of flow through an earth dam; those in which two or more soils are involved, as when the core of an earth dam is of one material and the outer shell of another; and cases in which the coefficient of permeability is greater in one direction than in another. Details of the sketching method are beyond the scope of this manual, particularly for the more complicated cases. Flow lines and equipotentials intersect at right angles to form squares (or rectangles). Some areas do not resemble squares, but can be subdivided further to form approximate squares. Usually it is not convenient to divide the entire section into squares and a row of rectangles will remain. The ratio of the lengths of the sides of each rectangle should be the same throughout. Even crude approximations of the flow net may provide a reasonable estimate of seepage quantities and give some motion of the seepage control measures which may be needed.

## 268. Uses of Flow Net

The flow net is useful in estimating the quantity of seepage, the hydrostatic head at any point in a cross section, the average hydraulic gradient for any element, and seepage forces.

  *a*. The quantity of seepage per unit of width and unit of time can be estimated from the following relationship, which is based on Darcy's law; $q=kh$ $(N_f/N_p)$, in which $k$ is the coefficient of permeability, $h$ is the total hydrostatic head, $N_p$ is the number of squares between the adjacent flow lines, and $N_f$ is the number of flow channels (number of squares between adjacent equipotentials).

  *b*. The equipotentials represent contours of equal head, so that open-end standpipes installed at different points along an equipotential will register the same height of water, after equilibrium has been established. Thus the hydrostatic head at any point in a cross section can be determined, as follows: $h_1=h(n_p/N_p)$, where $n_p$ is the number of equipotential drops between the point under consideration and zero potential. This computation may be useful in the determination of uplift pressures on the bottom of an impervious layer or a masonry dam.

  *c*. The average hydraulic gradient may be determined for any element, as follows: $i=h_2/L$, in which $h_2$ is the average difference in head between the approaching and receding faces of any element, and $L$ is the average length of the flow path. If the element is a square of the flow net, $h_2$ is $\Delta h$, the head loss between the two equipotential boundaries of the square, which is a constant fraction of the total head loss, $h$. If the element under consideration is at the discharge face, then the gradient is called the discharge or *escape gradient*. Maximum gradients occur where the squares are the smallest. Thus it is possible to determine by a quick glance at the flow net critical points in the section, where gradients and seepage forces are a maximum. Flow nets must be carefully drawn to permit accurate determinations of escape gradients. Escape gradients are very important, since they permit determination of the safety against boiling and piping (par. 91) and may indicate the necessity for filters or other drainage treatment in the zone where seepage emerges.

  *d*. The seepage force which acts on a segment of unit thickness may be calculated as follows: $F=\Delta ha\gamma_w$, where $a$ is the average width of the segment and $\gamma_w$ is the unit weight of water. The seepage forces which are obtained from a well-constructed flow net can be combined with gravity forces for stability analyses. Use of seepage forces in this fashion is beyond the scope of this manual, as is the construction of adequate flow nets for the more complicated cases. Reference should be made to standard textbooks in soil mechanics for this information.

## 269. Typical Flow Nets for Water-Retaining Structures

In figures 69 and 70 are shown flow nets for a number of water-retaining structures. These flow nets and related items are discussed in paragraphs 270 and 271. The flow nets shown are all for relatively simple cases involving homogeneous soils. In addition to providing typical cases for examination, their principal purpose is to highlight the need for the seepage control devices which are discussed in more detail in the next section.

## 270. Flow Beneath Sheet-Pile Wall

*a.* In figure 69 is shown the flow net for the condition of flow beneath a sheet-pile wall. On the drawing, solid lines within the flow net are flow lines, and dashed lines are equipotential lines. The flow net is symmetrical and conforms with boundary conditions previously listed.

*b.* In the example of figure 69 the number of equipotential drops, $N_p$, is 9 and the number of flow channels, $N_f$, is about 4.5. Assume that $k = 10^{-4}$ centimeters per second = 0.0001 centimeters per second

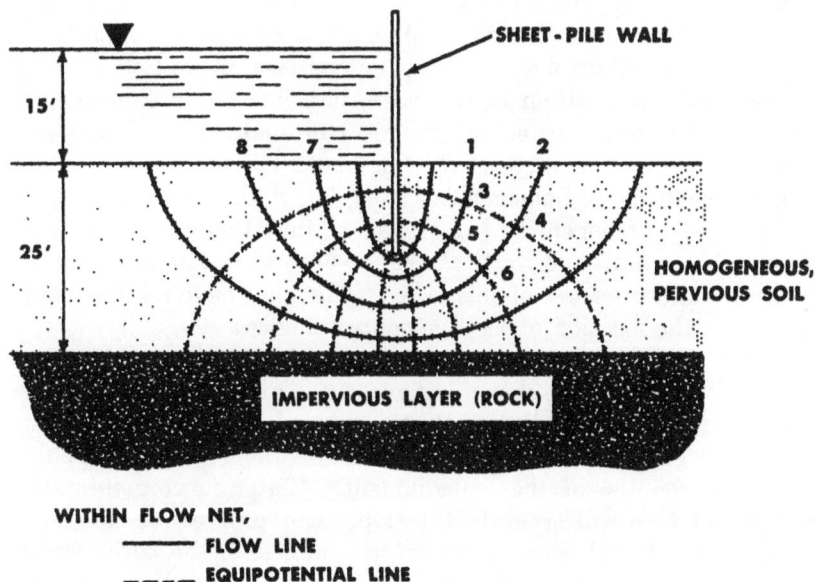

WITHIN FLOW NET,
——— FLOW LINE
————— EQUIPOTENTIAL LINE

*Figure 69. Flow net (flow beneath sheet-pile wall).*

for this soil. The quantity of seepage, $q$, per foot of length of the wall then is: $q=kh(N_f/N_p)=(0.0001/30.5)$ 15 $(4.5/9.0)=0.0000024$ cubic feet per second, since there are approximately 30.5 centimeters in 1 foot and $h=15$ feet. This is about 0.21 cubic feet per day per foot of wall.

c. At points 1 and 2 in figure 69 the hydrostatic head, $h_1=h_2=0$. At points 3 and 4, $h_3=h_4=(1/9)15=1\frac{2}{3}$ feet. At points 5 and 6, $h_5=h_6=(2/9)15=3\frac{1}{3}$ feet. At points 7 and 8, $h_7=h_8=15$ feet. Most of the figures within the flow net are squares, since their average dimensions are equal. For example, the average of distances 1–2 and 3–4 is the same, or nearly the same, as the average of distances 1–3 and 2–4. Thus, the element 1,2,3,4 is a square. The head loss across each square, $\Delta h$, is $(1/9)15=1\frac{2}{3}$ feet.

d. The average hydraulic gradient for the element 1,2,3,4 is $\Delta h/L=(1\frac{2}{3})/7$, since the average dimension of this square is about 7 feet. This value is much less than the critical hydraulic gradient (par. 91), and there should be no danger of boiling within this element. This is *not* the critical square, since it is not the smallest one. The calculations simply show the approach which can be used. Escape gradients are only approximate, since further subdivision of squares would produce somewhat larger gradients.

## 271. Flow Through Embankment of Homogeneous Soil

a. For cases of flow through embankments, the topmost flow line is a free water surface which forms the upper boundary of the flow net. It is called the *saturation line* and is defined as the line above which atmospheric pressure exists and below which there is hydraulic pressure. Above this line the soil may be saturated or partially saturated by capillary action. Seepage analyses are complicated by the fact that the position of the saturation line is not known, and must be determined by either graphical or mathematical methods which are beyond the scope of this manual. Discussions here given pertain principally to homogeneous soils, but it should be recognized that a lack of homogeneity , in the sense that there is a difference between horizontal and vertical permeabilities, is characteristic of practically all structures of this type. This may have an important effect upon the location of seepage control devices.

b. Flow nets for pervious embankments which are located on impervious foundations are shown in figure 70. Figure 70 ① represents flow through a simple rolled-earth dam of uniform section on an impervious base. The seepage line for such a section emerges at a point well above the toe of the embankment. Water flowing from the downstream face will saturate the slope, and progressive sloughing will occur which will cause an ultimate weakness of the entire structure. Complete failure may take place eventually. Even though such conditions may not lead to ultimate failure, continual maintenance and remedial measures will be required, even though the quantity of water flowing through the embankment is relatively small. Flattening of the downstream face of the embankment may tend to alleviate a potentially dangerous situation, but a simpler and more effective solution will result from the use of drains. The function of the drains is to lower the saturation line so that seepage does not appear on the downstream face, but is controlled and conducted safely to the desired points of emergence. Longitudinal interior drains, toe drains, and drainage blankets may be used to control seepage. Figure 70 ② illustrates the use of a toe drain for this purpose, while figure 70 ③ represents the use of a drainage blanket.

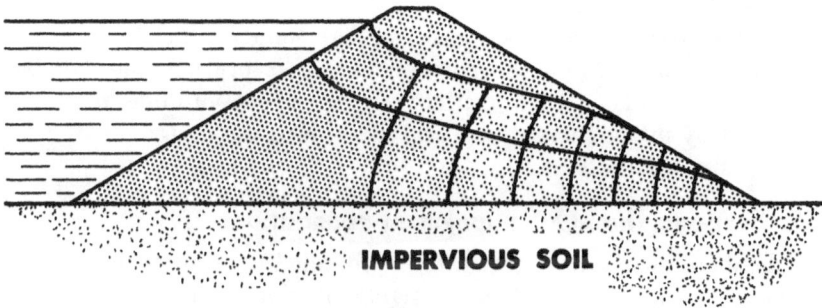

① Simple section

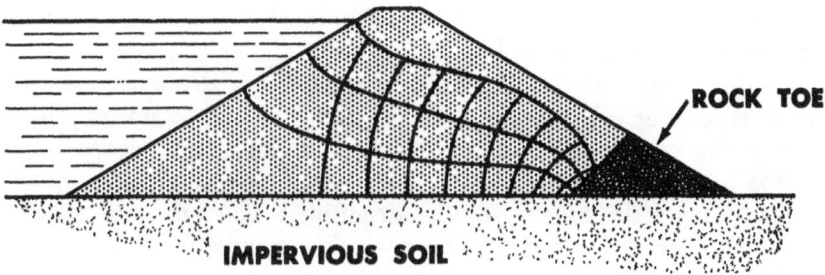

② Use of toe drain to control seepage

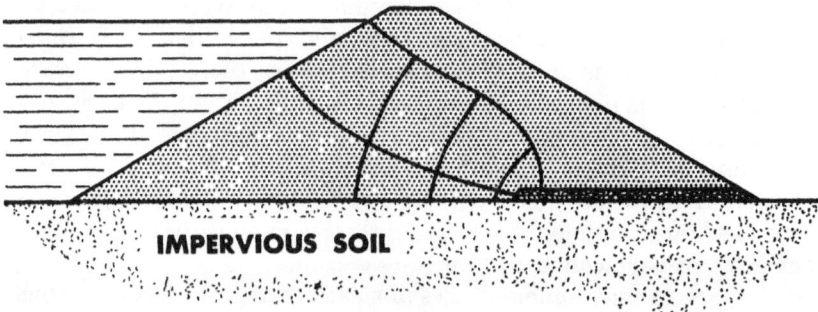

③ Use of drainage blanket to control seepage

Figure 70. Flow nets for homogeneous embankments on impervious foundations.

*c*. The introduction of drainage facilities into a structure creates a problem of potential internal erosion of soil from the dam or foundation. It is necessary that drains be designed with protective filters of granular material sufficiently pervious to permit free flow of water, and properly graded to filter out fine soil particles which would impregnate and eventually clog the filter and drain. Requirements relative to filter materials have been given previously (par. 93).

*d*. Where large quantities of free-draining granular material are available, it is frequently practical to build dams or levees with large pervious zones in conjunction with more impervious zones. Such a design may not only provide for economical use of available soils, but provides an effective means of controlling seepage. A design of this type is illustrated in figure 77.

*e*. If the foundation is pervious, so that some of the flow passes through the foundation, the position of the saturation line will be affected materially. The position of the saturation line may be controlled by introducing drains and filters into the design, or by the use of composite sections.

## 272. Seepage Through Embankment Foundations

*a*. Embankments may frequently be founded upon pervious strata, such as an alluvial deposit of relatively pervious sands and gravels at or near the surface, with an impervious layer at greater depth. If the foundation is many times as pervious as the structure, the construction of a flow net may be simplified by considering flow through the foundation only. If the permeability approaches that of the dam, then the combined flow through both elements must be considered. For all cases, it must be recognized that uncontrolled outcropping of seepage and the development of uplift pressures at the toe are conducive to piping, boils, and quick conditions which may endanger the safety of the structure. The effect of stratification is important, and flow nets which are drawn for these conditions must take this factor into consideration.

*b*. Unstable conditions due to seepage through an embankment foundation exist when subsurface flow has a relatively free inlet upstream or riverside of a dam or levee, and the outlet is restricted downstream or landside of the structure. Since the vertical permeability of stratified deposits is likely to be much less than the horizontal permeability, an important factor tending to produce detrimental restriction of outflow is natural stratification. Natural pervious deposits can rarely be considered homogeneous with respect to permeability, and are commonly stratified to the extent that potentially dangerous seepage concentration can develop.

*c*. Important and common cases of stratification are those in which pervious sands and gravels are overlain by a continuous, relatively impervious, natural blanket. The greatest uplift pressure would be created beneath the top stratum if it is assumed that no outflow or seepage can occur downstream through the blanket, and there are no other sources of natural pressure relief. Then there can be no loss in head through the pervious foundation and the full pressure head is exerted against the blanket. In such a case, and in order to prevent rupture, boils, or piping, the thickness of the blanket must be such that its effective weight is equal to or greater than the total head. Normally, it is impossible for the full pressure head to be exerted against the blanket, since pressure relief is realized from seepage through the blanket, by lateral flow to natural channels, breaks in the blanket, or other sources. Thus, an impervious blanket, to be effective in preventing loss of stability, may require an effective weight of only 50 to 75 percent of the total head. Full advantage of natural blankets can be taken for control of seepage under relatively low heads. For higher heads, the usual depth of impervious blanket can not be depended upon to confine and prevent outcropping of seepage. If the blanket is continuous and extends upstream or riverward for a considerable distance, much benefit in the reduction of seepage can result, provided that the blanket is not disrupted by borrow excavation. Artificially constructed impervious blankets are used for the control of seepage, as indicated in paragraphs 273 through 279.

## Section III. CONTROL OF SEEPAGE

### 273. General

In this section are described various methods for the control of seepage through and beneath water-retaining structures. Some of these have been mentioned previously and additional information is presented here. The choice of the type of seepage control to be used in a given case is dependent on a number of factors, most important of which is the character of the foundation. An adequate exploration program is necessary to insure that the method adopted is suited to local conditions and will fulfill the intended purpose. In many cases a combination of methods will be best. This is particularly true when seepage through an embankment and underseepage are interrelated, since a solution which is satisfactory for the embankment may not necessarily provide safety against foundation underseepage. Availability of materials and cost must be given consideration in selecting the method to be used.

### 274. Cutoffs

a. Whenever it is economically feasible seepage should be cut off. Two ways of doing this are shown in figure 71. It should be realized that no cutoff is completely impervious, and that the reduction of seepage is a relative matter.

*b.* Seepage can be cut off by connecting an impervious portion of the structure with an impervious stratum in the foundation by backfilling a cutoff trench with an impervious soil. In some cases the impervious layer of the foundation may be sandwiched in between an upper and lower pervious layer; a cutoff to such an impervious layer would reduce seepage only through the upper pervious layer. However, where the combined thickness of the impervious layer and the upper pervious layer is sufficient, they may be able to resist the upward seepage pressures existing in the lower pervious layer, and thus remain stable.

*c.* In cases where earth cutoffs are not practical a cutoff may be obtained by the use of sheet piling. Steel sheet piling is generally preferred because of its high strength. To be effective, the pile cutoff should be as tight and continuous as possible. Steel sheet piling is not completely watertight, since leakage may occur at the interlocks, splices, and through torn interlocks. The more pervious the soil the more effective is the piling in reducing seepage. The longer the length of the seepage path the less effective is piling, as compared with other methods. Steel piling is susceptible to cor-

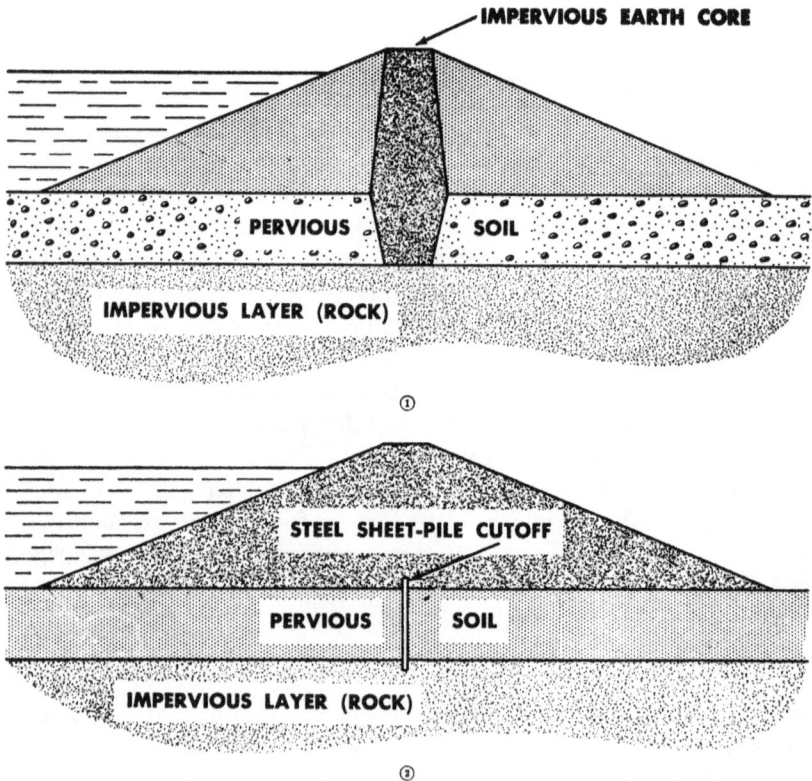

*Figure 71. Seepage cutoffs.*

rosion, although this should not be a matter of concern except in permanent structures.

## 275. Drainage Blankets and Horizontal Drains

Drainage blankets have previously been described in relation to the control of seepage through embankments. This type of control is also useful in controlling underseepage when the structure is built on fairly homogeneous sands and gravels. The blanket will prevent a concentrated flow of water at the toe and direct the flow to a controlling outlet. Horizontal interior drains offer about the same advantages and are feasible in cases in which a pervious, reasonably homogeneous stratum is overlain by an impervious top stratum of moderate depth; the drain must extend through the top layer and contact the pervious layer. When seepage through the structure is simultaneous with underseepage, control of both may be accomplished by filters and drains. However, even moderate stratification of the structure or the foundation will decrease the efficiency of blankets and drains. When pronounced stratification is apparent, seepage may bypass drains and emerge in the vicinity of the toe, or even some distance beyond the toe, thus defeating the purpose of the drain. For such cases other methods of seepage interception may prove more reliable and efficient.

## 276. Pressure Relief Wells

Pressure relief wells may be used to relieve foundation seepage pressures. Wells of this sort can be sunk to penetrate the pervious layer, including any minor impervious layers, thus providing an additional means of seepage escape and reducing the uplift pressures in the vicinity of the wells. These relief wells are nothing more than controlled artificial springs which prevent removal of soil, such as occurs in natural boils. The design of such well installations is beyond the scope of this manual. A group of wells connected to a common header pipe, or with each well being pumped individually, is often used to lower the ground-water level for construction purposes, as indicated in paragraph 330.

## 277. Impervious Upstream Blankets

Impervious upstream or riverside blankets which overlie a pervious foundation are effective in reducing the quantity of seepage, and to some extent will reduce uplift pressures and escape gradients downstream or landside. The benefits derived from the blanket are due to the dissipation of a part of the reservoir or river head through the blanket, the proportion being dependent upon the thickness, length, and effective permeability coefficient of the blanket. The construction of impervious blankets is particularly adaptable to the treatment of exposed areas adjacent to the structure, especially if excess impervious material which is unsuitable for embankment construction is available. If a natural blanket exists, full advantage should be taken of this fact, and borrow operations should be controlled to prevent excessive excavation of the natural impervious top blanket. Conversely, reinforcement of areas which are thin and weak may sometimes be justified. Blankets frequently will give adequate control of seepage for structures which are subjected to low heads; for higher heads, they should be supplemented by downstream drainage systems. Seepage sometimes can be controlled by "puddling," which means agitating the soil and water in the reservoir in order to cause a filter blanket to be deposited on top of the pervious soil in the bottom of the reservoir.

## 278. Landside Berms

On levee projects, in particular, landside berms are sometimes used to reinforce the landside portion of the levee. These berms generally are built up to the height at which seepage breaks out on the landside slope of the levee; the berm then extends landward for some distance with diminishing thickness. The berm serves to reinforce the landside area to resist upward seepage pressure by its weight. It also forces the seepage to travel a greater distance before emerging. Boils that may develop are then farther from the levee slope. The berm is a potential source of borrow which may not be completely saturated and may be valuable for emergency repairs to the levee. A landside berm is illustrated in figure 76 ①.

## 279. Sublevees

Sublevees are levees of lesser height set landward of the main levee. In some cases a sublevee may enclose a single weak area back of the main levee. A sandbag ring (fig. 76 ②) may be used for this purpose, particularly under emergency conditions. In other cases, the main levee may be connected to a main sublevee by a number of connecting levees to form separate pool areas. The pool areas are permitted to flood, either by seepage inflow or by pumping or siphoning from the river. The purpose of the sublevees is to provide a landside pool to counterbalance partially the upward seepage pressure.

# Section IV.  STABILITY OF EARTH MASSES

## 280. General Considerations

*a.* Cut slopes, embankments, and similar earth masses, and the underlying foundations of embankments are subject to shear stresses which are caused by the action of gravity. Shear stresses may also be caused by seepage and outside forces, such as earthquakes. It is essential that earth masses of this general character remain stable and retain an unchanged position, and that they not undergo detrimental settlement or distortion. The analysis of the stability of a deep cut, a high fill, or a large earth dam may be exceedingly complex. Only the general nature of the problems involved, and stability analyses which may be made, can be presented here. However, definite suggestions for the design of shallow cuts, low embankments, and low dams and dikes are given in the sections which follow.

*b.* The discussion which is contained in this chapter deals with unretained slopes. Unretained slopes are not supported or contained by any structure, other than the soil itself. They are dependent for stability upon the strength of the soil comprising the slope and the underlying material. Unretained slopes may be created by geologic means, as in natural riverbanks and hills, or by the construction of cuts and fills. Failure of an unretained slope may result when either the soil within the slope itself or the soil within the foundation is overstressed, due to increased loading or decreased strength.

*c.* As indicated above, the structural strength of the foundation will frequently govern the design of the cross section of an embankment. Additional discussion relative to embankment foundations is given in paragraph 291.

*d.* Retained slopes are discussed in chapter 12.

## 281. Slope Failures, Cohesive Soils

*a.* A failure in a relatively homogeneous cohesive soil often occurs in which the unstable portion of the mass is separated from the stable portion by a failure surface, or a narrow failure zone. The shape of the failure surface is approximately circular. The general appearance of a failure surface of this type is shown in figure 72. A failure in nonhomogeneous soils, especially if they are highly stratified, may be

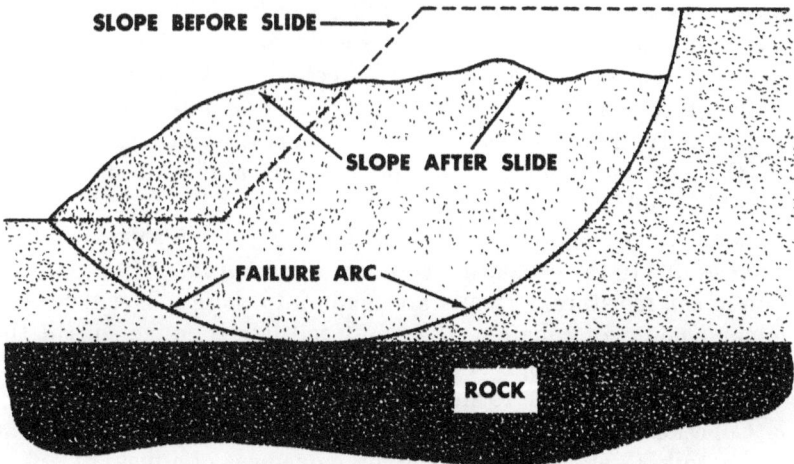

*Figure 72. Deep-seated slide in homogeneous soil.*

much more complex, and the failure surface or zone may approximate a combination of circular arcs and planes.

*b.* In homogeneous cohesive soils a failure surface generally emerges at the toe of the slope, if the angle of the slope with the horizontal is more than 53°. If the slope is flatter, the failure surface usually will emerge beyond the toe. The point at which it emerges will be influenced by any firmer material which may be present in the foundation. If a firm layer is present in the slope, or slightly beneath a horizontal plane through the toe of the slope, the failure surface may emerge above the toe. Failures of these types are, in general, known as *deep-seated slides*. The character of the slide will depend upon the type of soil present.

*c.* In soft, plastic, unfissured clays the sliding mass deforms in a progressive manner. If the clay is very soft there may not be a well-defined failure surface, but instead a zone of plastic movement in the soil. Even on very flat slopes, this movement may take place in certain soils and is called *creep*. Creep can be very injurious to structures which are founded upon a slope of this sort. Brittle, unfissured clays tend to move as a mass and will generally have a well-defined failure surface. Because of their sensitivity, clays of this type frequently experience a very marked reduction in strength in any overloaded zone. The failure will usually be progressive, although at times it may occur suddenly and with very little warning. Slides in fissured clays are difficult to analyze and predict, and may occur long after construction is complete due to progressive softening.

## 282. Slope Failures, Cohesionless Soils

*a.* In general, homogeneous cohesionless soils which are not subject to seepage are not susceptible to failure by deep-seated slides. If the slope angle is greater than the angle of internal friction of a cohesionless soil, the surface will fail by sliding parallel to the slope. The angle of internal friction concerned should be that of the soil in a loose condition, since the surface of a cohesionless soil mass is frequently quite loose. This value of the angle of internal friction is approximately the same as the *angle of repose*, which is the angle with the horizontal formed by the sides of a pile of dry, loose, cohesionless soil.

*b.* Some essentially cohesionless soils may exhibit some cohesive properties due to the cementing action of lime or iron, or the capillary action of soil moisture. If cuts are made into these soils which have a minor amount of cohesion, they may fail like cohesive soils if the excavated slope is too high or steep. If the small amount of cohesion is destroyed by exposure to the air, the failure will be superficial.

*c.* Cohesionless soils which are subjected to seepage may fail like cohesive soils—a slide failure involving movement of a mass of soil.

*d.* The effects of volume change on the shearing strength of loose cohesionless soils have been described previously (par. 134). The temporary liquefaction of fine-grained cohesionless soils may result in a flow failure, which may take place very quickly and result in a very flat final slope.

## 283. Stability Analyses

*a.* Detailed stability analyses of earth slopes are beyond the scope of this manual. Only the general nature of such analyses will be indicated here.

*b.* Most methods of slope stability analyses involve the assumption of a failure surface. In the majority of cases, the failure surface is assumed to be circular in cross section and infinite in width. Methods of analysis then involve the determination of forces acting upon the soil mass above an assumed sliding surface, and equating the forces which tend to produce rotation of the mass to those which tend to resist movement. Forces involved usually are the weight of the soil mass, seepage or water pressure forces, and the strength of the soil (friction and cohesion) along the sliding surface. Factors of safety are determined by assuming the soil mass to be in equilibrium, computing the soil strengths required for this condition, and comparing them with the available soil strengths. Another approach to the analysis is to use the design strengths of the soil in the analysis, and then compare the forces tending to resist movement with those which tend to cause movement. In other cases, the moments tending to cause movement may be compared with the resisting moments. In any case, it is apparent that an infinite number of trial failure surfaces could be investigated. Thus an important part of the analysis is the selection of the "critical" failure surface, the one which has the lowest factor of safety. Considerable information is available in soil mechanics literature to aid in the selection of the critical surface, particularly for homogeneous soils.

*c.* The general nature of slope stability analyses is indicated in figure 73. This represents a trial circular arc failure surface for a homogeneous soil mass which is not subjected to seepage. The soil has constant shearing strength, $s=c$, along the failure arc. In this case the resisting moment about the center of the circular arc is simply, $M_r=cRL$, where $c$ is the cohesion, $R$ is the radius of the circular arc, and $L$ is the length of the arc. The only force tending to cause movement is the weight of the soil mass, $W$, and the moment tending to cause failure, $M_f=Wx$, where $x$ is the perpendicular distance from the line of action of $W$ to the center. The factor of safety for this particular arc then would be F. of S.$=M_r/M_f$. The arc shown may not be the critical one. If seepage is present, then seepage forces must be included in the analysis; this normally is done by the use of a flow net. In case that the factor of safety is not ample to insure satisfactory performance of the structure, the design would be changed.

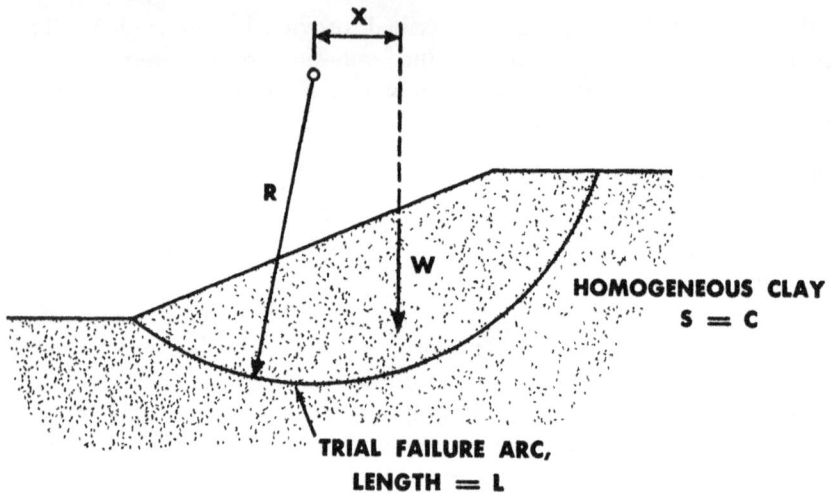

MOMENT TENDING TO CAUSE FAILURE, $M_F = Wx$

RESISTING MOMENT, $M_R = CRL$

*Figure 73.   Circular arc analysis of a slope failure in a homogeneous clay soil.*

## Section V.  CUT SLOPES

### 284. Standard Slopes

*a.* The discussion in this section is principally concerned with open cuts made as a part of a highway or railroad location.  It may also be applicable to wide, open cuts which are made for the purpose of obtaining borrow excavation or, occasionally, for structures.  Excavations which require bracing are discussed in chapter 12.

*b.* The standard slope for highway cuts in most soils is 1½ (horizontal) to 1 (vertical).  This slope should be satisfactory for cuts in most soils if the depth of the cut does not exceed about 20 feet. Because of erosion, slopes in sandy or loamy soils are sometimes reduced to 2 to 1, or even 3 to 1 in unusual cases.  Many cuts deeper than 20 feet behave satisfactorily with slopes of 1½ to 1 where favorable soils exist.  In deeper cuts in questionable soils the slope should be designed upon the basis of experience, a stability analysis, or the excavation of trial slopes.  Cuts in unfavorable soils are briefly discussed in paragraph 385.

*c.* Slopes steeper than the standard should be planned only in rock, dense sandy soil which is interspersed with boulders, and in loess. Emergency conditions may sometimes force the use of steeper slopes in any soil.  Steeper slopes should be excavated with the full knowledge that they may be subject to disastrous slides or may require very heavy maintenance.

## 285. Slopes in Difficult Soils

*a. General.* Cohesive sandy or gravelly soils, or cohesionless soils which are dry or moist, constitute favorable soils for the use of standard slopes. Certain other soils present difficulties and may require flatter slopes, extensive drainage treatments, or the construction of retaining walls or similar structures. These troublesome soils should be avoided whenever it is feasible to do so in the interest of safety and economy.

*b. Loose Sand.* Difficulties which may be presented by loose, saturated sands have been emphasized previously. Slides which occur in these deposits are caused by liquefaction and may be produced by a shock, or rapid change in the water table. In some situations it may be feasible to increase the density of a very loose sand deposit and eliminate the danger of a slide.

*c. Loess.* True loess which is located above the water table is a stable material, except that it is highly susceptible to erosion. For this reason, highway cuts in loess frequently are made vertical. Vertical slopes will be stable over a long period of time. However, in most situations it is desirable to make the bottom of the cut somewhat wider than normal, since there will be occasional falls of material as time passes. A vertical slice will fall and leave a vertical surface which again will be stable. Slides of this type in loess are generally due to progressive weakening of the material at the toe of the slope, because of percolating rainwater. Loess located below the water table is likely to be highly unstable, because of its high void ratio and the leaching action of water in removing the cementing agents.

*d. Clays.* Medium and dense homogeneous clays usually present no particular problems in stability for shallow cuts. Soft clays, however, cannot be expected to stand at the standard slope of 1½ to 1 to heights of more than 10 feet, or even less. Saturated sensitive clays, fissured clays, and clays which contain lenses or layers of water-bearing silt or sand can not be expected to be stable at the standard slope and should be treated as special cases.

*e. Detritus.* It is sometimes necessary to cut slopes in detritus, which is a loose accumulation of relatively sound pieces of rock intermingled with completely weathered ones. Detritus may exist as a blanket covering a gentle slope, or it may occur as talus at the foot of a steep rock cliff. If detritus is dry or well drained, it may be very stable and easily able to withstand slopes as steep as 1 to 1. However, detritus which is derived from some types of rock, particularly some types of schist and laminated shale, is highly unstable, even on very flat slopes. Slides may occur with little or no warning when these soils become saturated. Difficulties encountered in these soils may be prevented by adequate drainage. A deep drain may be placed at the top of the slope in order to cut off the water and prevent saturation.

## 286. Surface Protection

The surface of earth slopes must normally be protected against the erosive effects of surface water. The selection of the type of protection to be used depends upon the soil, climatic conditions, and condition of exposure.

*a. Turf Culture.* One of the easiest and most effective ways of reducing erosion on cut slopes is through the cultivation and development of a firm turf. Every effort should be made in areas which are subject to erosion, except under emergency conditions, to encourage the growth of native grasses on exposed slopes. Formation of a firm turf may be accomplished by seeding, sprigging, or sodding the slope with suitable native grasses. The details of selecting, planting, and cultivating these grasses will not be presented here. Reference should be made to TM 5–630 for a complete discussion of the subject. In general, seeding will not be too successful on slopes which are steeper than about 2 to 1, and these slopes must be sodded. In most cases erosion may be successfully controlled by turf culture in areas of moderate rainfall and on moderate slopes.

*b. Intercepting Ditch.* Under expedient conditions or in areas of easily eroded soils or severe rainfall, erosion of cut slopes may be successfully controlled by the construction of an intercepting ditch at the top of the slope. The purpose of the ditch is to intercept surface water. The water which is collected must be carried along the ditch and discharged down the slope at selected intervals into a

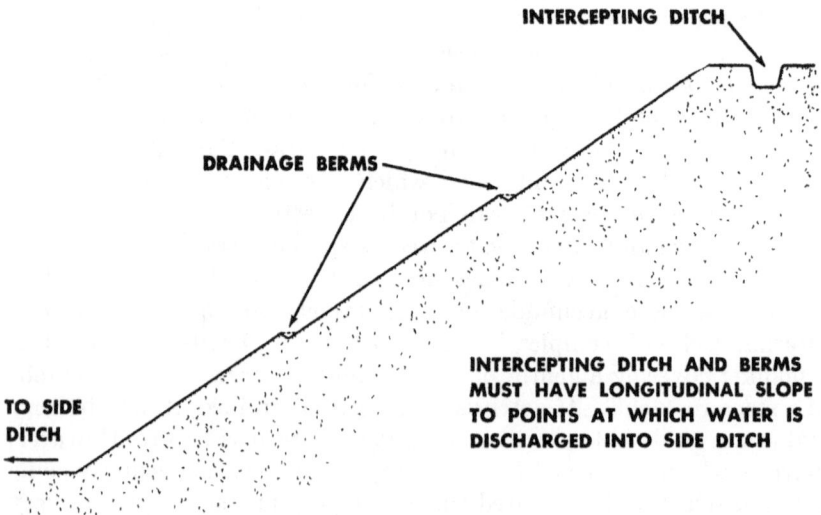

INTERCEPTING DITCH

DRAINAGE BERMS

TO SIDE DITCH

INTERCEPTING DITCH AND BERMS MUST HAVE LONGITUDINAL SLOPE TO POINTS AT WHICH WATER IS DISCHARGED INTO SIDE DITCH

*Figure 74. Intercepting ditch and drainage berms used to control erosion on a cut slope.*

side ditch. In deep cuts or under severe erosion conditions, the intercepting ditch may be supplemented by the use of drainage berms, as indicated in figure 74.

*c. Severe Erosion Conditions.* In some areas measures described above may not be sufficient to control erosion on a steep cut slope. This is particularly true in mountainous regions of heavy rainfall, where slopes may frequently be 1½ to 1, 1 to 1, or even steeper in the interest of economy. In such situations, the slope may have to be protected by the placing of rip-rap, or by the use of retaining walls or cribbing for short distances.

## Section VI.   EMBANKMENTS

### 287. General

This section is principally concerned with rolled-earth fills which are used for the support of highways. The material is also generally applicable to railroad fills, although railroad fills frequently are not as carefully compacted as highway fills, since rough surface may be prevented by proper maintenance of the ballast. This discussion is also primarily concerned with fills which are less than 50 feet high, and which are not subjected to prolonged innundation. Hydraulic fills are beyond the scope of this manual, since the equipment necessary for their construction normally would not be available to the military engineer in the field.

### 288. Cross Section

Standard slopes used in embankments vary from 1½ to 1, to 3 to 1, depending upon the soil used, height of fill, and the condition of exposure. Information which may be used as a guide in the selection of slopes to be used under given conditions is contained in table XII, appendix II, which is based upon civil practice in the design of highway fills. The width of the top of the embankment will depend upon the design of the road. In ideal conditions, the minimum width of the embankment at the top would be about 40 feet, which would provide for two 12-foot lanes and two 8-foot shoulders. Lesser widths may be both feasible and necessary.

### 289. Materials

The suitability of materials for use in embankments is indicated in column (7) of table VII, appendix II. Reference is also made to paragraph 215.

### 290. Construction

The construction of rolled-earth fills is covered in detail in chapter 9.

## 291. Embankment Foundations

*a.* As has been indicated, an embankment may fail because of failure of the soil upon which it rests. The stability of an embankment foundation is fully as important as the stability of the body of the embankment itself, since it is obviously of little avail to construct a stable, well-compacted embankment, if the foundation can not support the load.

*b.* Three principal types of failure may occur in embankment foundations. One of these is a failure by shear, which may mean that a deep-seated slide takes place in the foundation, which also may involve the embankment, or that the soil simply flows laterally, allowing the embankment to subside. In this type of failure the bearing capacity of the soil is exceeded by the weight of the embankment. Approximate methods given in chapter 8 may be used to estimate the possibility of this occurence, if the weak layer of soil is at least half as thick as the embankment is wide. Otherwise, the analysis must be based upon the assumption of a failure surface. Another way in which an embankment may fail is through excessive settlement. In such a case the embankment may remain structurally intact, but excessive settlement due to consolidation may cause the failure of a permanent pavement placed on the fill. Both of these types of failure are most likely to occur when major fills must be constructed over weak soils, such as the highly organic soils which may be encountered in a swamp or soft, compressible clay layers. Remedies which are suggested in paragraph 292 may generally apply to weak soils which may be subject to either one of these types of failures.

*c.* The third type of failure which may occur is that due to piping, which has been discussed previously. This type of failure is usually a possibility, if the fill must be founded over thin cohesionless strata which may be subject to an increase in neutral stress (water pressure). This occurrence is not a usual one in relation to highway fills but, if such a situation does exist, seepage control measures described in paragraphs 273 through 279 may be necessary.

## 292. Special Treatment of Embankment Foundations

*a. General.* Several general measures may be undertaken to permit the construction of an embankment over questionable soils. One obvious step is to flatten the slopes, thus spreading the load over a larger area. Another method which may be useful is to construct a gravel blanket or berm at the toe of the embankment slope, in order to provide additional weight to aid in the confinement of the underlying soil and prevent a lateral bulge from occurring. The fill may be constructed very slowly, so that the underlying soil can consolidate as the load is applied. This method is frequently advantageous if time permits, since the shearing strength of most soils increases with consolidation.

*b. Gravity Subsidence.* In some instances, a fill may simply be placed upon the surface of an unsatisfactory foundation soil and allowed to settle as it will, with no special treatment of the underlying soil. This method is obviously unsuitable for anything except emergency conditions, or for a very lightly traveled road, since the fill may settle rapidly during or shortly after construction, and settlement due to consolidation may continue for a long time. In either case, the satisfactory placing of a wearing surface on the fill is virtually impossible, and a long period of maintenance must be anticipated, as more and more soil must be added to the fill as time passes, until the embankment eventually comes to a stable condition.

*c. Total or Partial Excavation.* If the unsuitable soil is of shallow depth, the most common and satisfactory solution to the problem is simply to excavate the unsuitable soil and replace it with a suitable granular soil as indicated in figure 75. This method is very economical if the depth of the undesirable soil is less than 5 or 6 feet. Under expedient conditions, excavation and replacement to a depth of 10, or even 12, feet may be feasible in some situations. As the unsuitable material is removed, suitable material is dumped in to take its place. Conditions are frequently favorable to the use of a drag line in these operations. If the undesirable soil is not too deep, a solution may be achieved by partially excavating the unsuitable material and backfilling. This method should be used with caution, since the removal of a shallow portion of a deep, weak soil will not necessarily insure against a shear failure or excessive settlement.

*d. Displacement.* As indicated in *b* above, the undesirable soil may simply be displaced by the weight of the embankment. The process of displacement may be accelerated by blasting or jetting. These methods are sometimes used when the unsuitable soil has a depth up to about 20 feet, or a little more. When explosives are used, the embankment is built to an elevation considerably higher than the finished grade, is possible. Then deep charges of explosives are detonated in order to displace some of the underlying soil and liquefy the remainder, thus permitting it to be displaced more readily by the weight of the fill. Jets of water may be pumped into the underlying soil in a separate, somewhat similar, method to liquefy the soil and accelerate displacement. Both methods are effective if properly done, but require experienced personnel.

*e. Vertical Sand Drains.* A very effective approach to the problem of settlement of an embankment founded on a compressible soil is the use of vertical sand drains. In this method vertical holes, generally from 18 to 24 inches in diameter and spaced from 15 to 20 feet center-to-center, are driven through the compressible soil and backfilled with sand. The fill is then built at a controlled rate. The drains

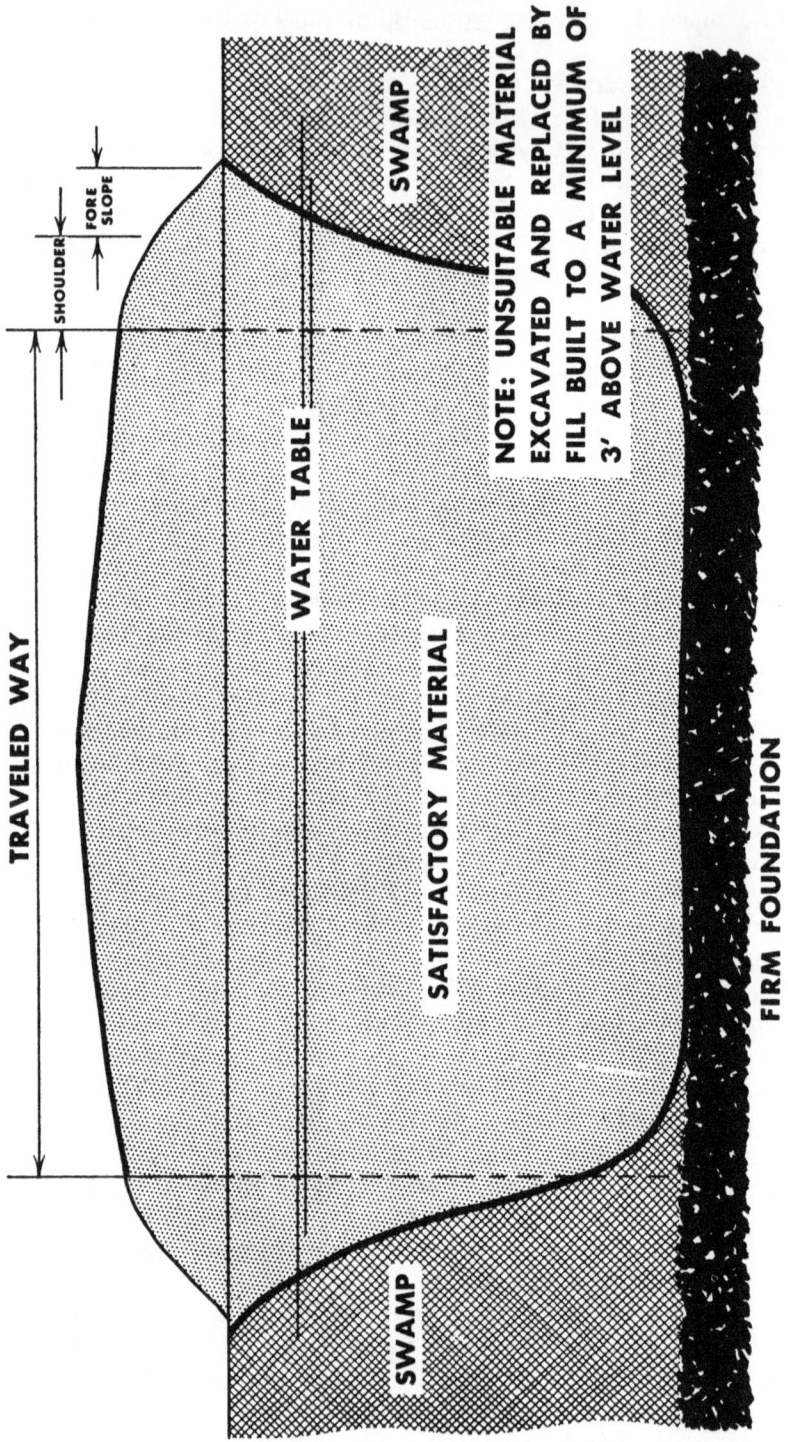

TRAVELED WAY

SHOULDER

FORE SLOPE

SWAMP

SWAMP

WATER TABLE

SATISFACTORY MATERIAL

NOTE: UNSUITABLE MATERIAL EXCAVATED AND REPLACED BY FILL BUILT TO A MINIMUM OF 3' ABOVE WATER LEVEL

FIRM FOUNDATION

Figure 75. Typical cross section of fills in swampy areas.

provide additional paths for the water to escape as it is forced from the voids of the soil and greatly accelerate consolidation. The method requires careful supervision and elaborate construction equipment.

## 293. Surface Protection

*a.* As with cut slopes, embankment slopes may require protection from the erosive effects of surface water.

*b.* Turf is very effective in controlling erosion on embankment slopes in areas of moderate rainfall (par. 286).

*c.* Drainage dikes and gutters constructed along outside shoulder lines are effective in protecting embankment slopes from erosion. The dikes are low, being generally from 4 to 8 inches in height and of trapezoidal or rounded cross section. Simple earth dikes are used, as are dikes made wholly or partially of a bituminous plant mixture. Surface water is confined inside the dikes and then is discharged down the slope at selected intervals into the side ditch. Shallow gutters may be used in the same fashion as dikes. Both work best when the shoulder is surfaced.

*d.* Elaborate protective devices may be needed under severe exposure conditions. For example, in a sidehill location which parallels a river, rip-rap, a paved revetment, steel sheet or timber piling, or cribbing may be required to protect the bank against erosion, particularly during flood stages.

## Section VII.  EARTH DIKES AND DAMS

## 294. General

This section is concerned with levees, which are comparatively small, long, earth dikes which serve to protect low-lying land against periodic inundation because of floods, high water, or high tides; and with small earth dams. The principal use of the earth dams described in this section is in the storage of water needed for military purposes. As has been emphasized in previous discussions, the design of large earth dams is beyond the scope of this manual. Understanding of the principles which are stressed in this section and in previous discussions will permit the satisfactory design and construction of small earth dams, although there is no substitute for experience in this field.

## 295. Levees

*a.* Levees, as commonly constructed, vary from earth dams in several important respects. First, they are frequently called upon to retain water during only a portion of a year. Second, their location is determined by the necessity of fulfilling their primary function, regardless of foundation conditions; they are frequently founded upon very weak soils. Another difference is that levees are usually built from materials available by shallow excavation near the site, while earth dams are frequently constructed of select materials which may be hauled a considerable distance. Finally, settlement of a levee is generally not an important factor.

*b.* In the United States, levees are commonly built with flat slopes and from materials easily available, which are frequently poor soils. Common levee sections have slopes of 3 to 1 on the land side and 6 to 1 on the river side. Steeper slopes may be used where good materials are available, the foundation is suitable, the levee is carefully compacted or is in a restricted space. The slopes are usually protected against erosion by turf cultivation, although more elaborate methods may be required in critical situations.

*c.* In some circumstances, particularly when the levee is founded on pervious soils, protection against boiling may be necessary. A landside berm (par. 278) or ring levee (par. 279) may be used for this purpose. A sandbag ring levee is illustrated in figure 76.

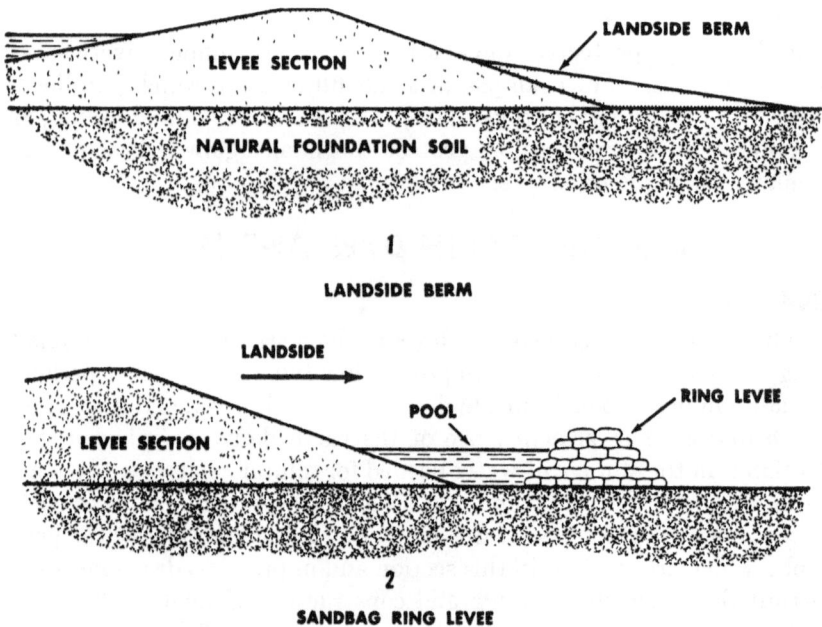

*Figure 76. Measures used to protect levees against the effects of seepage.*

*d.* Levees in the United States are frequently constructed by a dragline working on the top of the completed levee section. Borrow must then be obtained from areas within the working range of the dragline. Little or no compaction is required, other than that obtained by the passage of equipment on the top of the fill. The width of the top is frequently established by construction requirements.

## 296. Earth Dams

Earth dams are very widely used to create storage reservoirs, which may be required for water supply, irrigation, flood control, hydroelectric power, or a combination of purposes. They are frequently the most economical type for use in a given situation, particularly when located on an earth foundation. Masonry dams are also widely used, particularly for greater heights.

## 297. Location of Earth Dams

The proper location of an earth dam for storage purposes frequently involves a long and exhaustive investigation which is beyond the scope of this manual. Where water supply is the principal objective, consideration must be given to the water-producing characteristics of the watershed and their relation to the desired capacity of the reservoir. The dam must be high enough to provide the desired storage capacity, it should be as close as possible to the area to be served in order to reduce the cost of conduits for transmitting the water, and the reservoir should be high enough to permit gravity flow of the water to the area to be served. Topography is obviously an important factor, and careful and exhaustive level and topographic surveys are frequently necessary. In an ideal situation, the reservoir would be located in a place where a valley forms a comparatively broad, level area bounded by steep hills, with a narrow downstream and providing a good dam site. The geology of the site is important, since more than one dam has been built in an area underlaid by subterranean sinks or channels, which carried away the water behind the dam faster than it could be impounded. Detailed information about the geology of water supply is contained in TM 5–545. Ideally, the reservoir site should be underlain by an impervious layer at shallow depth. An adequate soil exploration program of the dam site is particularly important. Adequate sources of good fill material may be an important consideration. Swamp locations are not generally desirable for reservoirs, since the organic matter may have an undesirable effect on the quality of water, particularly during periods of low water level.

## 298. Design of Earth Dams

a. Principal elements of an earth dam for storage generally are the embankment section, the spillway, and outlet works. Most of the discussion presented here is concerned with the embankment section itself, with only brief mention being made of spillways and outlets. It must be emphasized that overtopping is a common cause of failure in earth dams, and that the provision of adequate spillway capacity is absolutely essential.

b. Principal items of concern in the design of an embankment include the stability of the embankment itself, stability of the foundation, and the control of seepage. Information presented previously is applicable to earth dams; some additional facts are contained in this section. Also discussed are freeboard requirements, crown widths, and surface protection.

## 299. Embankment Section

*a.* Stability of the embankment section may be investigated by the methods discussed in paragraphs 280 through 283.

*b.* There are two basic types of earth dam sections, i. e. homogeneous and nonhomogeneous sections. A homogeneous section may be used when only one material is readily available; steps must be taken to control seepage through the downstream face of the dam. If various types of soils are available, they may be utilized effectively in a nonhomogeneous section, such as those shown in figure 77. The impervious material is used to reduce seepage and the pervious

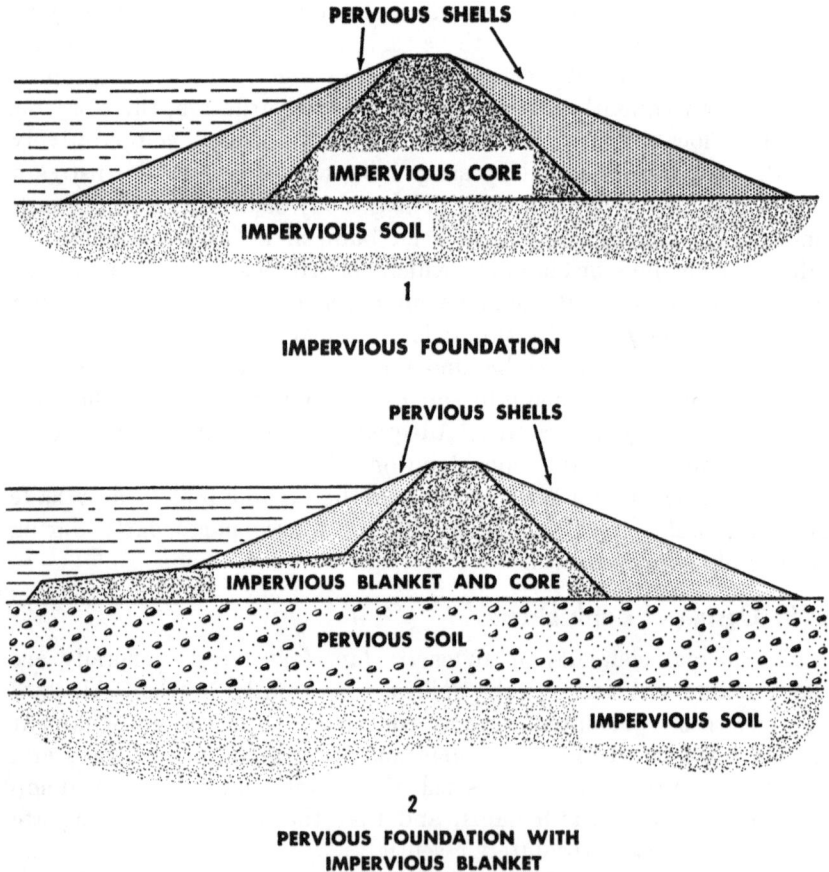

PERVIOUS SHELLS

IMPERVIOUS CORE

IMPERVIOUS SOIL

1

IMPERVIOUS FOUNDATION

PERVIOUS SHELLS

IMPERVIOUS BLANKET AND CORE

PERVIOUS SOIL

IMPERVIOUS SOIL

2

PERVIOUS FOUNDATION WITH
IMPERVIOUS BLANKET

*Figure 77.   Nonhomogeneous earth dam sections.*

material for stability and drainage. Generally, the impervious material is placed in a central core, with pervious materials used in the outer sections or "shells," ranging from the least pervious at the core to the most pervious at the outer faces.

*c.* It is undesirable to generalize about slopes for large earth dams, because of the many factors involved. On small dams, downstream slopes frequently are from 2 to 1 to 2½ to 1; on upstream slopes, from 2½ to 1, to 3 to 1. Flatter slopes may be needed if fine-grained soils are used in a homogeneous section, if the foundation is weak, or for other reasons.

*d.* Because of overtopping, the vertical distance between the top of the water in the reservoir and the top of the dam, or freeboard, is important. For small dams, a distance of 5 feet above the maximum height of water to be expected over the spillway should be adequate, although this distance may be increased under severe exposure conditions, particularly for a large reservoir in an area where high winds occur.

*e.* The top width (crown width) of an earth dam should not be less than 8 or 10 feet because of maintenance requirements. It may be much wider, if the crown is to be used as a roadway, or if it is anticipated that wave erosion will take place during flood stage in the reservoir.

*f.* Slope protection is required on an earth dam to protect the upstream face against erosion. Common practice calls for the use of rip-rap from the level which corresponds to low water elevation in the reservoir to a few feet above normal maximum water surface. Both dumped and hand-placed rip-rap are used, and must be protected by a filter layer, if they are placed against a fine-grained soil. The thickness of rip-rap varies from 18 to 60 inches for dumped rock, depending on exposure, and from 12 to 30 inches for hand-placed rock. Hand-placed rock should be compactly placed with the longest dimensions of the stones perpendicular to the face. Turf may be used in the freeboard section on the upstream slope, and is commonly used on the downstream slope to protect against surface erosion. Rip-rap may be required on the downstream slope, particularly if seepage occurs through the face. Sand bags may be used under expedient conditions.

## 300. Materials

Soils which are available in the area where an earth dam is to be built frequently must be utilized in construction in the interests of economy. If a choice of soils is possible, a more economical and satisfactory design may be achieved. Information relative to the value of the soils of the groups of the Unified Soil Classification System is given in column (7) of table VII, appendix II. The gravelly and sandy soils with little or no fines (GW, GP, SW, SP) are stable, pervious, and readily compacted with suitable equipment. The well-graded materials are a little more desirable than the poorly graded ones, but all are suitable for use in the pervious sections of embankments. Poorly graded sands are a little more difficult to utilize and, in general, should have flatter embankment slopes than the SW soils. The gravels and sands with fines (GM, GC, SM, SC) have variable characteristics, depending on the nature of the finer fraction and the gradation of the entire sample. These materials are often sufficiently stable and impervious to be used for the impervious sections of embankments. Soils in these groups should be carefully examined to be sure that they are properly zoned with relation to other materials in an embankment. Of the fine-grained soils, the CL group is best adapted to embankment construction. The soils of the ML group require careful compaction in the field to attain the desired strengths. CH soils have fair stability when used on flat slopes, but have detrimental shrinkage characteristics which make them difficult to use. The highly organic soils are not suitable for use in embankments, but may be used in applications where strength is not necessary.

### 301. Earth Dam Foundations

*a.* Embankment foundations are discussed in paragraph 291. Statements previously made are applicable to foundations of earth dams as well.

*b.* Information relative to the value for foundations of the groups of the Unified Soil Classification System is contained in column (11) of table VII, appendix II. It is believed that the material contained therein is self-explanatory, particularly in view of the discussion given in chapter 8.

### 302. Control of Seepage

*a.* Information relative to the control of seepage is given in paragraphs 273 through 279.

*b.* Information relating to the control of seepage in embankments constructed from soils of the groups of the Unified Soil Classification System is contained in column (12) of table VII, appendix II. Additional discussion relative to these soils is given below.

*c.* In the control of seepage through embankments, it is the relative permeability of adjacent materials, rather than the actual permeability, which governs their use in a given location. The more impervious soils (GW, GC, SW, SC, CL, CH) may be used in core sections or in homogeneous embankments to retard the flow of water. The coarse-grained, free-draining soils (GW, GP, SW, SP) are best suited for use in pervious shells, when these are required on the downstream slope, or on the upstream slopes where the possibility of drawdown exists. Care should be taken in the arrangement of materials to prevent piping within the section. Dams have been successfully constructed entirely of sand (SW, SP, SM) or of silt (ML), with the section made large enough to reduce seepage to an allowable value without the use of an impervious core. In general, free-draining sands and gravels are preferred for use in drains and toe sections, but a silty sand (SM) may effectively drain a clay and be entirely satisfactory.

*d*. With relation to seepage through dam foundations, the free-draining gravels (GW, GP) may carry considerable quantities of water and some means of positive control, like a cutoff trench, may be needed. Clean sands may be controlled by a cutoff or by an upstream impervious blanket. Slightly less pervious materials, such as GM, SM, or ML soils, may require a minor amount of seepage control, such as given by a toe drain; if they are sufficiently impervious, no control may be necessary. The relatively impervious soils generally require no seepage control, since only small volumes of water pass through them. The degree of control of uplift pressures which may be encountered because of seepage through the foundation is similar to that indicated above. That is, the more pervious soils may require control measures while the more impervious ones do not. Control of uplift pressures should not be applied indiscriminately just because certain soils are encountered. Rather, the use of control measures should be based upon a careful evaluation of existing conditions.

## 303. Construction

The construction of embankments, including compaction, is discussed in chapter 9. This information is applicable to the embankment section of earth dams constructed by rolling. The construction operations required for spillways and outlet works are beyond the scope of this manual.

## 304. Spillways

The provision of adequate spillway capacity for an earth dam is essential to insure against failure of the dam by overtopping. It is very desirable to provide the spillway for an earth dam at a location which is some distance away from the structure, as through a separate saddle in the rim of the reservoir. Experience has shown that putting the spillway channel over the embankment has frequently resulted in failure, and this practice is not recommended in situations where it can be avoided. A trough or chute spillway, which consists of an open conduit which conveys the water from the reservoir to the waterway downstream from the dam, is most commonly used with an earth dam. The spillway is preferably constructed in rock and lined with concrete; concrete founded on earth, and channels and tunnels in unlined rock have also been used. Spillway channels in earth cannot be expected to function satisfactorily, except possibly in emergency situations and for temporary structures of short life. Spillways are best designed with uniform cross sections and straight alinement throughout.

Adequate design of a spillway depends upon an estimate of the maximum flow which must be accommodated and the hydraulic design of the channel, neither of which is discussed here.

## 305. Outlets

Properly designed outlet works must normally be provided as a part of the design of an earth dam in order that the impounded water may be drawn off as desired. It is desirable that outlets be tunneled through the dam abutments as far from the structure as is practical, or located in separate concrete structures which are founded on rock. If this can not be done, the best approach is to locate the outlet conduits in open cut in rock in the dam abutments or foundations. Concrete conduits founded on earth beneath the embankment may present many difficulties and must be carefully designed if they are to function properly. Improperly designed outlet works through the dam have caused numerous failures. If conduits are passed through the embankment, they normally must be designed with gates at the intake end, with provisions for controlling the flow of seepage between the masonry and the earthen material of the dam, and inverse filters to protect the fill where the conduit emerges on the downstream side of the dam. Outlet works such as pipe must be watertight. Pipe should be tested under pressure before back-filling. Both outlets and spillways may require stilling basins or other protective works at points of discharge, in order to avoid the danger of destruction of the embankment because of scour.

# CHAPTER 12

# EARTH-RETAINING STRUCTURES

## Section I. LATERAL EARTH PRESSURE

### 306. Introduction

*a.* This chapter is concerned with the design and construction of earth-retaining structures—structures which must be used to restrain a mass of earth which will not stand unsupported. Such structures are commonly required when a cut is made or an embankment formed with slopes which are too steep to stand alone. Structures included in the discussion are retaining walls, bracing systems used in temporary excavations, and briefly, bulkheads and anchors.

*b.* Earth-retaining structures are commonly subjected to lateral thrust from the earth masses which they support. The pressure of the earth on such a structure is commonly called *lateral earth pressure.* The lateral earth pressure which may be exerted by a given soil on a given structure is a function of many variables. It must be estimated with a reasonable degree of accuracy before an earth-retaining structure may be properly designed. In many cases the lateral earth pressure may be assumed to be acting in a horizontal direction, or nearly so.

*c.* Three types of lateral earth pressure are commonly encountered in the design of engineering structures. These are the *earth pressure at rest* (at-rest pressure), the *active earth pressure* (active pressure), and the *passive earth pressure* (passive pressure).

### 307. At-Rest Pressure

*a.* As has been indicated previously (par. 89), the vertical pressure acting at a point in a horizontal plane which is located $z$ feet below the surface of an extensive mass of soil with a level surface is $p=\gamma z$, where $\gamma$ is the total unit weight of the soil. If an imaginary vertical plane were introduced in this soil mass through the point in question, the lateral (horizontal) earth pressure on the vertical plane at this point would be $p_0=K_0\gamma z$, where $p_0$ is the (unit) earth pressure at rest. The pressure at this point would remain the same if a thin, frictionless wall were introduced into the soil mass to replace the vertical plane and the soil on one side of the wall removed, provided that the wall were introduced in such a manner that the earth mass did not move; i. e. *no deformation of the earth mass were permitted.*

*b.* An accurate determination of the value of $K_0$ which is applicable to a given soil in a given condition involves extensive laboratory or field testing, and is not ordinarily done. The value of $K_0$ for sands varies from about 0.4 for a loose condition to 0.6 or more for a dense condition. With most clay soils, the value of $K_0$ is close to 1.0.

*c.* Since the value of $p_0$ varies directly with depth below the surface, the pressure distribution on the back of the assumed frictionless wall is triangular or "hydrostatic", as indicated in figure 78. If the height of the wall is $H$ feet, then the total at-rest pressure acting on the back of the wall, $P_0$, per foot of wall is $\frac{1}{2} K_0 \gamma H^2$. With the assumption of a level surface, the line of action of $P_0$ is horizontal and is located $H/3$ above the bottom of the wall.

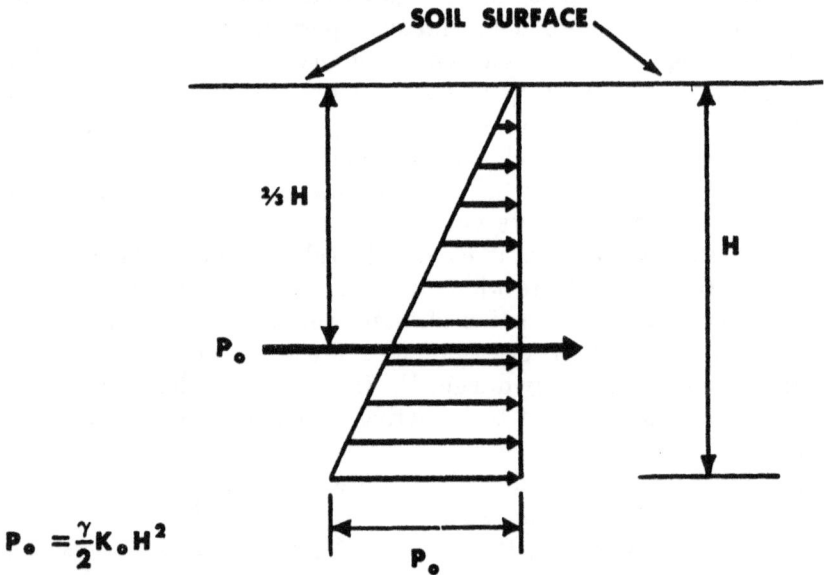

$$P_0 = \frac{\gamma}{2} K_0 H^2$$

*Figure 78.    Earth pressure at rest.*

## 308. Active Earth Pressure

*a.* The expressions for active earth pressure which are given in this paragraph, as well as those for passive earth pressure which are given in paragraph 309, are based upon the assumptions of a level surface of the soil behind the wall, no friction between the wall and the soil, and a plane failure surface. They are very widely used and are known as the *Rankine formulas for earth pressure.* Limitations which are applicable to their use will be emphasized as the discussion proceeds.

*b.* If the wall which was introduced into the discussion in paragraph 307 is allowed to move outward slightly, away from the soil mass which it supports, the horizontal thrust on the wall will be reduced.

If the movement is sufficiently great, the horizontal thrust will reach a minimum value which corresponds to a state of shear failure in the soil behind the wall. This *minimum* value of the earth pressure is the *active pressure*. Further movement of the wall will just continue the shear failure with very little change in the pressure.

c. Under the assumptions which have been listed, it may be shown that the coefficient of active earth pressure, $K_A$, for a cohesionless soil above the water table is given by the expression, $K_A = \dfrac{1 - \sin \phi}{1 + \sin \phi}$, where $\phi$ is the angle of internal friction. The value of $\phi$ may be determined by laboratory testing or may be estimated from table IV. Similarly, the unit weight may be determined by field measurements or estimated from experience. In figure 81 is given a plot of the values of $K_A$ for different values of $\phi$.

d. For a cohesionless soil, the distribution of active earth pressure on the back of a wall is similar to that previously explained for at-rest pressure. The distribution of active earth pressures, and the location (and direction) of the total active pressure, $P_A$, per foot of wall are shown in figure 79. For cohesionless soil, $P_A = \frac{1}{2} K_A \gamma H^2$. As a numerical example, assume that a long retaining wall is 15 feet high and supports a dry granular backfill, for which $\phi = 40°$ and $\gamma = 110$ lbs.

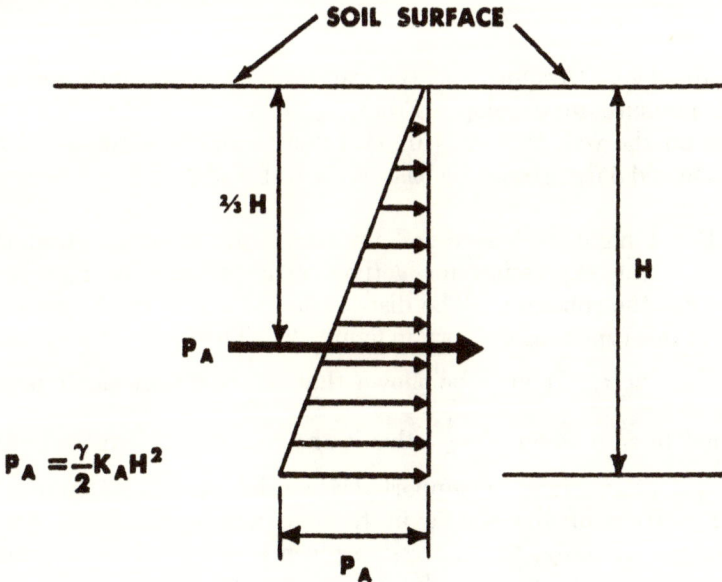

**SOIL SURFACE**

⅔ H

H

$P_A$

$P_A = \dfrac{\gamma}{2} K_A H^2$

$P_A$

**1**

**COHESIONLESS SOIL**

*Figure 79. Rankine active earth pressures.*

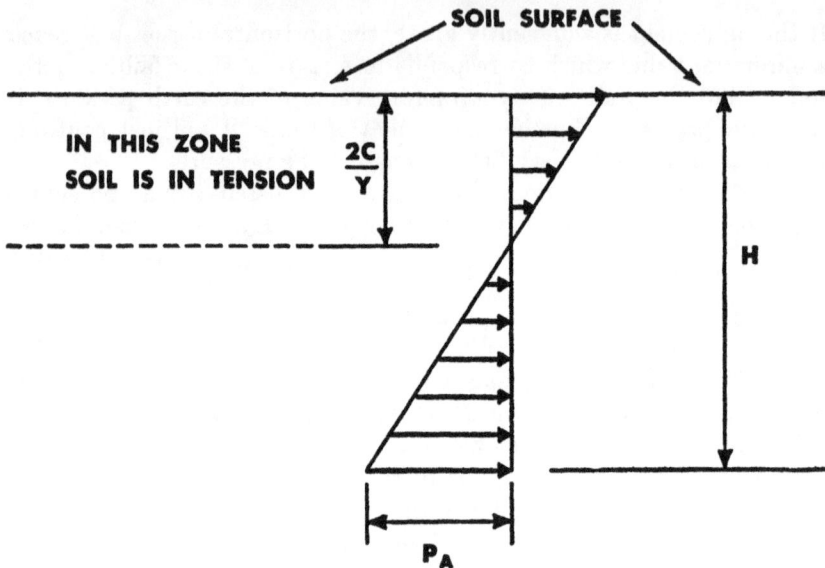

**IN THIS ZONE SOIL IS IN TENSION** $\dfrac{2C}{\gamma}$

SOIL SURFACE

H

$P_A$

2

**HIGHLY COHESIVE SOIL**

*Figure 79. Rankine active earth pressures*—Continued.

per cubic foot. Assume that the wall can yield sufficiently for active earth pressure to develop. From figure 81, $K_A=0.21$. The total thrust on the wall then is $\frac{1}{2}(0.21)110(15)^2=2600$ pounds per foot of wall, located 5 feet above the base of the wall and acting in a horizontal direction.

*e.* For a highly cohesive soil for which the shearing strength, $s$, equals $c$, the expression for active earth pressure is $P_A=\gamma z-2c$, where $c$ is the cohesion. The distribution of active earth pressure in a soil of this type is also shown in figure 79. The total active pressure, $P_A=\dfrac{\gamma H^2}{2}-2cH$. It may be shown that the resultant earth pressure is equal to zero when $H=\dfrac{4c}{\gamma}$; this is the reason why vertical cuts in clay will often stand unsupported for brief periods of time. The upper portion of the soil is in tension, which will cause vertical cracking. Actually, the expression given above is of little practical consequence, because soils of this sort do not remain in the active (or passive) state for a very long period of time. This is because of *creep* or a slow yield of the soil which tends to return the pressure to the at-rest condition. This is the reason that permanent retaining walls which support a clay backfill must be designed to withstand the at-rest pressure.

## 309. Effect of Submergence on Active Pressure in Cohesionless Soils

In the numerical example given above, the soil was located above the water table. Assume that the water table rises to the surface of the ground and, because of saturation, the total unit weight of the soil is increased to 125 pounds per cubic foot. The effective unit weight, $\gamma'$, then is $125-62.5$ equals 62.5 pounds per cubic foot. The soil pressure on the wall is then reduced to $\frac{1}{2}(0.21)$ $62.5$ $(15)^2=2100$ pounds per foot, assuming that $\phi$ remains unchanged. However, to this must be added the water pressure acting on the wall. This will be $\frac{1}{2}(62.5)(15)^2=7030$ pounds per foot of wall. Thus, the combined pressure of soil and water on the wall will be $2100+7030=9130$ pounds per foot, more than three times as great as before. This example serves to emphasize that, although it is possible to design a retaining wall which supports a cohesionless soil for combined soil and water pressure, it is generally much less costly to provide adequate drainage of the back-fill to insure against this eventuality.

## 310. Passive Earth Pressure

*a.* If the wall introduced into the discussion in paragraph 307 is moved toward the soil mass, the pressure on the wall increases above the value of the at-rest earth pressure. If the movement against the soil is sufficiently great, the horizontal thrust will reach a maximum value, which again corresponds to a state of shear failure in the soil behind the wall. This *maximum* value of the earth pressure is the *passive pressure*. Further movement of the wall in this direction will simply continue the shear failure with little change in pressure. Equations presented below for passive earth pressure are based upon the Rankine assumptions previously discussed.

*b.* For a cohesionless soil above the water table, the value of the coefficient of passive earth pressure, $K_p$, is $\dfrac{1+\sin \phi}{1-\sin \phi}$. A plot of the relationship between $K_p$ and $\phi$ is shown in figure 81. The distribution of passive earth pressure, and the direction and location of the total passive pressure per foot of wall, $P_p$, are shown in figure 80 for the Rankine assumptions. For a cohesionless soil, $P_p=\frac{1}{2}K_p \gamma H^2$, where the symbols have meanings previously given. It will be noted that $K_p$ is numerically larger than $K_0$, and is ordinarily several times as great as $K_A$, the ratio depending on the value of $\phi$.

*c.* For a highly cohesive soil for which $s=c$, the value of $p_p$ may be shown to be $\gamma z + 2c$. The distribution of passive earth pressure is trapezoidal in form, as indicated in figure 80. The total passive earth pressure per foot of wall for a highly cohesive soil under the Rankine assumptions is $P_p=\dfrac{\gamma H^2}{2}+2cH$. As noted previously, this

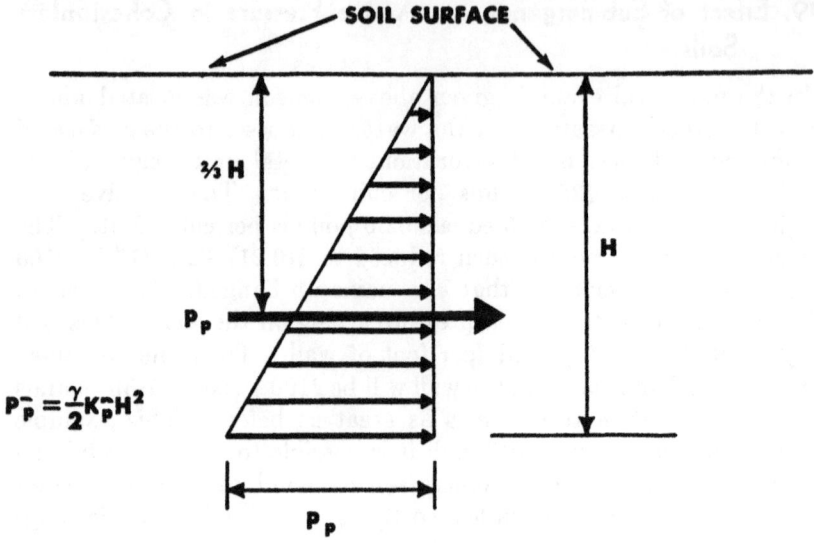

**SOIL SURFACE**

⅔ H

$P_P$

$$P_P^- = \frac{\gamma}{2} K_P^- H^2$$

$P_P$

**1**

**COHESIONLESS SOIL**

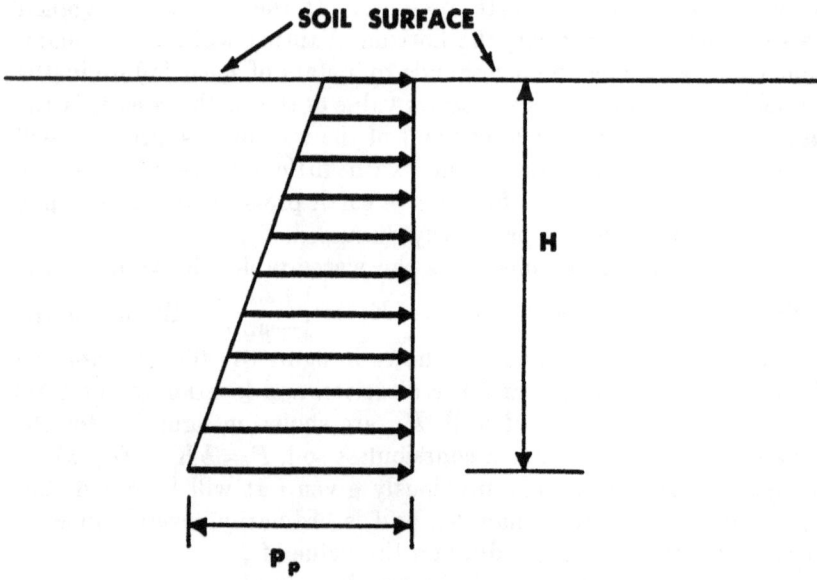

**SOIL SURFACE**

H

$P_P$

**2**

**HIGHLY COHESIVE SOIL**

*Figure 80. Rankine passive earth pressures.*

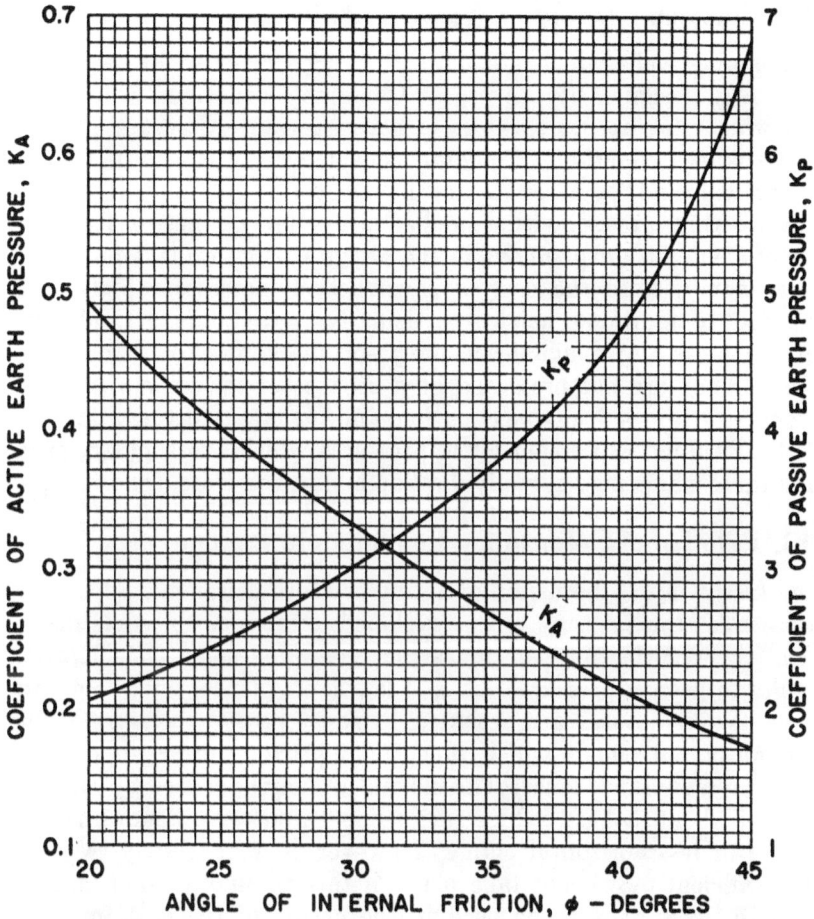

*Figure 81. Relationships between Rankine coefficients of active and passive earth pressure for cohesionless soils and the angle of internal friction.*

value may not be retained over a long period of time in a clay soil because of creep.

## 311. Pressures Exerted by Other Soils

Expressions for active and passive pressures exerted by soils which do not fall into the classification of either cohesionless or highly cohesive are available in most soil mechanics textbooks. They are not presented here because they are dependent upon an accurate determination of $c$ and $\phi$; this determination is beyond the scope of testing methods normally available to the military engineer in the field. In the active state, pressures exerted by these soils will be greater than those exerted by cohesionless soils under comparable conditions, but considerably less than those exerted by typical highly cohesive soils.

## 312. Coulomb's Theory of Active Earth Pressure

Retaining walls which are more than 20 feet or so in height are frequently designed by the use of Coulomb's equations, which give somewhat more accurate values of active pressure than are afforded by Rankine's equation. Coulomb considered the effects of friction on the wall and, as a result, developed expressions for active earth pressure which give somewhat smaller values than do those of Rankine. Use of Coulomb's expressions in the design of high walls is prompted by economy, since the cost of construction increases rapidly with each foot of additional height, particularly above 20 feet. Coulomb's equations are also available in many soil mechanics texts and references. They are not presented here because of their complexity, as compared with the Rankine equation, and the fact that the proper use of the Rankine expression gives pressures which are slightly larger than those which may actually develop and is, therefore, conservative.

## 313. Relation of Deformation Conditions to Pressure

*a.* In the initial discussion of active pressure it was indicated that, in order for this pressure to be developed, there must be some movement or "deformation" of the structure. The same consideration applies to passive pressure, but is of less consequence, since comparatively few structures are designed to resist passive pressure. The amount of movement which must take place in order to develop active pressure is not great. The movement may be caused by rotation of a wall about its base or by straight-line "translation", movement in a more or less horizontal plane. However, it is absolutely essential that sufficient movement take place, if this minimum value (active) of earth pressure is to be used in the design of the structure and developed in the field. If sufficient movement can not take place, then a higher value of earth pressure must be used in the design.

*b.* The only structures which can move sufficiently for the pressure to decrease to the active value are retaining walls founded on earth. These walls are normally and properly designed to resist active earth pressure.

*c.* If the structure is rigidly held in place, as in the case of a concrete retaining wall on a rock foundation, the required amount of deformation can not occur, and the structure should be designed to resist at-rest pressures. Similarly, some structures, such as the abutments for some types of bridges, must not be permitted to deform enough to produce active pressure and should be designed for at-rest pressures.

*d.* Certain types of structures are flexible enough to deform somewhat under lateral earth pressure, but not enough to permit active pressure to develop. The behavior of a structure of this type is sometimes called a case of incomplete deformation. Behavior of this sort is typical of bracing systems used in earth excavations. Neither the magnitude nor distribution of pressure in these cases can be safely calculated by the Rankine equations. The distribution of pressure in such cases is very complex. Their design should, at the present time, be based upon the use of empirical pressure diagrams derived from observations made on actual structures. Additional information relative to structures of this type is contained in paragraphs 325 through 331.

## Section II. RETAINING WALLS

### 314. Introduction

*a.* A *retaining wall* is a structure which is intended to provide support for a mass of soil; the wall holds the soil in a fixed position. In civil practice, a retaining wall is ordinarily regarded as a permanent structure while in military construction it may be either temporary or permanent in nature.

*b.* Retaining walls are used in many applications. For example, a structure of this sort may be used in a highway or railroad cut in order to permit the use of a steep slope and avoid excessive amounts of excavation. They are similarly used on the embankment side of side-hill sections in order to avoid excessive volumes of fill. Bridge abutments and the headwalls of culverts frequently function as retaining walls. Many times in the construction of buildings and various industrial structures retaining walls are used to provide support for the sides of deep, permanent excavations.

*c.* Permanent retaining walls are generally constructed from plain or reinforced concrete; stone masonry walls are also used occasionally. In military construction timber crib retaining walls are important; their design is emphasized in later paragraphs in this section.

### 315. Lateral Earth Pressures To Be Used in Design

*a.* Retaining walls which support level and well-drained backfills, are founded on earth, and have vertical back surfaces which are designed to withstand the Rankine active pressure given in paragraph 308. Concrete retaining walls on rock should be designed to resist earth pressures at rest, as should permanent walls which support clay backfills. If the back of the wall is not vertical, the top of the fill is inclined with the horizontal, or a surcharge is to be placed on the top of the fill, the pressures given by the Rankine equations must be modified. Approximate methods of taking into account the effects of these variables are presented below.

*b.* Several combinations of variables which are frequently encountered in determining the design loads on retaining walls are shown in figure 82. Approximate values of the pressures on the wall in each case may be determined as follows:

(1) If the wall has a sloping back and supports a level backfill,

the active pressure is assumed to act at a point on a vertical plane which passes through the heel of the wall. The pressure, $P_A$, is computed in the regular way. The weight of the triangular mass of soil above the back of the wall, $W$, may be considered as acting vertically downward through its center of gravity. Since it contributes to the stability of the wall, it is conservative to neglect this force, and this is frequently done.

(2) When the surface of the backfill is inclined at an angle with the horizontal, the computation of the pressure is more complicated than in other cases. For this case the Rankine

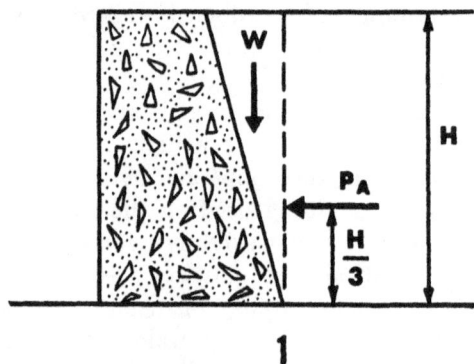

**1**

**SLOPING BACK, LEVEL BACKFILL**

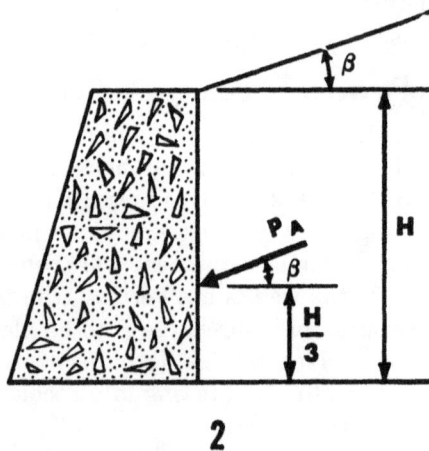

**2**

**VERTICAL WALL, INCLINED BACKFILL**

*Figure 82. Lateral earth pressures to be used in design of retaining walls which support cohesionless backfills.*

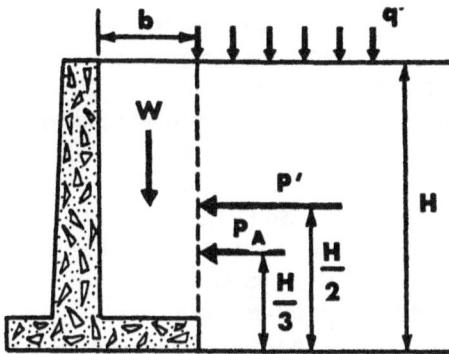

**3**

**LEVEL BACKFILL, UNIFORM SURCHARGE**

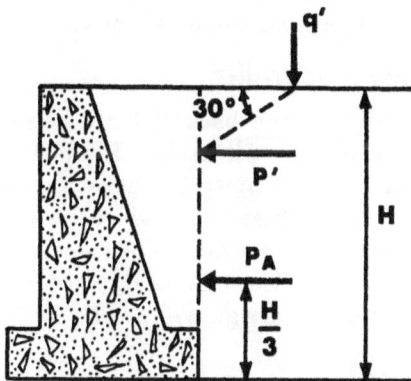

**4**

**LEVEL BACKFILL, LINE LOAD**

*Figure 82*—Continued.

formula, derived in the same way as for a horizontal surface, can be expressed as follows:

$$P_A = \frac{\gamma H^2}{2} \cos \beta \left( \frac{\cos \beta - \sqrt{\cos^2 \beta - \cos^2 \phi}}{\cos \beta + \sqrt{\cos^2 \beta - \cos^2 \phi}} \right)$$

in which $\beta$ is the angle between the surface of the backfill and the horizontal. In this case, $P_A$ acts in a direction which is parallel to the surface of the backfill and at a point H/3 above the base.

(3) If the surface of the backfill is horizontal and supports a uniformly distributed surcharge, then the pressure on the wall at any depth is increased by an amount equal to $K_A q'$,

**239**

where $q'$ is the intensity of the uniform surcharge. The total force on the wall due to the surcharge, $P'$, is equal to $K_A q' H$; it acts at a distance $\frac{1}{2}H$ above the base of the wall, as indicated in figure 82. If the surcharge extends over the portion of the backfill above the base of the wall, the downward force is increased from $W$ to $W+q'b$, where $b$ is the width of the backfill above the base of the wall.

(4) If a line load, $Q'$, or a concentrated load acts on the surface of the backfill, this also contributes to the pressure on the wall. The additional lateral pressure on the wall, $P'$, is equal to $K_A Q'$. The point of application of this force may be found approximately by the procedure indicated in figure 82.

(5) Other variables may be encountered. For example, the slope of the back of the wall or the slope of the backfill may not be uniform. Information relative to the pressures exerted in these cases is beyond the scope of this manual.

## 316. Backfills for Retaining Walls

*a. General.* The design of the backfill for a retaining wall is as important as the design of the wall itself. If the Rankine formulas are used in design, the backfill must be reasonably clean, granular material which is essentially cohesionless, which is easily drained, and which is not susceptible to frost action. The use of backfill materials which meet these requirements is economical, since it insures that the active pressure assumed in the design will not be exceeded during the life of the structure.

*b. Materials.* Materials which are best for backfills behind retaining walls are clean sands, gravels, and crushed rock. In the Unified Soil Classification System the GW and SW soils are to be preferred, if they are available, and GP and SP soils are also satisfactory. These granular materials require compaction, so as to be stable against the effects of vibration. Compaction also generally increases the angle of internal friction, which is desirable, in that it decreases the lateral pressure exerted on the wall. Materials of the GM, GC, SM, and SC groups may also be used for backfills behind retaining walls, but must be protected against frost action and may require elaborate drainage provisions. Fine-grained soils are not desirable as backfills if their use can be avoided, because they are difficult to drain. If clay soil must be used, the wall should be designed to resist earth pressures at rest.

*c. Compaction.* Backfills behind retaining walls are commonly put in place after the structure has been built. The method of compaction depends upon the soil, equipment available, and working space. Since most backfills are essentially cohesionless they are best compacted by vibration. Rolling equipment suitable for use with these soils has been discussed in chapter 9. Common practice calls for the backfill to be placed in layers from 3 to 4 inches in thickness; each layer is compacted to a satisfactory degree of density. In areas which are inaccessible to rollers or similar compacting equipment, compaction may be done by the use of mechanical air tampers or hand tools.

## 317. Drainage Provisions for Backfills

Drainage of the backfill is essential in order to keep the wall from being subjected to water pressure and to prevent frost action. Common drainage provisions used on concrete walls are shown in figure 83.

*a.* When the backfill is composed of clean, easily-drained materials, it is customary to provide for drainage by providing *weep holes* through the wall. Weep holes are commonly made by embedding pipes of from 4 to 6 inches in diameter in the wall; they are spaced from 5 to 10 feet center-to-center both horizontally and vertically. It is best to provide a filter of granular material around the entrance to each weep hole, in order to prevent the soil from washing out or the drain from becoming clogged. If possible, this material should conform to the requirements previously given for filter materials (par. 93).

*b.* Weep holes have the disadvantage of discharging the water seeping through the backfill at the toe of the wall, where the soil pressures are greatest. The water may weaken the soil at this point and cause the wall to fail. A more effective solution, which is also more expensive, is to provide a longitudinal back drain along the base of the wall, as shown in figure 83. A regular pipe drain should be used, surrounded with a suitable filter material. The drainage may be discharged away from the ends of the wall.

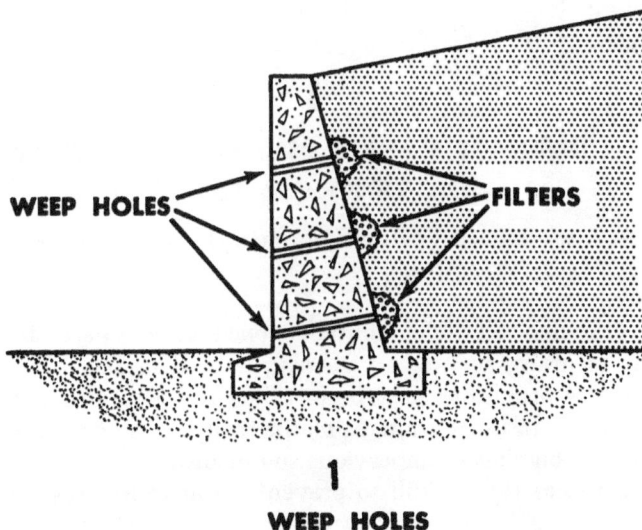

**WEEP HOLES**

*Figure 83.  Drainage provisions for retaining walls.*

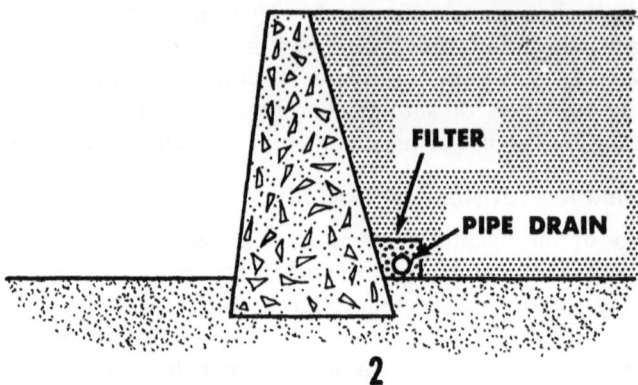

**LONGITUDINAL BACK DRAIN**

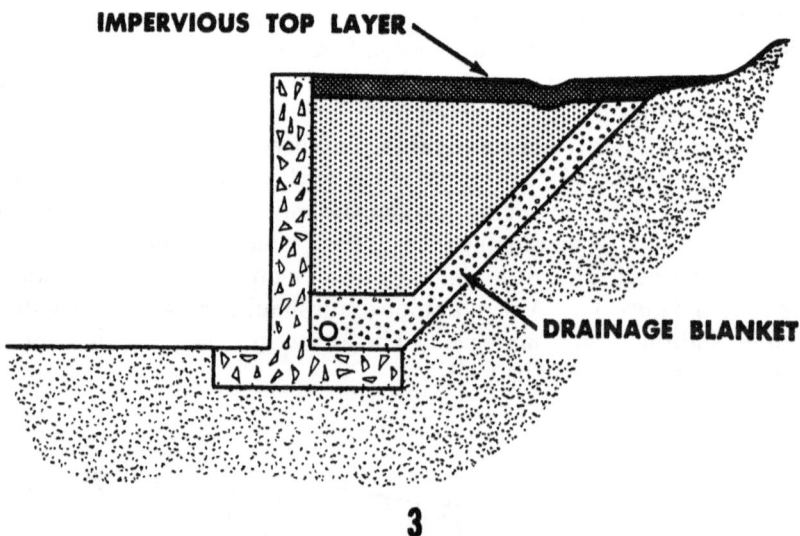

**DRAINAGE BLANKET USED IN COMBINATION
WITH IMPERVIOUS TOP LAYER**

*Figure 83. Drainage provisions for retaining walls*—Continued.

*c.* If a granular soil which contains considerable fine material and is difficult to drain, such as an SC soil, is used in the backfill, then more elaborate provisions may be installed to insure drainage. One such approach is to use a drainage blanket, such as shown in figure 83. If necessary, a blanket of impervious soil or bituminous material may be used on top of the backfill to prevent water from entering the fill from the top. Such treatments are relatively expensive.

## 318. Frost Action in Backfills

If the conditions for detrimental frost action are present in the backfill, including a frost-susceptible soil, availability of water, and freezing temperatures, steps must be taken to prevent the formation of ice lenses and the resultant severe lateral pressures which may be exerted against the wall. The usual way of preventing frost action is to substitute a thick layer of clean, granular, nonfrost-susceptible soil for the backfill material immediately adjacent to the wall. The width of the layer should be as great as the maximum depth of frost penetration in the area. This method is shown in figure 84. As with other structures, the bottom of a retaining wall should be located beneath the line of frost penetration.

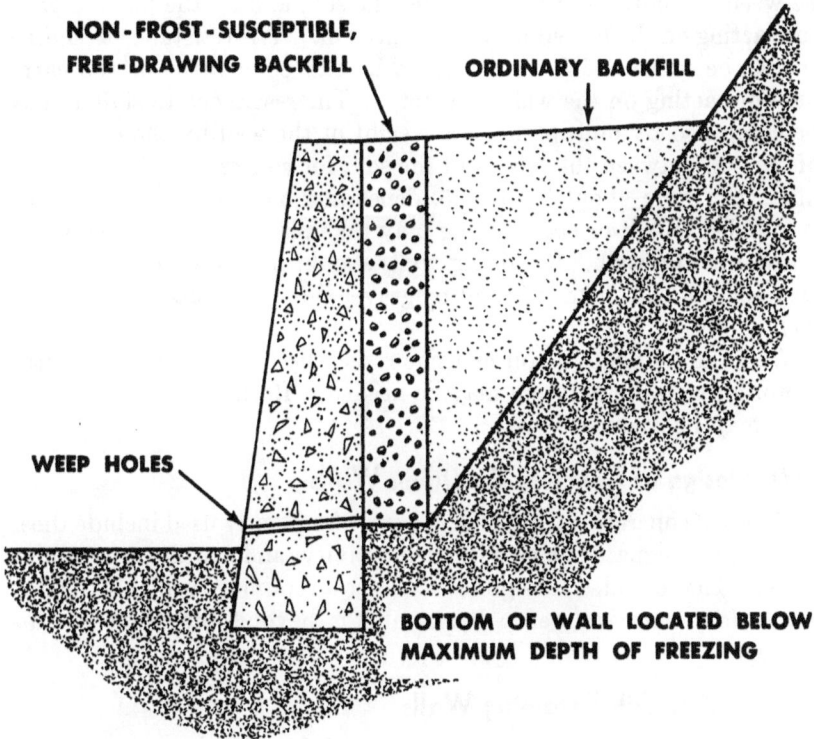

*Figure 84. Recommended treatment to eliminate frost action behind retaining walls.*

## 319. Principles of Design of Retaining Walls

From a structural standpoint, any retaining wall must be designed so that it is structurally capable of withstanding the forces which act upon it. From the standpoint of soil mechanics, the foundation upon which the wall rests must also be capable of supporting the loads which are applied to it, including the weight of the wall. In order for the foundation to be satisfactory the bearing capacity of the soil must not be exceeded, so that the wall will not fail by overturning; the wall must not be subjected to excessive settlement; and the wall must not fail by sliding. Important considerations relative to each of these factors are as follows:

*a.* With relation to the soil pressure beneath the wall, the wall should be designed so that the resultant of the earth pressure and weight of the wall falls within the middle third of the base. The maximum pressure on the soil (usually beneath the toe of the wall) must not exceed the allowable bearing capacity, as presented in chapter 8.

*b.* Methods of settlement presented in chapter 8 may be applied to retaining walls.

*c.* The sliding of a retaining wall is prevented by frictional resistance between the bottom of the wall and the soil, and by the passive pressure acting on the buried portion of the wall. The resistance to sliding should be at least 1.5 times the horizontal component of the earth pressure acting on the wall for safety. The resistance to sliding may be computed by multiplying the weight of the wall by the coefficient of friction between the base and the soil; for concrete walls this value may vary from about 0.3 to 0.5, depending on the soil. If the wall is founded on silt or clay, sliding may take place in the soil below the bottom of the wall. Retaining walls should not be built directly on clay soils; a thin layer of sand or gravel should be placed before the wall is built.

*d.* If the foundation soil can not resist the loads of the wall satisfactorily, a pile foundation may be needed. Both vertical and batter piles may be required.

## 320. Design of Concrete Retaining Walls

Types of concrete walls which are most frequently used include those which are designated as gravity, semigravity, and cantilever retaining walls. The detailed design of concrete retaining walls is readily available in many reference books and is beyond the scope of this manual.

## 321. Timber Crib Retaining Walls

*a.* A very useful type of retaining wall for military purposes, especially where the construction is to be temporary in nature, is constructed in the form of cribbing, using timber. The crib or cells are filled with earth, preferably with clean, coarse, granular material. A wall of this sort gains its stability through the weight of the material used to fill the cells, along with the weight of the crib units themselves. The longitudinal member in a timber crib is called a *stretcher*, while a transverse member is a *header*.

*b.* One of the principal advantages of a timber crib retaining wall is that it may be constructed with unskilled labor, and with a minimum of equipment. Suitable timber is available in many military situations, and the timbers may be salvaged for later use. Little foundation excavation is generally required and may be limited to shallow trenching for the lower part of the crib walls. The crib may be built in short sections, one or two cribs at a time, and where the amount of excavation is sufficient, and is suitable, it may be used for filling the cells. A crib of this sort may be used on foundation soils which are weak and might not be able to support a heavy wall, since the crib is fairly flexible and able to undergo some settlement and shifting without distress. However, this should not be misunderstood, as the foundation soil must not be so soft as to permit excessive differential settlement which would destroy the alinement of the crib.

## 322. Design of Timber Crib Walls

*a.* For important timber crib walls, the base width must be determined in the fashion indicated in paragraph 319 for solid retaining walls, particularly with regard to the position of the resultant pressure within the middle third of the base and the magnitude of maximum soil pressure. Experience has indicated that, in general, a satisfactory design will be achieved if the base width is a minimum of 50 percent of the height of the wall, provided that the wall does not carry a surcharge and is on a reasonably firm foundation. If the wall carries a heavy surcharge, the base width should be increased to a minimum of 65 percent of the height. In any case, the width of the crib at top and bottom should be not less than 4 feet.

*b.* Timber crib walls may be built with any desired batter or even vertical. The batter which is most generally used and recommended is 1 horizontal to 4 vertical, as indicated in figure 85. If less batter is used, the base width must be increased to insure that the resultant pressure will fall within the middle third of the base. The desired batter is generally achieved by placing the base on a slope which is equal to the batter. The toe may be placed on sills and this is frequently done with high walls. Sometimes double-cell construction is used to obtain the necessary base width of high walls. The wall is then decreased in width, or "stepped-back," in the upper portions of of the wall, above one-third height. Additional rows of bottom stretchers may be used to decrease the pressure on the soil, or to avoid detrimental settlement.

*c.* The front and rear walls of the crib should be connected at each panel point. The crib must be kept an essentially flexible structure and must be free to move somewhat in any direction, so as to adjust itself to thrusts and settlements.

*d.* The material which is used in filling the cells should be placed in thin layers and well compacted. Backfill behind the wall should also be compacted and should be kept close to, but not above, the level of the material in the cribs.

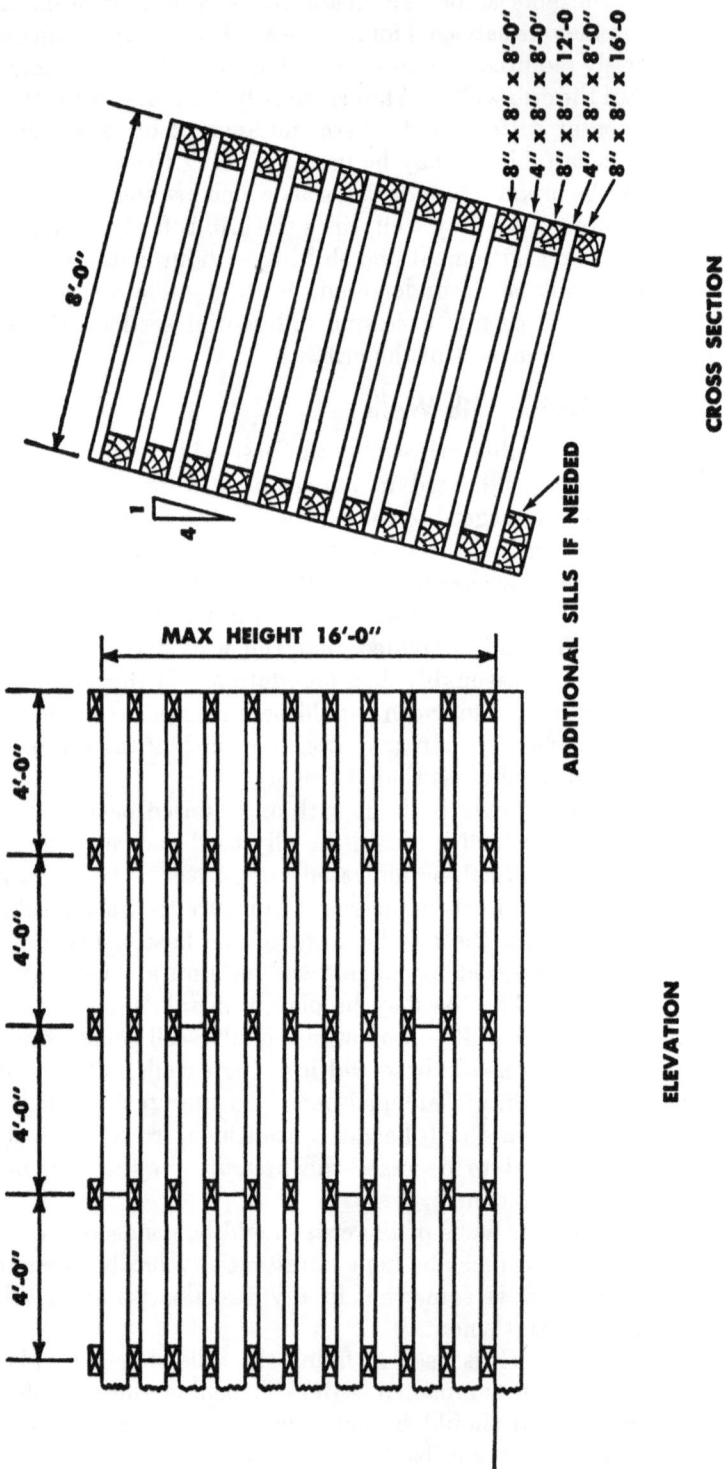

8'' x 8'' x 8'-0''
4'' x 8'' x 8'-0''
8'' x 8'' x 12'-0
4'' x 8'' x 8'-0''
8'' x 8'' x 16'-0

CROSS SECTION

8'-0''

1
4

ADDITIONAL SILLS IF NEEDED

MAX HEIGHT 16'-0''

4'-0''

4'-0''

4'-0''

4'-0''

4'-0''

ELEVATION

*Figure 85. Typical timber crib retaining wall.*

246

*e.* It is not customary to provide for drainage behind timber crib walls.

## 323. Typical Timber Crib Walls

In figure 85 are shown the elevation and cross section of a timber crib retaining wall which may be used to a maximum height of about 16 feet. A similar arrangement may be used for heights up to about 8 feet, with a minimum width of 4 feet. For heights above 16 feet, the headers are usually 6-inch by 12-inch timbers and the stretchers 12-inch by 12-inch timbers. Timbers are normally connected together by means of heavy (¾-inch diameter) drift bolts.

## 324. Other Timber Retaining Walls

Other types of timber retaining walls are used for low heights, particularly in connection with culverts and bridges. A wall of this sort may be built by driving timber posts into the ground and attaching thereto planks or logs. Details relative to retaining walls used in conjunction with bridge abutments are given in FM 5–9. Figure 86 illustrates two other types of timber retaining walls.

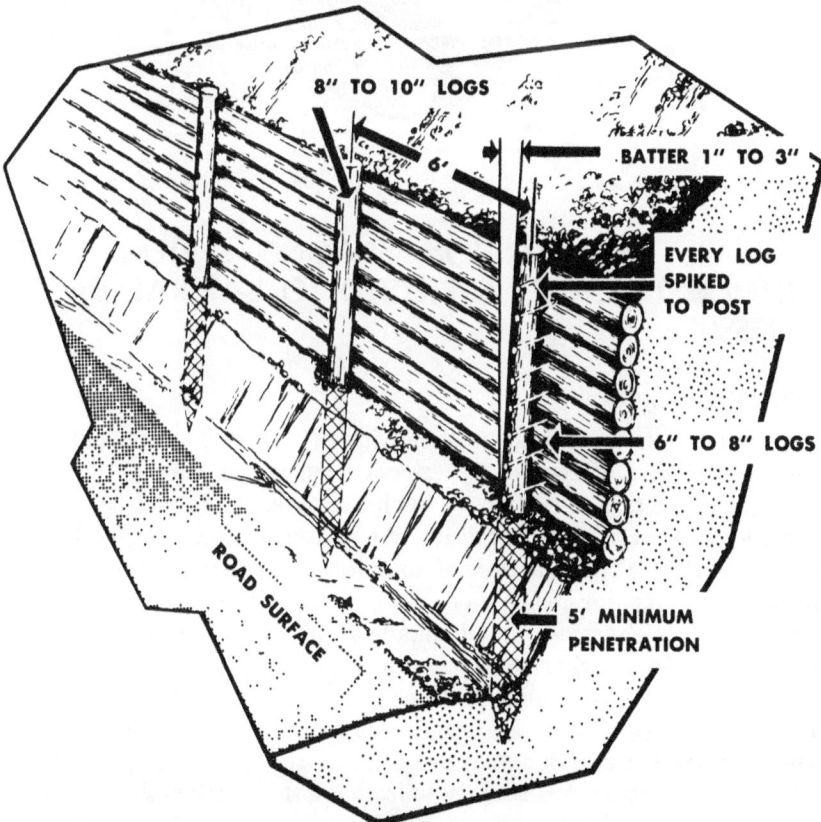

① Log retaining wall

*Figure 86. Other timber retaining walls.*

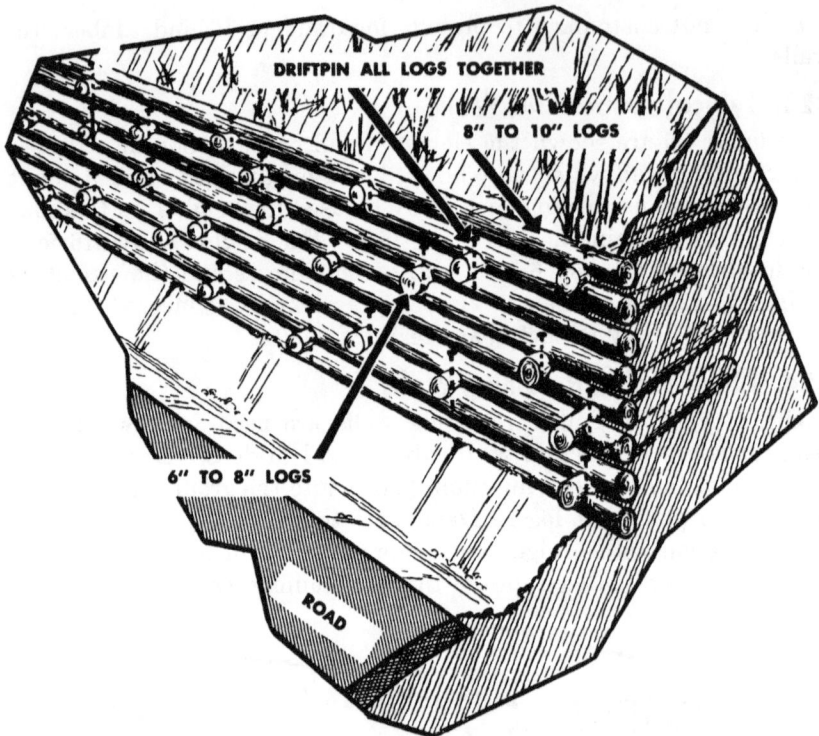

② Partially cribbed log retaining wall

*Figure 86. Other timber retaining walls*—Continued.

# Section III. EXCAVATION BRACING SYSTEMS AND OTHER STRUCTURES SUBJECTED TO LATERAL EARTH PRESSURE

## 325. Introduction

*a.* This section is primarily concerned with bracing systems used to protect the sides of temporary excavations, which may be required during construction operations. Such temporary excavations may be required for a variety of purposes, but are most often needed in connection with the construction of foundations for structures and the placing of utility lines, such as sewer and water pipes.

*b.* Shallow excavations may be made as open cuts with unsupported slopes, particularly when the excavation is being done above the water table. In general, recommendations previously given relative to safe slopes in cuts (par. 284) are applicable here, if the excavation is to remain open for any length of time. If the excavation is purely temporary in nature, most sandy soils above the water table will stand at somewhat steeper slopes, as much as ½ to 1 for brief periods, although some small slides may take place. Clays may be excavated to shallow depths with vertical slopes and will remain stable briefly. Generally speaking, it is much safer to brace cuts in clay which extend to depths of 5 feet or more below the surface, unless flat slopes are used.

*c.* Even for relatively shallow excavations, the use of unsupported cuts may be unsatisfactory for several reasons. Cohesive soils may stand on steep slopes temporarily, but bracing is frequently needed to protect against a sudden cave-in and accident. Required side slopes, particularly in loose, granular soils, may be so flat as to require an excessive amount of excavation. If the excavation is being done close to other structures, space may be limited or the consequences of the failure of a side slope may be very serious; considerable subsidence of the adjacent ground may take place, even though the slope does not actually fail. Finally, if the work is being done below the water table, the excavation may have to be surrounded with a temporary structure that will permit the excavation to be unwatered. Methods which may be used in draining excavations of this sort are discussed briefly in paragraph 330.

## 326. Bracing of Shallow Excavations

*a.* The term *shallow excavation* refers to excavations which are made to depths of from 12 to 20 feet below the surface, depending principally on the soil involved. The lower limit is applicable to fairly soft clay soils, while the upper limit is generally applicable to sands and sandy soils.

*b.* Several different schemes may be used to brace the sides of a narrow shallow excavation. Two of these arrangements are shown in figure 87. In the first of these procedures, timber planks are driven around the boundary of the excavation to form what is called *vertical sheeting*. The bottom of the sheeting is kept at or near the bottom of the pit or trench as excavation proceeds. The sheeting is held in place by means of horizontal beams called *wales*, which in turn are usually supported against one another by means of horizontal members called *struts*, which extend from one side of the excavation to the other. The struts may be cut slightly long, driven into place, and held by nailing or cleating. They may also be held in position by means of wedges or shims.

*c.* Another scheme which may be used involves the use of horizontal timber planks to form what is called *horizontal lagging*. The lagging in turn is supported by vertical *soldier beams* and struts. If the excavation is quite wide, struts may have to be biaced horizontally or vertically, or both.

## 327. Design of Shallow Bracing Systems

*a.* Bracing systems for shallow excavations are commonly designed on the basis of experience. It must be recognized that systems of this sort represent cases of incomplete deformation, since the bracing

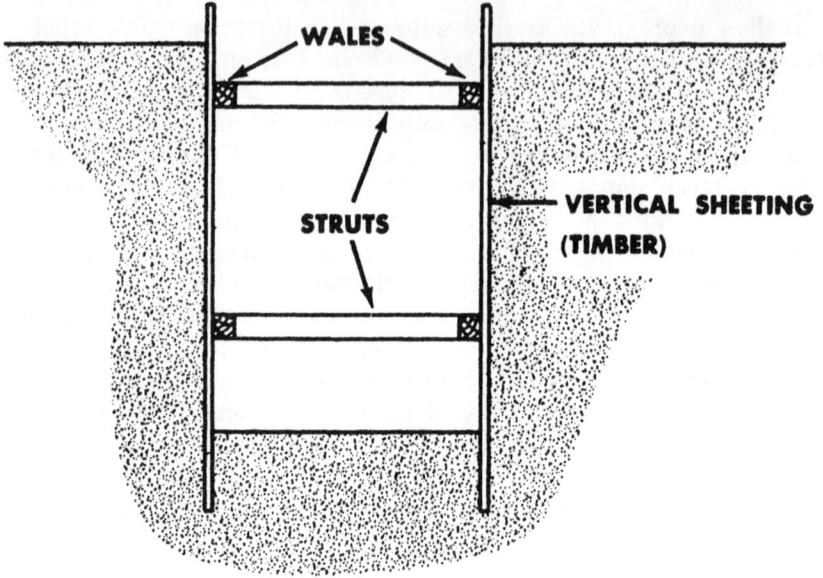

**1**

**VERTICAL SHEETING**

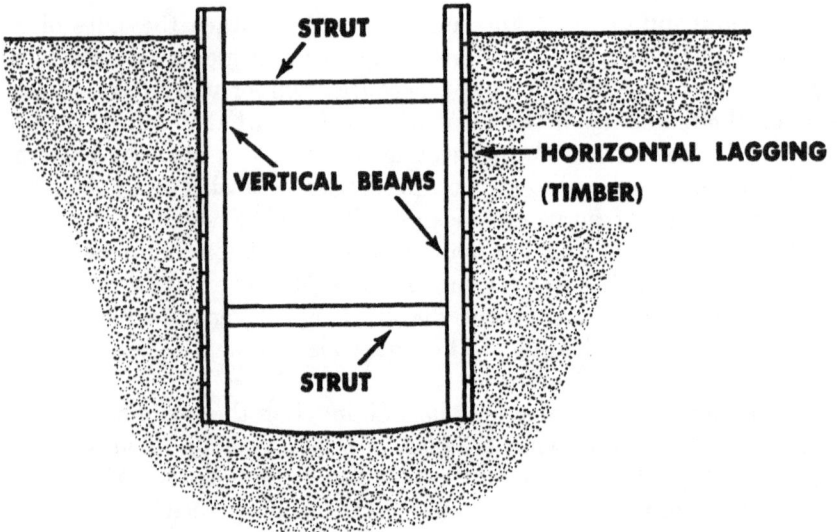

**2**

**HORIZONTAL LAGGING**

*Figure 87. Bracing systems for narrow shallow excavations.*

system prevents deformation at some points while permitting some deformation at others. Rankine's formulas for active earth pressure do not apply to this situation, since the distribution, magnitude, and point of application of the pressure are not the same as assumed in their derivation. The pressure distribution is more likely to be roughly parabolic or trapezoidal in shape than triangular, with the point of application of pressure being close to mid-height. If it is desired to use Rankine's formulas for active earth pressure, the intensity of pressure at the bottom of the excavation should be estimated with an ample factor of safety. Any effects of continuity in reducing the moment in the sheeting, lagging, wales, or beams should be neglected, as in computing the loads on the struts. It is not customary to reduce the size of the timbers used toward the top of the excavation.

*b*. Members used in bracing systems should be strong and stiff. In ordinary work, struts vary from 4-inch by 6-inch timbers for narrow cuts up to 8-inch by 8-inch timbers for excavations 10 or 12 feet wide. Heavier timbers are used if additional safety is desired. Struts are commonly spaced about 8 feet horizontally and from 5 to 6 feet vertically. Lagging or sheeting is customarily made from planks from 6 to 12 inches wide, with minimum thickness usually being about 2 inches.

## 328. Bracing of Wide Shallow Excavations

If the excavation is too wide to be cross braced by the use of struts, vertical sheeting may be used as indicated in figure 88. The wales are supported by inclined braces which are sometimes called *rakes;*

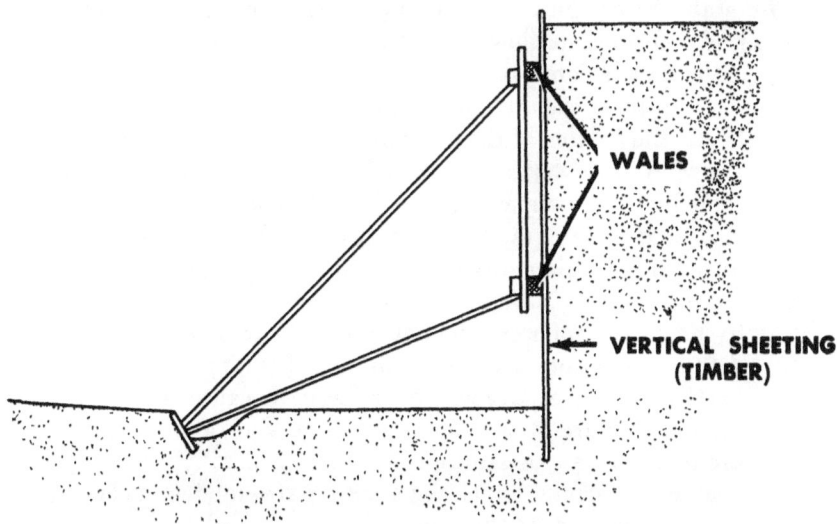

*Figure 88. Bracing of the side of a wide shallow excavation.*

the rakes in turn react against *kicker blocks* which are embedded in the soil. As the excavation is deepened, additional wales and braces may be added as necessary to hold the sheeting firmly in position. The use of this system is dependent upon the soil in the bottom of the excavation being firm enough to provide adequate support for the blocks.

### 329. Bracing Systems for Deep Excavations

*a.* For deeper excavations the use of timber becomes uneconomical, and use is generally made of steel sheet piling or steel H-beams. For example, steel sheet piling may be driven along the boundaries of the excavation, generally to a depth of several feet below the bottom of the proposed pit or trench. The purpose served in driving the piles below the bottom is to lessen the possibility of a failure by heaving, as indicated in *c* below. As the excavation is deepened, wales and struts are inserted as needed. Many times it is necessary to provide vertical support for the struts; this may be done by supporting the struts on vertical posts or by hanging them from beams which extend across the top of the cut. If the soil will stand unsupported briefly, it may be feasible to use steel H-beams driven with their flanges parallel to the side of the excavation instead of the steel sheeting. If this is done horizontal lagging is inserted between adjacent H-piles and wedged against the soil. Wales and struts are then added as before.

*b.* The design of bracing systems for deep excavations should be based upon semiempirical formulas for soil pressure which have been developed as a result of large-scale field observations and experience. Both safety and economy are essential on large projects. Such procedures are beyond the scope of this manual, but are available in most soil mechanics textbooks.

*c.* The stability of the bottom of the excavation must always be kept in mind. Two situations are apt to be dangerous.

(1) If the excavation is being made in sand below the water table, it is essential that the level of the water table be maintained below the bottom of the excavation. If this is not done, then the sand in the bottom of the excavation may become quick, sand boils may appear, and the whole bottom may rise. As indicated in paragraph 330, the water table may be lowered by pumping from ditches or sumps or by the use of well points. Either process must be carefully controlled in order to insure the safety of the bottom of the excavation.

(2) If the excavation is being made in soft clay, the soil in the bottom of the excavation may heave because of a bearing capacity failure of the soil adjacent to and in the bottom of the excavation. A failure of this sort generally may be prevented by driving steel sheet piling several feet below the proposed bottom of the excavation.

## 330. Drainage of Excavations

*a. Ditches and Sumps.* When an excavation is to be carried out to a level which is a few feet below the water table in pervious soils, it is frequently possible to drain the excavation by means of open ditches along the sides of the bottom of the excavation. The ditches channel the water into an open sump. A *sump* is a shallow pit into which the water flows by gravity. The water is then pumped out, frequently by the use of regular gasoline-operated pumps. If a loose, fine or silty sand is involved, there is some possibility that boils and piping may begin in the sump. This must be watched carefully while pumping is being carried on. In some cases it may be necessary to line the sides and bottom of the sump with a suitable filter material in order to prevent sand and silt from being washed away and pumped out with the water.

*b. Closed Drains.* In some situations, particularly when the effect of seepage on open ditches and sumps is troublesome, drain tiles or perforated pipes may be laid in trenches which are then protected with a suitable granular filter material.

*c. Well Points.* On large jobs involving excavation below the water table in pervious soils, the level of the water frequently is controlled by the use of well points. A *well point* is a perforated metal pipe which is generally from 2 to 4 feet in length and about 1½ inches in diameter. The well point, which is attached to the bottom of a riser pipe of similar size, is jetted into position; the end of the pipe below the surface is protected by a suitable granular filter. The water table may be lowered to a depth of about 15 feet below its original position by the use of a line of well points connected at intervals of from 2 to 6 feet in a straight line to a header pipe, which in turn is connected to a suction pump. The water table cannot be lowered more than about 15 feet by a single installation of well points at a given elevation because of the limitations of suction pumps. The water table may be lowered further by using several lines at successively lower elevations, so that the water table is lowered in stages. Well points are not effective in soils which have an effective size less than about 0.05 mm., unless gravity is supplemented by vacuum applied to the well points, in which case somewhat finer soils may be drained, although drainage usually requires a long time. Clays cannot be drained by well points because of their low permeabilities. The detailed design of well-point systems is beyond the scope of this manual.

## 331. Other Structures Subjected to Lateral Earth Pressures

Many other types of structures are subjected to lateral earth pressure. Among these are the flexible anchored bulkheads which may be used as a part of a water-front structure. Such a bulkhead is shown in figure 89. The embedded portion of a bulkhead of this

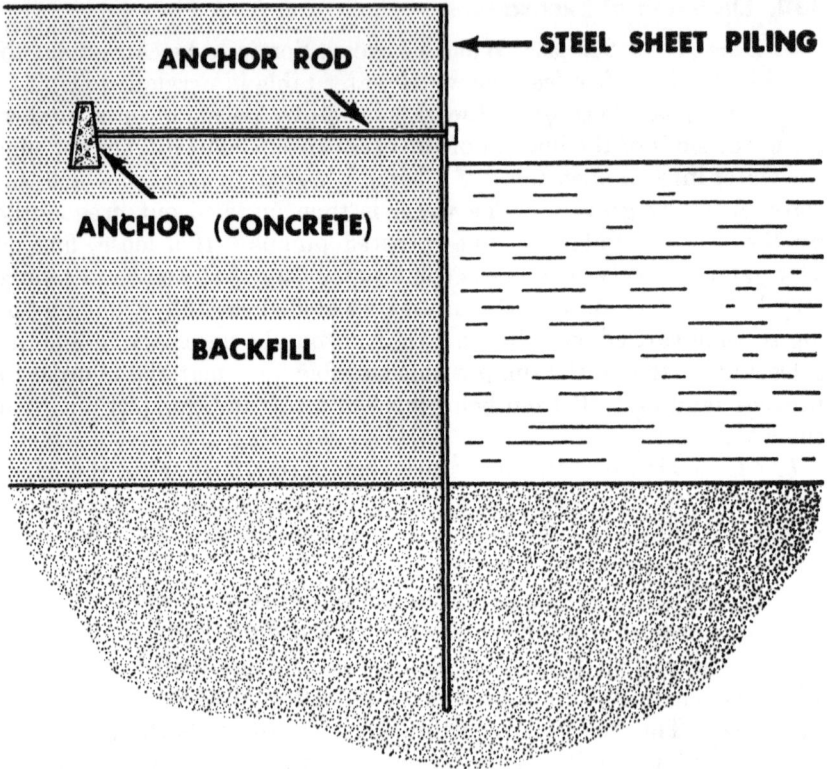

Figure 89.  Anchored bulkhead.

sort generally may be designed to take advantage of the passive earth pressure which may be developed below the ground line.  Anchorages may be similarly designed to resist passive pressure, provided that they are located far enough back from the structure that the assumed failure surface (a plane at an angle of $45+\phi/2$ with the vertical) may be developed.  Anchorages consisting of concrete blocks, timber, or two batter piles driven at opposite angles may be used to carry a portion of the load on a retaining wall or similar structure, if necessary. If the anchorage cannot move and prevents movement of the wall, the active earth pressure cannot be used in the design of the wall itself.

# CHAPTER 13

# SUBSURFACE SOIL EXPLORATION

## Section I. INTRODUCTION

### 332. General

The determination of subsurface soil conditions at the site of a proposed military construction project is a vital operation upon which the success or failure of the project may depend.  Information which is obtained by methods described in this chapter includes the nature, extent, and condition of soil layers; the position of the water table and the proximity of ledge rock which may affect design and construction; drainage characteristics; and sources of possible construction materials.  An adequate program of soil exploration at a particular location or in a given area requires that the military engineer be thoroughly familiar with the principles of soil action which have been stressed in the first twelve chapters of this manual.

### 333. Relation to Military Operations

Emphasis in this chapter is placed upon military construction, and particularly upon the engineer problem of subsurface soil exploration.  It is apparent that an estimate of soil conditions is an important factor in both the overall and detailed planning of military operations. Soils must be given consideration along with other factors which may be important in a given situation.  Soil conditions must be considered, for example, in the location of a road or the selection of the site of an airfield.  Information relative to road location and airfield site selection is contained in TM 5–545.  Information relative to airfields is also given in TM 5–255.  When time and the military situation permit, soil conditions may also be of vital importance in planning the location of a large structure.  The trafficability (ease with which vehicles and personnel may maneuver) of soils may be a vital factor in the planning of military operations, including the selection of beaches for amphibious landings.

### 334. Scope of This Chapter

Remaining material in this chapter is divided into three sections. The first of these deals with aids which may be used in the planning and execution of a subsurface soil exploration program.  Much of the information contained in this section is useful in broad planning functions, as well as in detailed planning, design, and construction. The second section is devoted to soil surveys for roads and airfields. In the final section attention is given to soil exploration for structures.

## Section II.  SOURCES OF INFORMATION OF VALUE IN PLANNING SOIL EXPLORATION PROGRAMS

### 335. Introduction

*a.* Many sources of information concerning soils may be available for the area in which an engineering construction project is to be carried out.  These sources include engineer intelligence reports, geologic and topographic maps and reports, agricultural soil maps and reports, and aerial photographs.

*b.* These sources of information may be the only ones available for objective areas held by the enemy and are of great value in strategic and tactical planning.  From the standpoint of engineering construction this information is of importance in two general ways. First, a study of this information, when available, will aid in securing a broad understanding of soil conditions and associated engineering problems which may be encountered in the area.  Second, such information is of great value in planning, conducting, and interpreting the results of detailed soil exploration programs which are necessary for design and construction.  Properly used, reports and maps discussed in this section are of value in holding soil sampling operations to a minimum.

*c.* Many of the corollary sources of information discussed are available for the more developed regions of the world.  All available sources of information should be carefully studied and correlated, where time and the military situation permit.

### 336. Intelligence Reports

*a.* Intelligence reports which include maps and studies of soil conditions are usually available for areas in which military operations have been planned.  Among the best and most comprehensive of these are the Terrain Intelligence Folios prepared by the Intelligence Branch of the Corps of Engineers, in cooperation with the U. S. Geological Survey.

*b.* Intelligence reports vary considerably in the amount of detail which is included, depending on the purpose for which they are intended.  For broad general planning, large areas are covered and relatively small-scale maps are prepared.  For detailed planning, soils are frequently delineated and described on large-scale maps. These reports are most useful in planning construction operations and are a source of information relative to geology, topography, terrain conditions, climate and weather conditions, sources of construction materials, and similar items.

*c.* Additional information relative to terrain intelligence reports and their preparation is contained in TM 5–545.

*d.* Normal engineer intelligence channels will produce much of the material discussed in the following paragraphs. In tactical situations additional sources of information will become available as our advance continues.

## 337. Geologic and Topographical Maps and Reports

*a.* It is apparent that there is a close relationship between geology and soil conditions. This fact has been emphasized by the discussion of the processes of soil formation presented in paragraphs 4 through 12. Information contained in geologic maps and reports may be either of general or specific usefulness to the engineer engaged in design and construction, depending on the type of map or report available. Advance geologic information will aid the engineer in planning, conducting, and interpreting the results of soil exploration programs and will help to avoid construction failures on unfavorable ground.

*b.* Specific information concerning both standard and special sources of geologic maps and literature is contained in TM 5–545, together with a discussion of the uses of such information.

*c.* Ordinary topographical maps may be of some usefulness in estimating soil conditions, particularly when reviewed in conjunction with geologic maps. Experience is required for such examinations, but a careful inspection of maps of known areas will provide a background for the recognition of similar features on maps of unknown areas. For example, such characteristic shapes as sand dunes, beach ridges, and alluvial fans may be detected on large-scale topographic maps, as may typical formations in glaciated areas.

## 338. Agricultural Soil Survey Maps and Reports

*a.* Information has been presented previously about the study of soils from an agricultural viewpoint, including that relating to the soil profile, soil horizons, and pedology (par. 11), and the classification of soils by agricultural soil scientists (par. 65).

*b.* Agricultural soil maps and reports are available for many of the developed agricultural areas of the world. As far as the maps are concerned, they are important sources of information relative to surface soils, since they show the areal extent of each of the soils encountered. Principal mapping units include the soil series, type, and phase (par. 11). Factors which are considered in the field surveys upon which the maps are based include the careful study of the soil horizons in test pits, highway and railway cuts, auger borings, and other exposed places. Each of the horizons is studied from the standpoint of color, texture, structure, organic content, and consistency. Other items of importance are drainage, topography, vegetation, agricultural land use, and geographical location. On the

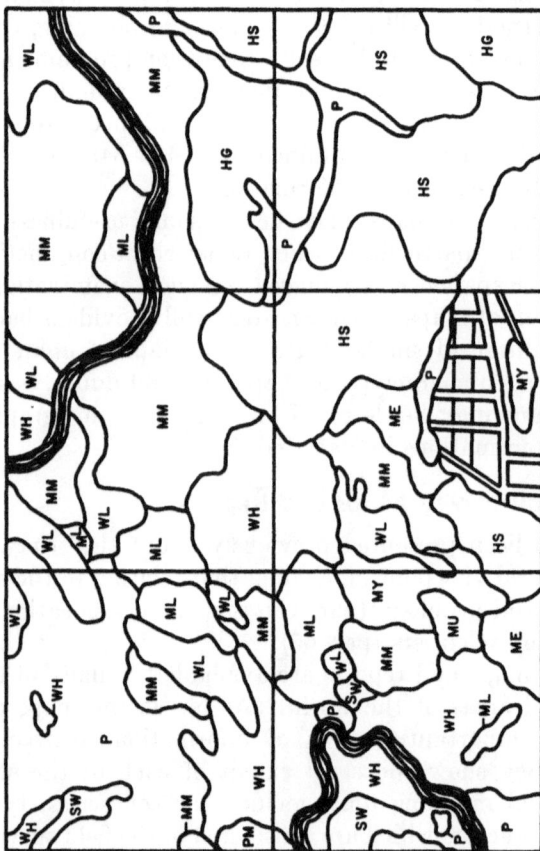

LEGEND

HG, HS – HOLYOKE

ME, ML, MM, MY – MERRIMAC

MU – MUCK

P – PAPAKATING

PM – PODUNK

SW – SWAMP

WH, WL – WHIPPANY

RIVER

UNCLASSIFIED CITY LAND

NOTE: LETTER DESIGNATION INDICATES SOIL SERIES AND TYPE. FOR EXAMPLE, HG IS HOLYOKE GRAVELLY LOAM; MM IS MERRIMAC LOAM, ETC. PHASE VARIATIONS NOT SHOWN. ON TYPICAL MAP COLOR IS USED TO INDICATE SOIL VARIATIONS. SCALE OF TYPICAL MAP IS 1 INCH = 1 MILE.

Figure 90.   A section of a simplified typical agricultural soil map.

basis of the field examination the soil is classified and placed in the proper mapping unit. The geographical location and extent of each soil is established by field measurements and the extensive use of aerial photographs. A section of a typical agricultural soil map is shown in figure 90.

c. Written reports generally accompany the maps and contain detailed information about the agricultural characteristics and uses of the soils encountered. Laboratory test results are generally minimized in these reports. The reports may be of both general and specific value for engineering purposes. Careful interpretation of the terminology used is necessary if they are to yield information of engineering value.

d. Agricultural studies are concerned primarily with surface soils, generally to a depth of about 6 feet below the natural ground surface. There value as aids in the engineering study of surface soils is apparent. For example, if the same soil is shown in two different areas on the map, it will be the same for agricultural purposes in the two locations. It very frequently will be the same for engineering purposes, also. If the soil is sampled and evaluated for engineering purposes in one area, then the amount of sampling and testing can be sharply reduced in the second area.

e. Caution should be exercised in the use of older agricultural soil maps in many regions, since the geographical location and areal extent of soils shown on the map may not be as accurate as is desirable for engineering purposes. This is not true of more modern soil maps.

## 339. Use of Aerial Photographs

Aerial photographs can be used to good advantage as aids in subsurface soil exploration, especially in areas which are inaccessible to other forms of examination. Two principal uses are found for aerial photographs in connection with an appraisal of soil conditions. Probably the most practical use for the photographs is in locating areas for field sampling in establishing the engineering soil map for a road or airport location. With experience and training in the interpretation of features shown on the photographs, they may also be used to predict the engineering characteristics of soils and to establish the boundaries between soils of unlike characteristics in unmapped areas. These uses are discussed briefly below.

## 340. Airphoto Interpretation

The use of aerial photographs in delineating and identifying soils is based upon the recognition of typical patterns formed under similar conditions of soil profile and weathering. Basic techniques for this procedure are discussed in TM 30–246. Principal elements which can be identified on a photograph and which provide clues to the identification of soils to a trained observer are land form, slopes, drainage patterns, erosional characteristics, soil color or "tone," vegetation and land use. Each of these is discussed briefly below; the items discussed are intended to serve only as examples of the information which may be derived from the examination of aerial photographs.

a. *Land Form.* The "form" or configuration of the land in different types of deposits is definitely characteristic and can be identified on aerial photographs. For example, glacial forms such as moraines, kames, eskers, and terraces have readily-identified forms. In desert areas characteristic dune shapes indicate areas which are covered by sands subject to movement by wind. In areas which are underlain by flat-lying, soluble limestone, the airphoto pattern is typified by sinkholes. Areas of muck and peat have characteristic nonconformities with any pattern and sharp boundaries.

b. *Slopes.* Prevailing ground slopes are generally representative of the texture of the soil. Steep slopes are characteristic of granular materials, while relatively flat and smoothly rounded slopes may indicate more plastic soils.

c. *Drainage Patterns.* The complete absence of surface drainage, or a very simple drainage pattern, is indicative of pervious soils. A highly integrated drainage pattern is frequently indicative of impervious soils, which in turn are frequently plastic. Drainage patterns are frequently indicative of the nature of the underlying rock in the areas. For example, in areas where alternately hard and soft layers of rock exist, major streams generally flow in valleys that have been cut in the softer rock. Also, in shale areas channels and gullies meander to a considerable extent because of the softness of the underlying material.

d. *Erosional Patterns.* Considerable information may be gained from the careful study of gullies. The cross section or shape of a gully is controlled primarily by the cohesiveness of the soil. Each abrupt change in grade, direction, or cross section is indicative of a change in the soil profile or rock layers. Short, V-shaped gullies which have steep gradients are typical of cohesionless soils. Gullies which are U-shaped and have steep gradients are indicative of deep, uniform silt deposits. Cohesive soils generally develop round, saucer-shaped gullies.

e. *Soil Color.* The true color of the soil is shown on the photograph by shades of gray, ranging from white to black. Soft, white colors are generally indicative of pervious, well-drained soils. Large flat areas of sands, for example, are frequently marked by uniform light gray color tones, a very flat appearance and no natural surface drainage. Clays and organic soils frequently appear as dark gray to black color tones. In general statement, sharp changes in the color tone represent changes in soil texture.

f. *Vegetation.* Vegetation may be indicative of surface soils, although its significance frequently is difficult of interpretation because of the effects of climate and other factors. With local experience, both cultivated and natural vegetative cover are reliable indicators of soil type.

*g. Land Use.* Ready identification of soils is frequently facilitated by observing agricultural land use. For example, orchards require well-draining soils and the presence of an orchard on level ground would imply a sandy soil. Wheat is frequently grown on loess-type soils. Rice is usually found in poorly draining soils underlain by impervious soils, such as clay. Tea grows in well-draining soils.

*h. Illustration.* In figure 91 is shown an aerial photograph and the corresponding engineering soil map. Pedologic and geologic information also was used in the preparation of the map. Sufficient information has not been included in this manual to permit the construction of such maps under all conditions. The illustration serves only to show what can be accomplished by the use of aerial photographs and associated information.

## 341. Miscellaneous Sources

*a.* Information frequently may be gained from conversations with local residents who are familiar with soils, such as civil engineers, contractors, surveyors, quarrymen, and the like. Information may also be gained in this way about unusual soil and foundation problems, local sources of construction materials, and the location and nature of soil and rock deposits. Such information must be carefully evaluated.

*b.* The use of foreign maps is described in detail in TM 5–248. In foreign countries topographical, geological, and agricultural soil maps are frequently obtainable through national, state, or provincial governments. Engineering agencies, such as a department of public works, may also be valuable sources of information.

## Section III. SOIL SURVEYS FOR ROADS AND AIR FIELDS

## 342. Purpose

Detailed soil surveys are made as an aid in the location, design, and construction of roads and airfields. The surveys are conducted in order to determine the location, nature, and condition of soil layers at the site; to obtain samples for examination, classification, and testing; to develop the soil profile (par. 350); to determine field drainage characteristics, including the location of the water table; and to determine the location of rock layers. For the purpose of this discussion three types of soil surveys are considered. Two of these are the *hasty* survey, which is made when time is very limited or under expedient conditions, and the *deliberate* survey, which is made when more time is available. Most of the discussion is devoted to the deliberate

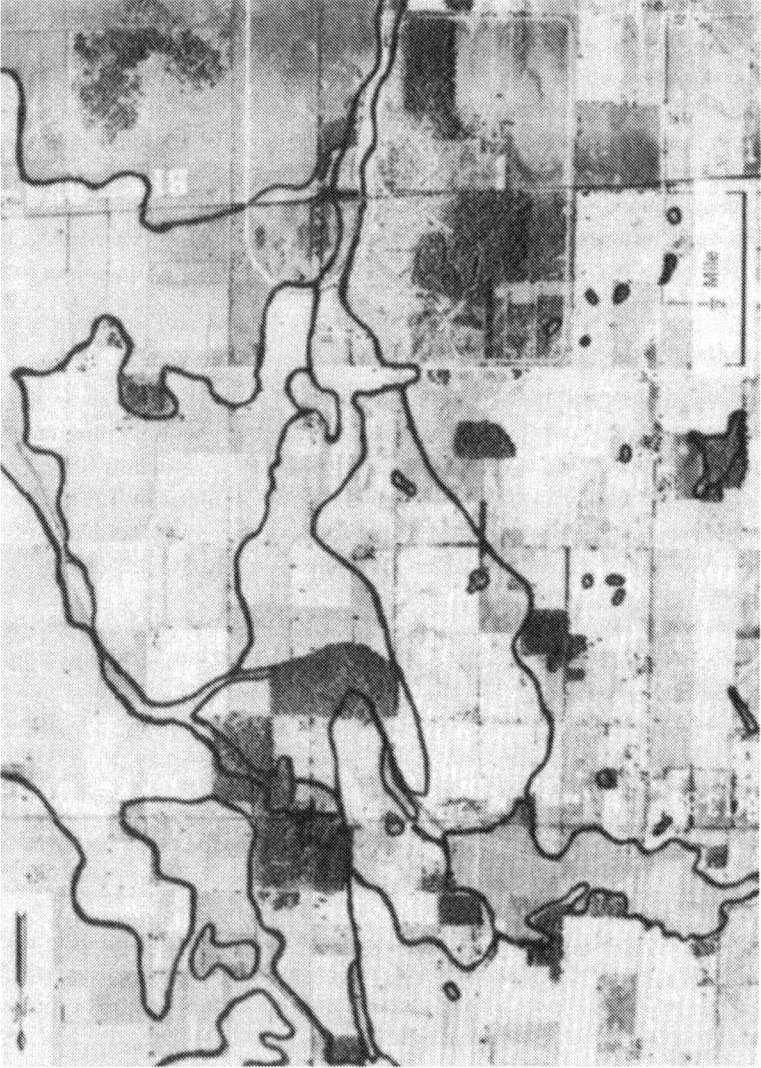

Figure 91. An aerial photograph and the corresponding engineering soil map

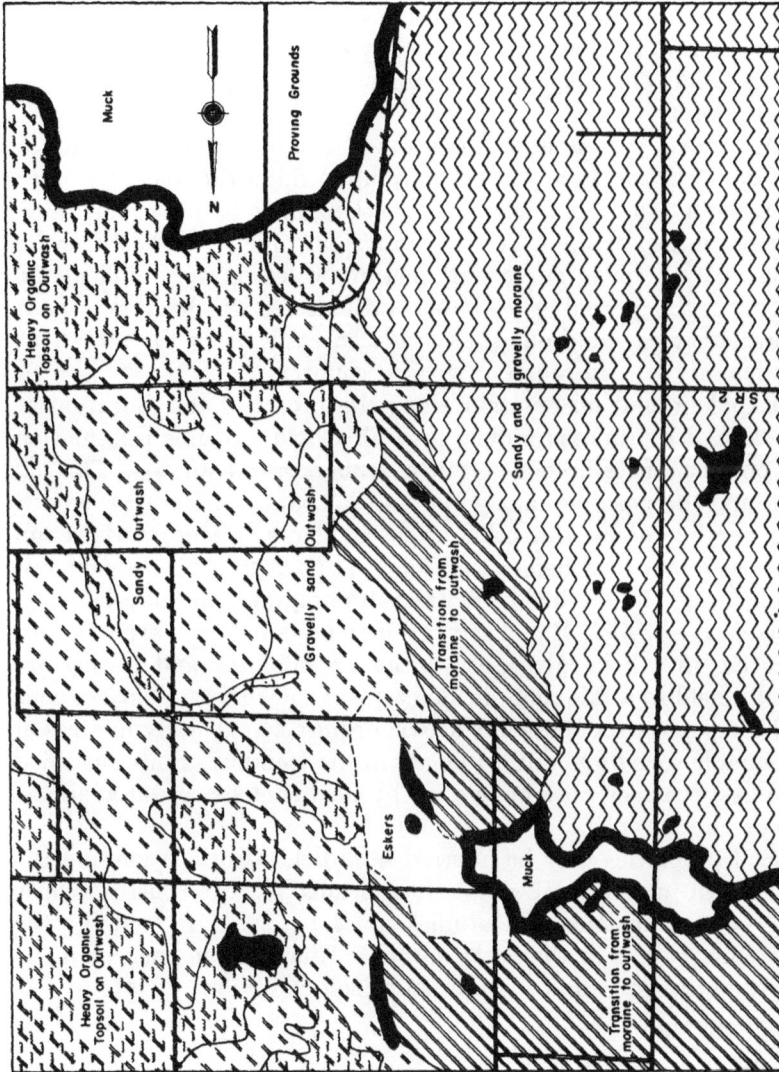

Figure 91—Continued.

263

survey, and this may be regarded as the standard type of soil survey for construction purposes. The third type of survey is concerned with the exploration of sources of borrow materials.

### 343. Hasty Soil Survey

*a.* The hasty soil survey is made when limitations of time and the military situation do not permit a more detailed examination of the soil conditions. It should be preceded by a careful study of the sources of information described in paragraphs 335 through 341, if circumstances permit.

*b.* A trained person may observe soil conditions at the planned site of construction from the air, keeping in mind the items discussed in paragraph 340. Significant observations may be noted mentally, in written notes, or upon available maps and photographs. Careful air observation gives an overall picture which is very difficult to secure from the ground, since important features which often cannot be detected from the ground may be seen easily from the air. This is particularly true in rough or wooded country.

*c.* Rapid ground observation along the proposed road location or at the site of a proposed airfield may also yield valuable information when circumstances do not permit a deliberate soil survey. The soil profile may be observed along the natural banks of streams, eroded areas, bomb craters, and other exposed places. Loose surface soil should be scraped off before the examination is made. Field identification, using the methods described in paragraph 56, may be made of the observed soils. Rapid samples may be taken from exposed soils for testing in a field laboratory, although sampling and testing are normally held to a minimum in this type of soil survey. Surface soils may be exposed by the use of a shovel or pick, particularly in areas of questionable soils or at critical points in the location. Soils identified in the hasty survey may be located by field sketches or on available maps or photographs.

*d.* Soil properties may be inferred from field identification and the values given in tables VI and VII, appendix II. As construction is carried out, additional soil studies must be made to amplify the information gained in the hasty survey. Changes can then be made in location, design, and construction as soil conditions dictate and as circumstances permit.

### 344. Deliberate Soil Survey

A deliberate soil survey is made when time permits in order to gain the detailed information needed for location, design, and construction. Time may be a factor in these surveys, also, and their scope may have to be limited in certain cases. A deliberate survey is required for major construction projects. It is frequently performed at about the same time that detailed topographical information is being obtained so that results of the soil survey may be integrated with other pertinent information. Sampling and testing which may be necessary to control the conduct of construction operations, e. g. compaction, are not considered a part of the soil survey.

## 345. Methods of Exploration

The principal method of exploration used in soil surveys for roads and airfields is by the use of the hand auger. Another method frequently used is to dig test pits. Other methods of exploration are used, as briefly described in *c* below.

*a. Hand Augers.* The method of obtaining samples by means of a hand auger is shown in figure 92. Samples obtained by the use of the hand auger are completely disturbed, but are satisfactory for determining the soil profile, classification tests, moisture content determinations, compaction tests, and similar purposes. By adding pipe extensions, they may be used up to depths of about 30 feet below the surface in some soils. They work best in clays and slightly plastic sands and silts, but may also be used in sampling cohesionless soils above the water table. They are not suitable for sampling clean sands below the water table and are difficult to use in gravelly soils and cemented layers, like hardpan.

*b. Test Pits.* A *test pit* is an open excavation which is large enough for a man to enter and study the soil in its undisturbed condition. The digging of a test pit is commonly done by hand and can be a long

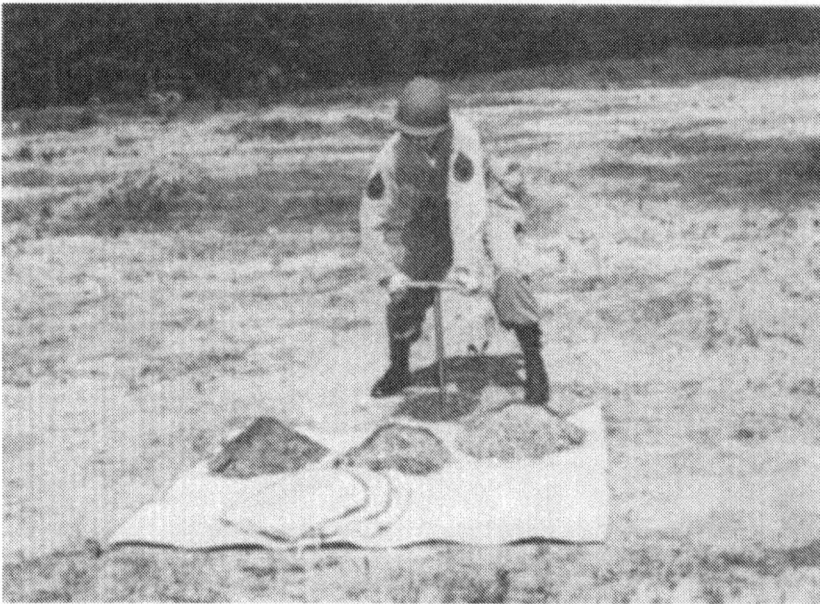

*Figure 92. Obtaining soil samples with hand auger.*

and tedious process. Both disturbed and undisturbed samples can be taken from test pits, the latter from the sides or bottom of the pit.

 *c. Other Methods.*

  (1) Under expedient conditions excavations can be made with a dragline or clamshell, or possibly a bulldozer, to permit hasty examination of the underlying soil.

  (2) Power-driven earth augers may sometimes be available and are excellent tools for subsurface soil exploration. Some of these will dig a hole 24 inches in diameter, which may be large enough for a small man to enter and make a visual inspection.

  (3) Sounding rods improvised from steel reinforcing rods or pipe may be used to provide valuable information in some situations. Their principal usefulness is in determining the depth to rock which underlies a soft layer. They may also be used, for example, in determining the depth of muck or peat which overlies a compact soil. Experience is required for interpretation of the results of soundings.

## 346. Procedure

 *a.* On airfield projects, auger holes or test pits are dug at 1,000-foot intervals along two lines 75 feet from the runway centerline. This procedure, in addition to providing a longitudinal profile, will provide a cross-sectional view of soil conditions beneath the runway. The 1,000-foot spacing should be regarded as the maximum spacing permissible; additional holes may be needed if the soil profile is not uniform, which is frequently the case. In road locations, auger holes or test pits should be dug along the centerline in areas where large cuts or fills are to be made.

 *b.* In planning the layout of test holes particular attention must be given to critical points, i. e. high and low points on the existing ground profile, and to points at which breaks occur in the surface profile, since such changes are frequently indicative of changes in subsurface soils.

 *c.* In cut areas all holes should extend 4 feet *below final subgrade elevation.* In fill areas, they should extend 4 feet below the natural ground elevation. All main holes should extend to the water table, if practicable. Depths of holes may well be increased in situations in which heavy loads are expected. This situation may occur, for example, when a high fill is to be built over a soft natural soil.

 *d.* It should be remembered that it is always better to dig too many holes than too few. The assumption of uniform conditions where they do not exist may have disastrous consequences and is not a risk worth saving a few test holes.

## 347. Locating, Numbering, and Recording

The engineer in charge of the soil survey is responsible for properly surveying, numbering, and recording each auger boring, test pit, or other exploratory investigation. A log is kept of each test hole which shows the elevation (or depth below the surface) of the top and bottom of each soil layer, the field identification of each soil encountered, and the number and type of each sample taken. Other information which may be included in the log is that relating to compactness of each soil, changes in moisture content, depth to ground water, and depth to rock.

## 348. Soil Samples

*a. General.* Bag, composite, moisture content, and undisturbed samples may be taken during a soil survey for a road or an airfield, as required. Details relative to methods to be used in obtaining and preserving samples are contained in TB 5–253–1.

*b. Bag Samples.* Bag samples are taken by the use of a shovel, pick, auger, or any other convenient hand tool. They are placed in bags without attempting to keep the soil in an undisturbed condition. These samples are used for mechanical analysis, plasticity, specific gravity, compaction, and laboratory-compacted CBR tests. The size of the sample taken will depend upon the tests to be performed.

*c. Composite Samples.* A composite sample is a bag sample which is taken in order to obtain a representative mixture of all soil layers within a profile to be investigated, or of the material contained in a stockpile or windrow. In securing this type of sample care must be taken to insure that the sample is truly representative.

*d. Moisture Content Samples.* Determination of the natural moisture content of a soil is frequently necessary and is accomplished by taking field samples which are sealed in containers to prevent the loss of moisture by evaporation. For fine-grained soils, about 100 grams of soil is generally enough to determine the moisture content. Much larger samples may be needed if the soil contains gravel.

*e. Undisturbed Samples.* An *undisturbed sample* is one which is cut, removed, and packed with the least possible disturbance. Every effort is made to preserve the void ratio, moisture content, and structure of the natural soil when undisturbed samples are taken. Undisturbed samples are used to determine the unit weight of the soil in place, or the shearing strength of the soil in the laboratory by means of the CBR or unconfined compression tests. For major projects, undisturbed samples may sometimes be required for shipment to a laboratory which has equipment for determining permeability, shearing resistance under varying conditions, and consolidation characteristics. Types of undisturbed samples which are frequently taken in shallow soil exploration programs include chunk samples cut by hand with a shovel and knife, cylinder samples which may be taken by forcing a CBR mold or special sampler into the soil, and box samples.

## 349. Water Table

The elevation of the water table is determined during the soil survey by observing the level at which free water stands in the test hole, usually an auger boring. To get an accurate determination, holes should be covered and inspected 24 hours after being dug, in order to allow the water to reach its maximum level.

## 350. Soil Profile

In engineering, the *soil profile* is defined as a graphical representation of a vertical cross section of the soil layers from the surface of the earth downward. Where heavy structures are involved, the soil profile may be determined to a depth of 100 feet or more below the surface. For roads and airfields the depth of the profile is generally much less, as described in paragraph 342. A definition of the soil profile from an agricultural viewpoint has been given previously (par. 11). The agricultural definition is slightly different from that used in engineering, but the difference is not great enough to cause any practical difficulty. The trend in soil engineering is toward the broad meaning of the term used by the agricultural soil scientist, in the sense that the soil profile is visualized as a product of the parent material and its environment, rather than just a drawing showing the results of a few isolated test holes.

## 351. Plotting the Soil Profile

*a.* A detailed field log is kept of each auger boring or test pit made during the soil survey. When the survey has been completed the information contained in the separate logs is consolidated, as shown in figure 93. In addition to the information shown, it is desirable to show the natural water contents of fine-grained soils along the side of each log, when this information is available. Other descriptive abbreviations may be used, as deemed appropriate.

*b.* The soil profile is drawn from the consolidated field data. It shows the location of test holes, profile of the natural ground to scale, location of any ledge rock encountered, field identification of each soil type, thickness of each soil stratum, profile of the water table, and the profile of the finished grade line. In some cases it may be possible to show on the profile the results of laboratory tests on significant soil strata; more frequently this information is given in a separate report. A soil profile based on the field data of figure 93 is shown in figure 94. Both figures represent part of an airfield soil survey, but the same procedure is used to record the results of soil surveys on road locations.

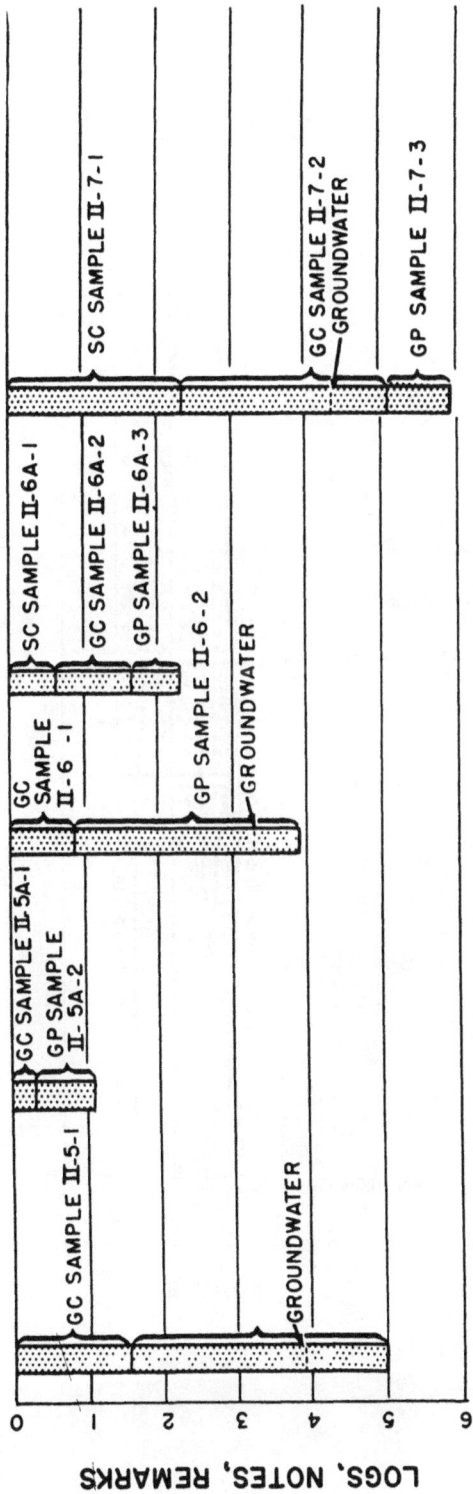

*Figure 93. Consolidated field data of a soil survey.*

269

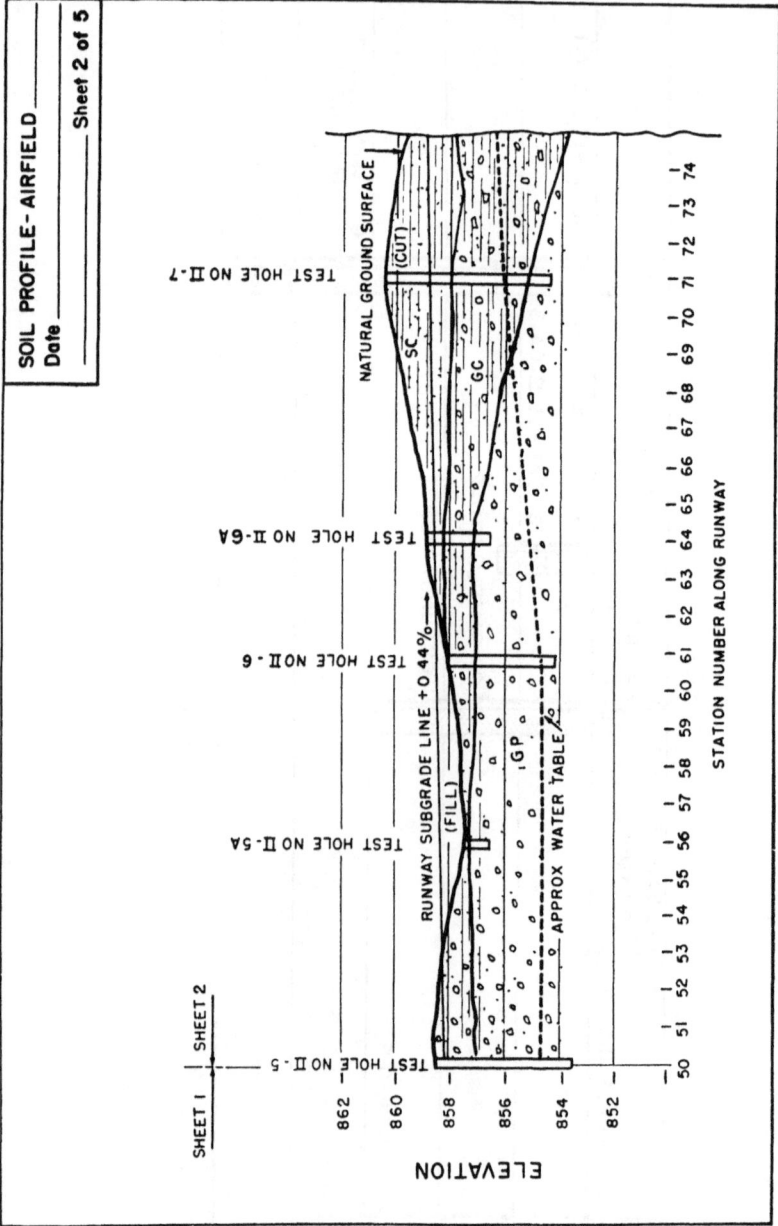

Figure 94. Section of a typical soil profile sheet.

## 352. Alternative Graphical Presentation

As shown in figures 93 and 94, the soil group of the Unified Soil Classification System is shown on the log or profile by letter designation, as GP. As an alternative the hatching symbols shown in column (4) of table VI, appendix II, may be used. In certain special instances the use of color to delineate soil types on maps and drawings is desireable. A suggested color scheme to show the major soil groups is shown in column (5) of table VI.

## 353. Uses of the Soil Profile

The soil profile has many practical uses in the location, design, and construction of roads and airfields. It has a great influence on the location of the grade line, which should be placed so that full advantage is taken of the best soils which are available at the site. The profile will show whether soils to be excavated are suitable for use in embankments, or if borrow soils will be required. It may show the existence of undesirable soils, such as peat or muck, or ledge rock close to the surface, which will require special construction measures. It will aid in the planning of drainage facilities so that advantage may be taken of the presence of well-draining soils. It may indicate that special drainage installations will be needed with soils which are more difficult to drain, particularly in areas where the water table is high. Considerations relative to capillarity and frost action may be particularly important when frost-susceptible soils are shown on the profile.

## 354. Sources of Borrow Materials

The location of borrow materials which will be suitable for use in fills, soil stabilization, bases, and surfaces is frequently an important function of the soil engineer on construction. Detailed information concerning the occurrence and use of soil and rock, including special localized materials such as tuff, caliche, and coral, is contained in TM 5–545. Information also is included therein relative to finding, evaluating, and working such deposits.

## 355. Borrow Exploration (Soils)

After a potential source of borrow material has been tentatively selected, the location may be explored in much the same way as for roads and airfields, provided that the material is soil. If the material is rock, special methods and equipment are required. In soils, samples may be taken from banks by excavating a groove from top to bottom of the stratum, thus securing a composite sample. If necessary, separate samples may be taken from separate layers. When the exploration is carried out by means of augers, holes should extend several feet below the probable limit of operations to determine if the borrow pit can be deepened, if necessary. The logs of borings are kept in the same way as before. Notes relative to natural moisture content and compactness of the soils encountered are important, since these factors may determine the type of equipment which can be used to work the deposit. The location of the water table is also important, since a high water table may seriously interfere with construction. If a large area is involved, borings are frequently arranged in a grid pattern, with recommended maximum spacings of holes generally being 100 feet center-to-center. The available yardage in a deposit may be estimated by determining the approximate average thickness and multiplying by the horizontal area. If the quantity of material is doubtful, alternate areas are located to prevent work stoppages.

## Section IV.  SOIL EXPLORATION FOR STRUCTURES

### 356. Introduction

The military engineer in the field rarely has to build massive structures. However, the occasional heavy load, such as that of a bridge pier, control tower, heavy machine or gun, dock, or dam will stress the soil to a much greater depth than the usual load on a road or airfield. For this reason it is sometimes necessary to determine the soil profile to a considerable depth below the surface. This may be necessary in order that a shallow foundation may be properly designed, or to determine if a pile foundation is necessary and the depth to which piles should be driven. The bearing capacity of soils beneath structures built on shallow and deep foundations is discussed in chapter 8. The effect of a weak layer of soil beneath the surface is particularly important. The methods described for deep exploration in this section generally will be satisfactory for loads encountered in military construction. Very large and heavy permanent structures may require additional study which is beyond the scope of this manual.

### 357. Preliminary Site Study

The preliminary study of soil conditions at the site of a proposed heavy structure is as important as it is for roads and airfields. Methods described in paragraphs 335 through 341 are applicable to structural sites, but are generally of less value than for roads and airfields since the area to be studied is usually much smaller. Geology of the site is important and is discussed in TM 5–545.

### 358. Depth and Spacing of Borings

*a. Spacing.* It is difficult to determine the proper spacing of borings for a structure before exploration begins, because the spacing is a function of the uniformity of soil conditions. As a general rule for structural foundations, one boring should be taken for each 2,500 square feet of area covered by the structure. If the foundation is to support a single heavy load, a boring at the center of the proposed foundation may be sufficient. In other cases, borings may be taken at the corners of the foundation site, with intermediate borings being made as needed to develop the soil profile. For earth dams and dikes borings are frequently spaced at 100-foot intervals along the proposed center line.

*b. Depth.* For heavy structures, a widely used general rule is that exploration should be carried to a depth which is equal to at least one and one-half (preferably two) times the width of the structure, or to bedrock. This rule may lead to excessive depths of borings if the structure is wide and low. If equipment is available to do so, it is highly desirable that rock encountered below the surface be drilled to a depth of at least 5 feet, in order to ascertain that bedrock has actually been reached and not an isolated boulder. It is important that borings be carried deep enough to detect weak or compressible soils which will be affected by the weight of the structure. No general rule is applicable to dam sites, the depth of borings depending on the type and height of the dam, and soil and geological conditions at the site. Borings must extend deep enough to detect pervious strata which may cause troublesome seepage through the foundation.

## 359. Exploration Methods

Three methods which have been described previously are sometimes used for soil exploration at structural sites. These are auger borings, test pits, and field plate loading tests. Auger borings and test pits were described in the previous section, while the field load test is described in paragraph 153. Auger borings are useful to relatively shallow depths, but have the disadvantage that the samples secured are completely disturbed and are not suitable for strength determinations. Test pits are limited to very shallow depths. The proper interpretation of the results of field plate loading tests requires that the soil profile be known.

## 360. Deep Borings

*a.* When it is necessary to determine soil conditions to depths beyond the range of the methods of the preceding paragraph, *deep borings* must be made. Many different methods, some of which require elaborate equipment, are used to make deep borings. Only two of the simpler methods are described here. Equipment for these procedures may be available to the military engineer in the field, or may be improvised.

*Figure 95.  Drive-sampling operation*

*b.* In some soils, samples can be taken to a considerable depth by simply driving a sampler (*d* below) into the ground. A simple rig for performing this operation is shown in figure 95. A 1-inch exploratory boring is shown in the figure, with the tube being driven into the ground by use of a light hammer. A winch mounted on the back of a jeep is used to raise the hammer. In this method the sampler, attached to the drill rod, is simply driven down by blows of a heavy weight impinging on a driving collar attached to the drill rod. A continuous *drive sample* is obtained from top to bottom of the hole. The method works best in plastic soils and may not be satisfactory in compact granular materials or below the water table, since the hole is not cased. Samples are disturbed, but are satisfactory for developing the soil profile. Carefully taken, they may be satisfactory for strength determinations by means of the unconfined compression test.

*c.* The best method of deep exploration for most structures is to advance the drill hole by means of suitable equipment and combine this operation with drive sampling of the soils encountered. Well-drilling equipment is satisfactory for advancing the hole, although the best equipment for the purpose is a rotary drilling rig. Wash borings may be used to advance the hole, but the soil which is brought to the surface in the wash water is useless as a sample. The hole is advanced to the desired depth and cleaned out. The sampler, attached to the drill rod, is then lowered to the bottom of the hole and driven into the soil as in *b* above. Drive samples generally are taken at 5-foot intervals and at intermediate depths when a significant change in soil occurs. Casing may be required below the water table to prevent the sides of the hole from caving in.

*d.* A device which is suitable for obtaining drive samples in many soils, particularly cohesive soils, which are usually the critical ones, is shown in the simplified sketch of figure 96. This is a *Shelby tube sampler*. Essential elements of this type are the tube itself, the ball

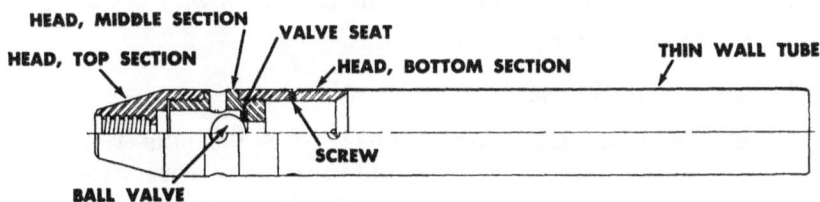

ASSEMBLY

*Figure 96. Simplified sketch of a Shelby tube sampler.*

check valve which permits the escape of water and air as the sampler is forced into the soil, and the drill rod connection. The sample tube is thin, seamless steel tubing, varying in outside diameter from as small as 1½ to as large as 5 inches. The usual size is 2 or 3 inches outside diameter and from 2 to 4 feet long. In civil practice, the assembly is brought to the surface after the sample is taken, the sample tube removed and the ends sealed, and the tube with the sample inside shipped to the laboratory for testing. The sample may be extruded from the sample tube in the field, if desired. If the samples are taken carefully, preferably by pushing the sampler into the soil rather than driving it, they will be essentially undisturbed.

*e.* If deep undisturbed samples are required for more extensive laboratory testing, as for the determination of additional data relative to shear strengths, consolidation characteristics, or permeability, larger and more elaborate sampling devices may be needed. Special equipment is required for core drilling in rock. The description of these sampling operations is beyond the scope of this manual.

## 361. Field Record and Soil Profile

A field log is kept of each deep boring in the fashion that has been indicated previously in paragraph 347. An important part of the log is the *penetration record*. The penetration record shows the number of blows required to force the sample tube a distance of one foot into the soil. The usefulness of the penetration record is explained in the next paragraph. A soil profile, or profiles, is constructed from the field logs, as indicated previously in paragraph 350. The field identification is frequently supplemented by the results of the liquid limit, plastic limit, natural moisture content, and unconfined compression tests performed on samples from critical soil layers. The natural moisture content and penetration record are particularly useful in correlating the information shown on the field logs. If the soil profile is nonuniform, additional borings may be needed to establish the location and extent of critical strata. After the profile has been established, methods described in chapter 8 may be used to design the foundation.

## 362. Penetration Record

A number of approximate correlations between the number of blows required to drive a sampler one foot and the relative density (par. 29) of cohesionless soils, or the consistency of cohesive soils, are in existence. The number of blows required is obviously a function of the amount of energy used in driving the sampler. The amount of energy applied is not standardized. One common procedure uses a 140-pound weight falling through a distance of 30 inches, while another uses a 300-pound weight with an 18-inch fall. Any correlation of this type is approximate, since the resistance to penetration is dependent not only upon the equipment used and the soil properties, but also upon the method of operation, depth below the surface, and other factors which have not yet been fully evaluated. An approximate correlation of this type is given in table XIV, appendix II.

# APPENDIX I

# BIBLIOGRAPHY

## 1. Military Publications

FM 5–9       Elementary Bridging.

TB 5–253–1     Soil Testing Set No. 1 and Expedient Tests.

TM 30–246     Interpretation of Aerial Photographs.

TM 5–248      Foreign Maps.

TM 5–252      Use of Road and Airdrome Construction Equipment.

TM 5–255      Aviation Engineers.

TM 5–545      Geology and Its Military Applications.

TM 5–630      Grounds Maintenance, Dust and Erosion Control, Repairs and Utilities.

## 2. Other Technical Publications

*a.* Airfield Pavement Design—Rigid Pavements, Engineering Manual for Military Construction, Part XII, Chapter 3, Corps of Engineers, Department of the Army, Washington, D. C., July, 1951.

*b.* The Appraisal of Terrain Conditions for Highway Engineering Purposes, Bulletin No. 13, Highway Research Board, Washington, D. C., 1941.

*c.* Base Course Drainage for Airport Pavements, *A. Casagrande* and *W. L. Shannon*, Proceedings Separate No. 75, American Society of Civil Engineers, New York, 1951.

*d.* Classification and Identification of Soils, *A. Casagrande*, Publication No. 432, Graduate School of Engineering, Harvard University, Cambridge, Mass., 1947.

*e.* Classification of Soils and Subgrade Materials for Highway Construction, Vol. 25, Proceedings, Highway Research Board, Washington, D. C., 1945.

*f.* Compaction of Embankments, Subgrades and Bases, Bulletin No. 58, Highway Research Board, Washington, D. C., 1952.

*g.* Foundation Engineering, *R. B. Peck*, *W. E. Hansen* and *T. M. Thornburn*, John Wiley & Sons, New York, 1953.

*h.* Foundations of Structures, *C. W. Dunham*, McGraw-Hill Book Company, New York, 1950.

*i.* Fundamentals of Soil Mechanics, *D. W. Taylor*, John Wiley & Sons, New York, 1948.

*j.* Handbook of Applied Hydraulics, *C. M. Davis et al.*, McGraw-Hill Book Company, New York, 1942.

*k.* Handbook of Culvert and Drainage Practice, Armco Drainage and Metal Products, Middletown, Ohio, 1950.

*l.* Highway Engineering, *L. J. Ritter, Jr.* and *R. J. Paquette*, The Ronald Press Company, New York, 1951.

*m.* Introductory Soil Mechanics and Foundations, *G. B. Sowers* and *G. F. Sowers*, The Macmillan Company, New York, 1951.

*n.* Pile Foundations and Pile Structures, Manual of Engineering Practice No. 27, American Society of Civil Engineers, New York, 1946.

*o.* Proceedings of the Conference on Soil Stabilization, Massachusetts Institute of Technology, Cambridge, Mass., 1952.

*p.* Proceedings of the American Railway Engineers Association, Vol. 43, Chicago, Illinois, 1942.

*q.* Procedures for Testing Soils, American Society for Testing Materials, Philadelphia, Pa., 1948.

*r.* Soil Mechanics Design, Part CXIX, Engineering Manual, Civil Works Construction, Corps of Engineers, Department of the Army, Washington, D. C., 1952.

*s.* Soil Mechanics, Foundations, and Earth Structures, *G. P. Tschebotarioff*, McGraw-Hill Book Company, New York, 1951.

*t.* Soil Mechanics in Engineering Practice, *K. Terzaghi* and *R. B. Peck*, John Wiley & Sons, New York, 1948.

*u.* Soil Mechanics Nomenclature, Manual of Engineering Practice No. 22, American Society of Civil Engineers, New York, 1941.

*v.* Soil Primer, Portland Cement Association, Chicago, Ill., 1950.

*w.* Soil Testing for Engineers, *T. W. Lambe*, John Wiley & Sons, New York, 1951.

*x.* Soil Tests for Military Construction, *G. E. Bertram*, Technical Bulletin No. 107, American Road Builders Association, Washington, D. C., 1946.

*y.* Subsurface Sewage Disposal, *J. E. Kiker, Jr.*, Bulletin No. 23, Florida Engineering and Industrial Experiment Station, Gainesville, Florida, 1948.

*z.* Standard Specifications for Highway Materials and Methods of Testing, Parts I and II, Sixth Edition, American Association of State Highway Officials, Washington, D. C., 1950.

*aa.* The Unified Soil Classification System, with Appendixes A and B, Technical Memorandum No. 3–357, Waterways Experiment Station, Vicksburg, Miss., 1953.

*ab.* Volcanic Ash and Lateritic Soils in Highway Construction, Bulletin No. 44, Highway Research Board, Washington, D. C., 1951.

# APPENDIX II

# TABLES

# TABLE V

## Table V. Unified Soil Classification System

**UNIFIED SOIL CLASSIFICATION**
(Including Identification and Description)

| Major Divisions | | | Group Symbols | Typical Names | Field Identification Procedures (Excluding particles larger than 3 inches and basing fractions on estimated weights) | | |
|---|---|---|---|---|---|---|---|
| **Highly Organic Soils** | | | Pt | Peat and other highly organic soils. | Readily identified by color, odor, spongy feel and frequently by fibrous texture. | | |
| **Fine-grained Soils** — More than half of material is smaller than No. 200 sieve size. The No. 200 sieve size is about the smallest particle visible to the naked eye. | Silts and Clays, Liquid limit greater than 50 | | OH | Organic clays of medium to high plasticity, organic silts. | *Identification Procedures on Fraction Smaller than No. 40 Sieve Size* — Dry Strength (Crushing characteristics): Medium to high | Dilatancy (Reaction to shaking): None to very slow | Toughness (Consistency near PL): Slight to medium |
| | | | CH | Inorganic clays of high plasticity, fat clays. | High to very high | None | High |
| | | | MH | Inorganic silts, micaceous or diatomaceous fine sandy or silty soils, elastic silts. | Slight to medium | Slow to none | Slight to medium |
| | Silts and Clays, Liquid limit less than 50 | | OL | Organic silts and organic silty clays of low plasticity. | Slight to medium | Slow | Slight |
| | | | CL | Inorganic clays of low to medium plasticity, gravelly clays, sandy clays, silty clays, lean clays. | Medium to high | None to very slow | Medium |
| | | | ML | Inorganic silts and very fine sands, rock flour, silty or clayey fine sands or clayey silts with slight plasticity. | None to slight | Quick to slow | None |
| **Coarse-grained Soils** — More than half of material is larger than No. 200 sieve size. | **Sands** — More than half of coarse fraction is smaller than No. 4 sieve size. (For visual classification, the 1/4-in. size may be used as equivalent to the No. 4 sieve size) | Sands with Fines (Appreciable amount of fines) | SC | Clayey sands, sand-clay mixtures. | Plastic fines (for identification procedures see CL below). | | |
| | | | SM | Silty sands, sand-silt mixtures. | Nonplastic fines or fines with low plasticity (for identification procedures see ML below). | | |
| | | Clean Sands (Little or no fines) | SP | Poorly-graded sands, gravelly sands, little or no fines. | Predominantly one size or a range of sizes with some intermediate sizes missing. | | |
| | | | SW | Well-graded sands, gravelly sands, little or no fines. | Wide range in grain size and substantial amounts of all intermediate particle sizes. | | |
| | **Gravels** — More than half of coarse fraction is larger than No. 4 sieve size. | Gravels with Fines (Appreciable amount of fines) | GC | Clayey gravels, gravel-sand-clay mixtures. | Plastic fines (for identification procedures see CL below). | | |
| | | | GM | Silty gravels, gravel-sand-silt mixtures. | Nonplastic fines or fines with low plasticity (for identification procedures see ML below). | | |
| | | Clean Gravels (Little or no fines) | GP | Poorly-graded gravels, gravel-sand mixtures. | Predominantly one size or a range of sizes with some intermediate sizes missing. | | |
| | | | GW | Well-graded gravels, gravel-sand mixtures, little or no fines. | Wide range in grain sizes and substantial amounts of all intermediate particle sizes. | | |

*TABLE V__Cont.*

## Table V.  Unified Soil Classification System

**UNIFIED SOIL CLASSIFICATION**
(Including Identification and Description)

| Information Required for Describing Soils | | Laboratory Classification Criteria |
|---|---|---|
| 6 | | 7 |

**Information Required for Describing Soils (Column 6)**

For undisturbed soils add information on stratification, degree of compactness, cementation, moisture conditions and drainage characteristics.

Give typical name; indicate approximate percentages of sand and gravel; max. size; angularity, surface condition, and hardness of the coarse grains; local or geologic name and other pertinent descriptive information; and symbol in parentheses.

Example:
Silty sand, gravelly; about 20% hard, angular gravel particles 1/2-in. maximum size; rounded and subangular sand grains coarse to fine; about 15% nonplastic fines with low dry strength; well compacted and moist in place; alluvial sand; (SM).

For undisturbed soils add information on structure, stratification, consistency in undisturbed and remolded states, moisture and drainage conditions.

Give typical name; indicate degree and character of plasticity, amount and maximum size of coarse grains; color in wet condition, odor if any, local or geologic name, and other pertinent descriptive information, and symbol in parentheses.

Example:
Clayey silt, brown; slightly plastic; small percentage of fine sand; numerous vertical root holes; firm and dry in place; loess; (ML).

---

Use grain-size curve in identifying the fractions as given under field identification.

**Laboratory Classification Criteria (Column 7)**

Determine percentages of gravel and sand from grain-size curve. Depending on percentage of fines (fraction smaller than No. 200 sieve size) coarse-grained soils are classified as follows:

| | |
|---|---|
| Less than 5% | GW, GP, SW, SP. |
| More than 12% | GM, GC, SM, SC. |
| 5% to 12% | Borderline cases requiring use of dual symbols. |

$$C_u = \frac{D_{60}}{D_{10}} \quad \text{Greater than 6}$$

$$C_g = \frac{(D_{30})^2}{D_{10} \times D_{60}} \quad \text{Between one and 3}$$

Not meeting all gradation requirements for GW

$$C_u = \frac{D_{60}}{D_{10}} \quad \text{Greater than 4}$$

$$C_g = \frac{(D_{30})^2}{D_{10} \times D_{60}} \quad \text{Between one and 3}$$

Not meeting all gradation requirements for SW

Atterberg limits below "A" line with PI less than 4

Atterberg limits above "A" line with PI greater than 7

Atterberg limits below "A" line or PI less than 4

Atterberg limits above "A" line with PI greater than 7

Above "A" line with PI between 4 and 7 are borderline cases requiring use of dual symbols.

Limits plotting in hatched zone with PI between 4 and 7 are borderline cases requiring use of dual symbols.

PLASTICITY INDEX

LIQUID LIMIT PLASTICITY CHART

For laboratory classification of fine-grained soils

Comparing Soils at Equal Liquid Limit
Toughness and Dry Strength Increase
with Increasing Plasticity Index

*TABLE V__Cont.*

*Table V.   Unified Soil Classification System*

**UNIFIED SOIL CLASSIFICATION**
(Including Identification and Description)

(1) <u>Boundary classifications:</u> Soils possessing characteristics of two groups are designated by combinations of group symbols. For example GW-GC, well-graded gravel-sand mixture with clay binder.

(2) All sieve sizes on this chart are U. S. standard.

FIELD IDENTIFICATION PROCEDURES FOR FINE-GRAINED SOILS OR FRACTIONS

These procedures are to be performed on the minus No. 40 sieve size particles, approximately 1/64 in. For field classification purposes, screening is not intended, simply remove by hand the coarse particles that interfere with the tests.

Dilatancy (Reaction to shaking)

After removing particles larger than No. 40 sieve size, prepare a pat of moist soil with a volume of about one-half cubic inch. Add enough water if necessary to make the soil soft but not sticky.
Place the pat in the open palm of one hand and shake horizontally, striking vigorously against the other hand several times. A positive reaction consists of the appearance of water on the surface of the pat which changes to a livery consistency and becomes glossy. When the sample is squeezed between the fingers, the water and gloss disappear from the surface, the pat stiffens, and finally it cracks or crumbles. The rapidity of appearance of water during shaking and of its disappearance during squeezing assist in identifying the character of the fines in a soil.
Very fine clean sands give the quickest and most distinct reaction whereas a plastic clay has no reaction. Inorganic silts, such as a typical rock flour, show a moderately quick reaction.

Adopted by Corps of Engineers and Bureau of Reclamation, January 1952

Dry Strength (Crushing characteristics)

After removing particles larger than No. 40 sieve size, mold a pat of soil to the consistency of putty, adding water if necessary. Allow the pat to dry completely by oven, sun, or air drying, and then test its strength by breaking and crumbling between the fingers. This strength is a measure of the character and quantity of the colloidal fraction contained in the soil. The dry strength increases with increasing plasticity.
High dry strength is characteristic for clays of the CH group. A typical inorganic silt possesses only very slight dry strength. Silty fine sands and silts have about the same slight dry strength, but can be distinguished by the feel when powdering the dried specimen. Fine sand feels gritty whereas a typical silt has the smooth feel of flour.

Toughness (Consistency near plastic limit)

After removing particles larger than the No. 40 sieve size, a specimen of soil about one-half inch cube in size, is molded to the consistency of putty. If too dry, water must be added and if sticky, the specimen should be spread out in a thin layer and allowed to lose some moisture by evaporation. Then the specimen is rolled out by hand on a smooth surface or between the palms into a thread about one-eighth inch in diameter. The thread is then folded and rerolled repeatedly. During this manipulation the moisture content is gradually reduced and the specimen stiffens, finally loses its plasticity, and crumbles when the plastic limit is reached.
After the thread crumbles, the pieces should be lumped together and a slight kneading action continued until the lump crumbles.
The tougher the thread near the plastic limit and the stiffer the lump when it finally crumbles, the more potent is the colloidal clay fraction in the soil. Weakness of the thread at the plastic limit and quick loss of coherence of the lump below the plastic limit indicate either inorganic clay of low plasticity, or materials such as kaolin-type clays and organic clays which occur below the A-line.
Highly organic clays have a very weak and spongy feel at the plastic limit.

# TABLE VI

Table VI.  *Characteristics Pertinent to Roads and Airfields*

| Major Divisions (1) | (2) | Letter (3) | Symbol Hatching (4) | Symbol Color (5) | Name (6) |
|---|---|---|---|---|---|
| COARSE GRAINED SOILS | GRAVEL AND GRAVELLY SOILS | GW | | Red | Well-graded gravels or gravel-sand mixtures, little or no fines |
| | | GP | | Red | Poorly graded gravels or gravel-sand mixtures, little or no fines |
| | | GM (d, u) | | Yellow | Silty gravels, gravel-sand-silt mixtures |
| | | GC | | Yellow | Clayey gravels, gravel-sand-clay mixtures |
| | SAND AND SANDY SOILS | SW | | Red | Well-graded sands or gravelly sands, little or no fines |
| | | SP | | Red | Poorly graded sands or gravelly sands, little or no fines |
| | | SM (d, u) | | Yellow | Silty sands, sand-silt mixtures |
| | | SC | | Yellow | Clayey sands, sand-clay mixtures |
| FINE GRAINED SOILS | SILTS AND CLAYS LL < 50 | ML | | Green | Inorganic silts and very fine sands, rock flour, silty or clayey fine sands or clayey silts with slight plasticity |
| | | CL | | Green | Inorganic clays of low to medium plasticity, gravelly clays, sandy clays, silty clays, lean clays |
| | | OL | | Green | Organic silts and organic silt-clays of low plasticity |
| | SILTS AND CLAYS LL > 50 | MH | | Blue | Inorganic silts, micaceous or diatomaceous fine sandy or silty soils, elastic silts |
| | | CH | | Blue | Inorganic clays of high plasticity, fat clays |
| | | OH | | Blue | Organic clays of medium to high plasticity, organic silts |
| HIGHLY ORGANIC SOILS | | Pt | | Orange | Peat and other highly organic soils |

*TABLE VI__Cont.*

*Table VI.   Characteristics Pertinent to Roads and Airfields*

| Value as Foundation When Not Subject to Frost Action (7) | Value as Base Directly under Bituminous Pavement (8) | Potential Frost Action (9) | Compressibility and Expansion (10) | Drainage Characteristics (11) |
|---|---|---|---|---|
| Excellent | Good | None to very slight | Almost none | Excellent |
| Good to excellent | Poor to fair | None to very slight | Almost none | Excellent |
| Good to excellent | Fair to good | Slight to medium | Very slight | Fair to poor |
| Good | Poor | Slight to medium | Slight | Poor to practically impervious |
| Good | Poor | Slight to medium | Slight | Poor to practically impervious |
| Good | Poor | None to very slight | Almost none | Excellent |
| Fair to good | Poor to not suitable | None to very slight | Almost none | Excellent |
| Good | Poor | Slight to high | Very slight | Fair to poor |
| Fair to good | Not suitable | Slight to high | Slight to medium | Poor to practically impervious |
| Fair to good | Not suitable | Slight to high | Slight to medium | Poor to practically impervious |
| Fair to poor | Not suitable | Medium to very high | Slight to medium | Fair to poor |
| Fair to poor | Not suitable | Medium to high | Medium | Practically impervious |
| Poor | Not suitable | Medium to high | Medium to high | Poor |
| Poor | Not suitable | Medium to very high | High | Fair to poor |
| Poor to very poor | Not suitable | Medium | High | Practically impervious |
| Poor to very poor | Not suitable | Medium | High | Practically impervious |
| Not suitable | Not suitable | Slight | Very high | Fair to poor |

*TABLE VI__Cont.*

Table VI.  *Characteristics Pertinent to Roads and Airfields*

| Name (6) | Value as Foundation When Not Subject to Frost Action (7) | Value as Base Directly under Bituminous Pavement (8) | Potential Frost Action (9) |
|---|---|---|---|
| Well-graded gravels or gravel-sand mixtures, little or no fines | Excellent | Good | None to very slight |
| Poorly graded gravels or gravel-sand mixtures, little or no fines | Good to excellent | Poor to fair | None to very slight |
| Silty gravels, gravel-sand-silt mixtures | Good to excellent | Fair to good | Slight to medium |
| | Good | Poor | Slight to medium |
| Clayey gravels, gravel-sand-clay mixtures | Good | Poor | Slight to medium |
| Well-graded sands or gravelly sands, little or no fines | Good | Poor | None to very slight |
| Poorly graded sands or gravelly sands, little or no fines | Fair to good | Poor to not suitable | None to very slight |
| Silty sands, sand-silt mixtures | Good | Poor | Slight to high |
| | Fair to good | Not suitable | Slight to high |
| Clayey sands, sand-clay mixtures | Fair to good | Not suitable | Slight to high |
| Inorganic silts and very fine sands, rock flour, silty or clayey fine sands or clayey silts with slight plasticity | Fair to poor | Not suitable | Medium to very high |
| Inorganic clays of low to medium plasticity, gravelly clays, sandy clays, silty clays, lean clays | Fair to poor | Not suitable | Medium to high |
| Organic silts and organic silt-clays of low plasticity | Poor | Not suitable | Medium to high |
| Inorganic silts, micaceous or diatomaceous fine sandy or silty soils, elastic silts | Poor | Not suitable | Medium to very high |
| Inorganic clays of high plasticity, fat clays | Poor to very poor | Not suitable | Medium |
| Organic clays of medium to high plasticity, organic silts | Poor to very poor | Not suitable | Medium |
| Peat and other highly organic soils | Not suitable | Not suitable | Slight |

# TABLE VI__Cont.

*Table VI.   Characteristics Pertinent to Roads and Airfields*

# Notes:

1.  Column 3, Division of GM,and SM groups into subdivisions of d and u are for roads and airfields only; subdivision is on basis of Atterberg limits; suffix d (e. g., GMd) will be used when the liquid limit is 28 or less and the plasticity index is 6 or less; the suffix u will be used when the liquid limit is greater than 28.

2.  Column 7, values are for subgrades and base courses except for base course directly under bituminous pavement.

3.  In column 8, the term "excellent" has been reserved for base materials consisting of high quality processed crushed stone.

4.  In column 9, these soils are susceptible to frost as indicated under conditions favorable to frost action described in the text.

5.  In column 12, the equipment listed will usually produce the required densities with a reasonable number of passes when moisture conditions and thickness of lift are properly controlled.  In some instances, several types of equipment are listed, because variable soil characteristics within a given soil group may require different equipment. In some instances, a combination of two types may  be necessary.
    a.  Processed base materials and other angular materials. Steel-wheeled rollers are recommended for hard angular materials with limited fines or screenings. Rubber-tired equipment is recommended for softer materials subject to degradation.
    b.  Finishing.  Rubber-tired equipment is recommended for rolling during final shaping operations for most soils and processed materials.
    c.  Equipment size.  The following sizes of equipment are necessary to assure the high densities required for airfield construction:
    Crawler-type tractor -- total weight in excess of 30,000 lb.
    Rubber-tired equipment -- wheel load in excess of 15,000 lb, wheel loads as high as 40,000 lb may be necessary to obtain the required densities for some materials (based on contact pressure of approximately 65 to 150 psi).
    Sheepsfoot roller -- unit pressure (on 6- to 12-sq-in. foot) to be in excess of 250 psi and unit pressures as high as 650 psi may be necessary to obtain the required densities for some materials. The area of the feet should be at least 5 per cent of the total peripheral area of the drum, using the diameter measured to the faces of the feet.

6.  Column 13, unit dry weights are for compacted soil at optimum moisture content for modified AASHO compactive effort.

# TABLE VII

*Table VII.  Characteristics Pertinent to Embankments and Foundations*

## Notes

1.

Values in columns 7 and 11 are for guidance only.  Design should be based on test results.

2.

In column 9, the equipment listed will usually produce the desired densities with a reasonable number of passes when moisture conditions ₱ 1 thickness of lift are properly controlled.

3.

Column 10, unit dry weights are for compacted soil at optimum moisture content for Standard AASHO (Standard Proctor) compactive effort.

| Major Divisions (1) | | Letter (3) | Symbol | | |
|---|---|---|---|---|---|
| | | | Hatching (4) | | Color (5) |
| COARSE GRAINED SOILS | GRAVEL AND GRAVELLY SOILS | GW | | | Red |
| | | GP | | | |
| | | GM | | | Yellow |
| | | GC | | | |
| | SAND AND SANDY SOILS | SW | | | Red |
| | | SP | | | |
| | | SM | | | Yellow |
| | | SC | | | |
| FINE GRAINED SOILS | SILTS AND CLAYS LL < 50 | ML | | | Green |
| | | CL | | | |
| | | OL | | | |
| | SILTS AND CLAYS LL > 50 | MH | | | Blue |
| | | CH | | | |
| | | OH | | | |
| HIGHLY ORGANIC SOILS | | Pt | | | Orange |

# TABLE VII__Cont.

Table VII.  Characteristics Pertinent to Embankments and Foundations

| Name (6) | Value for Embankments (7) | Permeability Cm Per Sec (8) |
|---|---|---|
| Well-graded gravels or gravel-sand mixtures, little or no fines | Very stable, pervious shells of dikes and dams | $k > 10^{-2}$ |
| Poorly-graded gravels or gravel-sand mixtures, little or no fines | Reasonably stable, pervious shells of dikes and dams | $k > 10^{-2}$ |
| Silty gravels, gravel-sand-silt mixtures | Reasonably stable, not particularly suited to shells, but may be used for impervious cores or blankets | $k = 10^{-3}$ to $10^{-6}$ |
| Clayey gravels, gravel-sand-clay mixtures | Fairly stable, may be used for impervious core | $k = 10^{-6}$ to $10^{-8}$ |
| Well-graded sands or gravelly sands, little or no fines | Very stable, pervious sections, slope protection required | $k > 10^{-3}$ |
| Poorly-graded sands or gravelly sands, little or no fines | Reasonably stable, may be used in dike section with flat slopes | $k > 10^{-3}$ |
| Silty sands, sand-silt mixtures | Fairly stable, not particularly suited to shells, but may be used for impervious cores or dikes | $k = 10^{-3}$ to $10^{-6}$ |
| Clayey sands, sand-silt mixtures | Fairly stable, use for impervious core for flood control structures | $k = 10^{-6}$ to $10^{-8}$ |
| Inorganic silts and very fine sands, rock flour, silty or clayey fine sands or clayey silts with slight plasticity | Poor stability, may be used for embankments with proper control | $k = 10^{-3}$ to $10^{-6}$ |
| Inorganic clays of low to medium plasticity, gravelly clays, sandy clays, silty clays, lean clays | Stable, impervious cores and blankets | $k = 10^{-6}$ to $10^{-8}$ |
| Organic silts and organic silt-clays of low plasticity | Not suitable for embankments | $k = 10^{-4}$ to $10^{-6}$ |
| Inorganic silts, micaceous or diatomaceous fine sandy or silty soils, elastic silts | Poor stability, core of hydraulic fill dam, not desirable in rolled fill construction | $k = 10^{-4}$ to $10^{-6}$ |
| Inorganic clays of high plasticity, fat clays | Fair stability with flat slopes, thin cores, blankets and dike sections | $k = 10^{-6}$ to $10^{-8}$ |
| Organic clays of medium to high plasticity, organic silts | Not suitable for embankments | $k = 10^{-6}$ to $10^{-8}$ |
| Peat and other highly organic soils | Not used for construction | |

## TABLE VII—Cont.

Table VII. *Characteristics Pertinent to Embankments and Foundations*

| Compaction Characteristics (9) | Std AASHO Max Unit Dry Weight Lb Per Cu Ft (10) | Value for Foundations (11) | Requirements for Seepage Control (12) |
|---|---|---|---|
| Good, tractor, rubber-tired, steel-wheeled roller | 125-135 | Good bearing value | Positive cutoff |
| Good, tractor, rubber-tired, steel-wheeled roller | 115-125 | Good bearing value | Positive cutoff |
| Good, with close control, rubber-tired, sheepsfoot roller | 120-135 | Good bearing value | Toe trench to none |
| Fair, rubber-tired, sheepsfoot roller | 115-130 | Good bearing value | None |
| Good, tractor | 110-130 | Good bearing value | Upstream blanket and toe drainage or wells |
| Good, tractor | 100-120 | Good to poor bearing value depending on density | Upstream blanket and toe drainage or wells |
| Good, with close control, rubber-tired, sheepsfoot roller | 110-125 | Good to poor bearing value depending on density | Upstream blanket and toe drainage or wells |
| Fair, sheepsfoot roller, rubber tired | 105-125 | Good to poor bearing value | None |
| Good to poor, close control essential, rubber-tired roller, sheepsfoot roller | 95-120 | Very poor, susceptible to liquefaction | Toe trench to none |
| Fair to good, sheepsfoot roller, rubber tired | 95-120 | Good to poor bearing | None |
| Fair to poor, sheepsfoot roller | 80-100 | Fair to poor bearing, may have excessive settlements | None |
| Poor to very poor, sheepsfoot roller | 70-95 | Poor bearing | None |
| Fair to poor, sheepsfoot roller | 75-105 | Fair to poor bearing | None |
| Poor to very poor, sheepsfoot roller | 65-100 | Very poor bearing | None |
| Compaction not practical | | Remove from foundations | |

Table VIII. Revised Public Roads Classification System

| General classification | Granular materials (35% or less of total sample passing No. 200) | | | | | | | Silt-clay materials (More than 35% of total sample passing No. 200) | | | |
|---|---|---|---|---|---|---|---|---|---|---|---|
| Group classification | A-1 | | A-3 | A-2 | | | | A-4 | A-5 | A-6 | A-7 (A-7-5, A-7-6) |
| | A-1-a | A-1-b | | A-2-4 | A-2-5 | A-2-6 | A-2-7 | | | | |
| Sieve analysis, percent passing: | | | | | | | | | | | |
| No. 10 sieve | 50 max. | | | | | | | | | | |
| No. 40 sieve | 30 max. | 50 max. | 51 min. | | | | | | | | |
| No. 200 sieve | 15 max. | 25 max. | 10 max. | 35 max. | 35 max. | 35 max. | 35 max. | 36 min. | 36 min. | 36 min. | 36 min. |
| Characteristics of portion passing No. 40: | | | | | | | | | | | |
| Liquid limit | | | | 40 max. | 41 min. | 40 max. | 41 min. | 40 max. | 41 min. | 40 max. | [1] 41 min. |
| Plasticity index | | 6 max. | N. P. | 10 max. | 10 max. | 11 min. | 11 min. | 10 max. | 10 max. | 10 max. | 10 max. |
| Group index [2] | 0 | | 0 | | 0 | | 4 max. | 8 max. | 12 max. | 16 max. | 20 max. |

[1] Plasticity index of A-7-5 subgroup is equal to or less than liquid limit minus 30. Plasticity index of A-7-6 subgroup is greater than liquid limit minus 30. Refer to figure 22 for graphical method of delineating A-7 subgroups.

[2] See figure 21 for group index formula and charts.

## Table IX. Civil Aeronautics Administration Soil Classification System

| Soil group | Retained on No. 10 sieve (percent) | Material passing No 10 sieve | | | Liquid limit (percent) | Plasticity index (percent) |
|---|---|---|---|---|---|---|
| | | Coarse sand (percent) | Fine sand (percent) | Combined silt and clay (percent) | | |
| E-1 | 0–45 | 40+ | 60– | 15– | 25– | 6– |
| E-2 | 0–45 | 15+ | 85– | 25– | 25– | 6– |
| E-3 | 0–45 | | | 25– | 25– | 6– |
| E-4 | 0–45 | | | 35– | 35– | 10– |
| E-5 | 0–45 | | | 45– | 40– | 15– |
| E-6 | 0–55 | | | 45+ | 40– | 10– |
| E-7 | 0–55 | | | 45+ | 50– | 10–30 |
| E-8 | 0–55 | | | 45+ | 60– | 15–40 |
| E-9 | 0–55 | | | 45+ | 40+ | 30– |
| E-10 | 0–55 | | | 45+ | 70– | 20–50 |
| E-11 | 0–55 | | | 45+ | 80– | 30+ |
| E-12 | 0–55 | | | 45+ | 80+ | |
| E-13 | Peat and muck (field identification) | | | | | |

*Note.* Classification is based upon the sieve analysis of the portion of the material which passes a No. 10 sieve. When a sample contains material coarser than a No. 10 sieve in an amount equal to or greater than the maximum limit shown in the table, a raise in classification may be permitted provided that the coarse material is reasonably sound and well graded. Coarse sand is material which passes a No. 10 sieve and is retained on a No. 60; fine sand passes a No. 60 but is retained on a No. 270; combined silt and clay is material finer than a No. 270 sieve.

*Table X.  Agricultural Soil Classification System*

## GRADUATION LIMITS OF TEXTURAL SOIL GROUPS [1]

First place the soil in Textural Group I, II or III, according to clay content (column 7), then according to silt content (column 6) or silt and clay content combined (column 5) and finally the sand content and gravel content (columns 1 to 4 inclusive). When clay content approaches the upper limit for that group, it is called "heavy," such as "heavy clay loam," etc., and a "light" soil when approaching the lower clay limit.

| Soil texture | (1) Fine and coarse gravel, (percent) | (2) Coarse sand and gravel, (percent) | (3) Sand and gravel, (percent) | (4) Fine sand, (percent) | (5) Silt and clay, (percent) | (6) Silt, (percent) | (7) Clay, (percent) |
|---|---|---|---|---|---|---|---|
| **Group IA.  SANDS AND GRAVELS** | | | | | | | |
| Gravel | 85 to 100 | | | | 0 to 15 | | 0 to 20 |
| Gravel and sand | 50 to 85 | | | | 0 to 15 | | 0 to 20 |
| Sand and gravel | 25 to 50 | | | | 0 to 15 | | 0 to 20 |
| Coarse sand | 0 to 25 | 50 to 100 | | | 0 to 15 | | 0 to 20 |
| Sand | 0 to 25 | 0 to 50 | | 0 to 50 | 0 to 15 | | 0 to 20 |
| Fine sand | 0 to 25 | | | 50 to 100 | 0 to 15 | | 0 to 20 |
| **Group IB.  LOAMY SANDS** | | | | | | | |
| Gravelly loamy coarse sand | 25 to 85 | 50 to 85 | | | 15 to 20 | | 0 to 20 |
| Gravelly loamy sand | 25 to 50 | 25 to 50 | | 0 to 50 | 15 to 20 | | 0 to 20 |
| Gravelly loamy fine sand | 25 to 35 | | | 50 to 60 | 15 to 20 | | 0 to 20 |
| Loamy coarse sand | 0 to 25 | 50 to 85 | | | 15 to 20 | | 0 to 20 |
| Loamy sand | 0 to 25 | 0 to 50 | | 0 to 50 | 15 to 20 | | 0 to 20 |
| Loamy fine sand | 0 to 25 | | | 50 to 85 | 15 to 20 | | 0 to 20 |

See footnote at end of table.

Table X.  *Agricultural Soil Classification System*—Continued

| Soil texture | (1) Fine and coarse gravel, (percent) | (2) Coarse sand and gravel, (percent) | (3) Sand and gravel, (percent) | (4) Fine sand, (percent) | (5) Silt and clay, (percent) | (6) Silt, (percent) | (7) Clay, (percent) |
|---|---|---|---|---|---|---|---|
| **Group IC.  SANDY LOAMS** | | | | | | | |
| Gravelly coarse sandy loam | 25 to 80 | 50 to 80 | --- | --- | 20 to 50 | --- | 0 to 20 |
| Gravelly sandy loam | 25 to 50 | 25 to 50 | --- | 0 to 50 | 20 to 50 | --- | 0 to 20 |
| Gravelly fine sandy loam | 25 to 30 | --- | --- | 50 to 55 | 20 to 50 | --- | 0 to 20 |
| Coarse sandy loam | 0 to 25 | 50 to 80 | --- | --- | 20 to 50 | --- | 0 to 20 |
| Sandy loam | 0 to 25 | 0 to 50 | --- | 0 to 50 | 20 to 50 | --- | 0 to 20 |
| Fine sandy loam | 0 to 25 | --- | --- | 50 to 80 | 20 to 50 | --- | 0 to 20 |
| **Group ID.  LOAMS AND SILT LOAMS** | | | | | | | |
| Gravelly loam | 25 to 50 | --- | --- | --- | 50 to 70 | 30 to 50 | 0 to 20 |
| Gravelly silt loam | 25 to 50 | --- | --- | --- | 50 to 100 | 50 to 80 | 0 to 20 |
| Loam | 0 to 25 | --- | --- | --- | 50 to 70 | 30 to 50 | 0 to 20 |
| Silt loam | 0 to 25 | --- | --- | --- | 50 to 100 | 50 to 80 | 0 to 20 |
| Silt | 0 to 25 | --- | --- | --- | 80 to 100 | 80 to 100 | 0 to 20 |
| **Group II.  CLAY LOAMS** | | | | | | | |
| Gravelly sandy clay loam | 25 to 80 | --- | 50 to 80 | --- | --- | --- | 20 to 30 |
| Gravelly clay loam | 25 to 50 | --- | 25 to 50 | --- | --- | 20 to 50 | 20 to 30 |
| Gravelly silty clay loam | 25 to 30 | --- | 0 to 30 | --- | --- | 50 to 55 | 20 to 30 |
| Sandy clay loam | 0 to 25 | --- | 50 to 80 | --- | --- | 0 to 30 | 20 to 30 |
| Clay loam | 0 to 25 | --- | 20 to 50 | --- | --- | 20 to 50 | 20 to 30 |
| Silty clay loam | 0 to 25 | --- | 0 to 30 | --- | --- | 50 to 80 | 20 to 30 |

## Group III. CLAYS

| | | | |
|---|---|---|---|
| Gravelly sandy clay | 25 to 70 | 50 to 70 | | 30 to 100 |
| Gravelly clay | 25 to 50 | 25 to 50 | 0 to 45 | 30 to 100 |
| Sandy clay | 0 to 25 | 50 to 70 | 0 to 20 | 30 to 100 |
| Silty clay | 0 to 25 | 0 to 20 | 50 to 70 | 30 to 100 |
| Clay | 0 to 25 | 0 to 50 | 0 to 50 | 30 to 100 |

[1] Basic concept for this textural classification is from the U. S. Bureau of Chemistry and Soils.

*Table XI.   Allowable Pressures for the Design of Shallow Foundations*

|  | Maximum allowable pressure, tons per square foot |
|---|---|
| Hard, sound rock | 40+ |
| Soft rock | 8–10 |
| Hardpan overlying rock | 10–12 |
| Compact gravel and boulder-gravel deposits; very compact sandy gravel | 10 |
| Loose gravel and sandy gravel; compact sand and gravelly sand; very compact inorganic sand-silt soils | 5–6 |
| Hard, dry consolidated clay | 5 |
| Loose coarse to medium sand; medium compact fine sand | 4 |
| Compact sand-clay soils | 3 |
| Loose fine sand; medium compact inorganic sand-silt soils | 2 |
| Firm or stiff clay | 1½–2 |
| Loose saturated sand-clay soils; medium soft clay | 1 |

*Note.*   Values are not applicable if foundation soil is underlain by a weaker soil.   Use of the tabular values for the design of shallow foundations of major structures is not recommended unless their use is justified by experience or additional investigation.

Table XII. *Recommended Requirements for Compaction of Highway Embankments*

| Revised public roads system | Approximate equivalent, unified system | Condition of exposure | | | | | |
|---|---|---|---|---|---|---|---|
| | | Condition 1 (not subject to inundation) | | | Condition 2 (subject to inundation) | | |
| | | Height of fill, feet | Side slope | Desired compaction, % AASHO maximum density | Height of fill, feet | Side slope | Desired compaction, % AASHO maximum density |
| A-1 | GW, GP, SW. Some GM or SM | Not critical | 1½ to 1 | 95+ | Not critical | 2 to 1 | 95 |
| A-3 | SP | Not critical | 1½ to 1 | 100+ | Not critical | 2 to 1 | 100+ |
| A-2-4 | Most GM and SM | Less than 50 | 2 to 1 | 95+ | Less than 10 | 3 to 1 | 95 |
| A-2-5 | | | | | 10 to 50 | ---- | 95-100 |
| A-2-6 or 7 | GC or SC | Less than 50 | 2 to 1 | 95+ | Less than 50 | 3 to 1 | 95-100 |
| A-4, A-5 | ML, MH | Less than 50 | 2 to 1 | 95+ | Less than 50 | 3 to 1 | 95-100 |
| A-6, A-7 | CL, CH | Less than 50 | 2 to 1 | 95-100 | Less than 50 | 3 to 1 | 95-100 |

NOTES

(1) Under Condition 2, higher fills of the order of 35 to 50 feet should be compacted to 100 percent, at least for portions subject to inundation. Major fills composed of unusual materials which have low shearing resistance should be analyzed by soil mechanics methods.

(2) For soils of the A-6 or A-7 groups, the lower compaction requirements shown obtain only for low fills (10 to 15 feet or less) not subject to inundation and not carrying large volumes of heavy traffic.

(3) Highly organic soils are not generally suitable for fill construction.

Table XIII. *Desired Gradation for Crushed Rock, Gravel, or Slag and Uncrushed Sandy and Gravelly Aggregates for Base Courses*

| Sieve designation | Percent passing each sieve, by weight | | | | |
|---|---|---|---|---|---|
| | Maximum aggregate size | | | | |
| | 3-inch | 2-inch | 1½-inch | 1-inch | 1-inch |
| 3-inch | 100 | ------- | ------- | ------- | ----- |
| 2-inch | 65–100 | 100 | ------- | ------- | ----- |
| 1½-inch | ------- | 70–100 | 100 | ------- | ----- |
| 1-inch | 45–75 | 55–85 | 70–100 | 100 | 100 |
| ¾-inch | ------- | 50–80 | 60–90 | 70–100 | ----- |
| ⅜-inch | 30–60 | 40–70 | 45–75 | 50–80 | ----- |
| No. 4 | 25–50 | 30–60 | 30–60 | 35–65 | ----- |
| No. 10 | 20–40 | 20–50 | 20–50 | 20–50 | 65  90 |
| No. 40 | 10–25 | 10–30 | 10–30 | 15–30 | 33–70 |
| No. 200 | 3–10 | 5–15 | 5–15 | 5–15 | 8- 25 |

Table XIV. *Approximate Correlation Between Resistance to Penetration of a Tube Sampler and the Relative Density of Cohesionless or Consistency of Cohesive Soils (New England Division, Corps of Engineers)*

| Relative density of cohesionless soils | Consistency of cohesive soils | Blows per foot |
|---|---|---|
| Very loose | Very soft | Less than 8. |
| Loose | Soft | 8–16. |
| Medium | Medium stiff | 16–55. |
| Compact | Stiff to medium hard | 55–110. |
| Very compact | Very hard | More than 110. |

*Note.* Sampler is 3.00 inches outside diameter and is driven by means of a 300-pound weight falling a distance of 18 inches.

# APPENDIX III

# NOMENCLATURE

The nomenclature used in this manual agrees essentially with that suggested in the publication "Soil Mechanics Nomenclature", which is Manual of Engineering Practice No. 22 of the American Society of Civil Engineers.

| | |
|---|---|
| $A$ | Area |
| $a$ | Average width of a segment of a flow net |
| $b$ | Width |
| $C$ | Coefficient |
| $C_c$ | Compression index |
| $C_g$ | Coefficient of gradation |
| $C_p$ | Coefficient of percolation |
| $C_u$ | Uniformity coefficient |
| $c$ | Cohesion per unit area |
| $D_d$ | Relative density |
| $D_f$ | Depth of foundation |
| $D_{10}$ | Grain size (diameter) corresponding to 10 percent on a grain-size distribution curve (Hazen's effective size). |
| $D_{15}$ | Grain size (diameter) corresponding to 15 percent on a grain-size distribution curve. |
| $D_{30}$ $D_{60}$ $D_{85}$ | Grain diameters, similar to $D_{15}$, above |
| $d$ | Deformation |
| $e$ | Void ratio |
| $e_O$ | Initial void ratio |
| $e_{max}$ | Void ratio of a granular soil in loosest possible condition |
| $e_{min}$ | Void ratio of a granular soil in densest possible condition |
| $G$ | Specific gravity of soil particles |
| $g$ | Acceleration due to gravity |
| $H$ | Height |
| $h$ | Hydraulic head |
| $h_c$ | Height of capillary rise above a free water surface |
| $I_p$ | Plasticity index |
| $i$ | Hydraulic gradient |
| $K_A$ | Coefficient of active earth pressure |
| $K_O$ | Coefficient of earth pressure at rest |
| $K_p$ | Coefficient of passive earth pressure |
| $k$ | Coefficient of permeability |
| $k_s$ | Subgrade modulus, corrected |
| $k_u$ | Subgrade modulus, uncorrected |
| $L$ | Length |
| $M_f$ | Moment tending to cause failure (stability analysis) |
| $M_r$ | Resisting moment (stability analysis) |
| $N_f$ | Number of flow channels in a flow net |

| $N_p$ | Number of squares between adjacent flow lines in a flow net |
|---|---|
| $n$ | Porosity |
| $n_p$ | Number of equipotential drops between a point in a flow net and zero potential. |
| $P$ | Normal force |
| $P_A$ | Active earth pressure |
| $P_O$ | Earth pressure at rest |
| $P_p$ | Passive earth pressure |
| $P'$ | Lateral force on the back of a retaining wall caused by a surcharge on the backfill. |
| $p$ | Normal stress (force per unit of area) |
| $p_A$ | Active earth pressure per unit of area |
| $p_O$ | Initial normal stress; also, at-rest earth pressure per unit of area |
| $p_p$ | Passive earth pressure per unit of area |
| $\bar{p}$ | Intergranular pressure |
| $Q$ | Total load |
| $Q_a$ | Allowable load on a bearing pile |
| $Q'$ | Line load (retaining wall design) |
| $q$ | Quantity of seepage per unit of width and unit of time |
| $q_O$ | Ultimate bearing capacity |
| $q_u$ | Unconfined compressive strength |
| $q'$ | Surcharge per unit of area |
| $R$ | Radius |
| $S$ | Shearing force |
| $S_r$ | Degree of saturation |
| $S_t$ | Sensitivity |
| $s$ | Shearing stress (or resistance) per unit of area; also, penetration of a pile under the final blow of the hammer. |
| $t_p$ | Time required for the water level to drop one inch in the percolation test |
| $u$ | Excess hydrostatic pressure |
| $u_w$ | Neutral pressure |
| $v$ | Velocity of flow (of water) |
| $W$ | Weight |
| $W_h$ | Weight of the ram (pile hammer) |
| $w$ | Moisture content |
| $w_L$ | Liquid limit |
| $w_p$ | Plastic limit |
| $x$ | Distance |
| $z$ | Depth |
| $\beta$ | Angle between the surface of an inclined backfill and the horizontal |
| $\gamma$ | Wet unit weight |
| $\gamma_d$ | Dry unit weight (dry density) |
| $\gamma_s$ | Unit weight of solids |
| $\gamma_w$ | Unit weight of water |
| $\gamma'$ | Submerged unit weight |
| $\Delta e$ | Change in void ratio |
| $\Delta H$ | Change in thickness of a soil layer under compressive load |
| $\Delta h$ | Head loss between two adjacent equipotential boundaries of a square in a flow net. |
| $\Delta p$ | Change in normal stress |
| $\phi$ | Angle of internal friction |

# INDEX

[AG 000.91 (3 Jun 54)]

By order of the Secretaries of the Air Force and the Army:

N. F. TWINING,
*Chief of Staff. United States Air Force.*

OFFICIAL:
E. E. TORO,
*Colonel, USAF,*
*Air Adjutant General.*

M. B. RIDGWAY,
*General, United States Army,*
*Chief of Staff.*

OFFICIAL:
JOHN A. KLEIN,
*Major General, United States Army,*
*The Adjutant General.*

www.ingramcontent.com/pod-product-compliance
Lightning Source LLC
Chambersburg PA
CBHW021550210326
41599CB00010B/384